Lie Algebras in Particle Physics

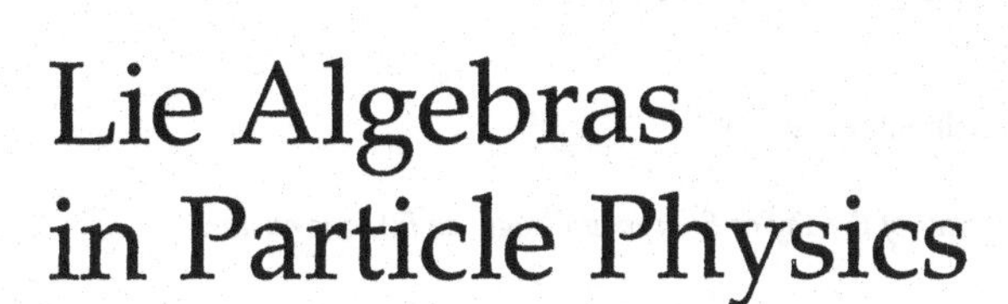

Lie Algebras in Particle Physics

Second Edition

Howard Georgi

CRC Press is an imprint of the
Taylor & Francis Group, an **informa** business

First published 1999 by Westview Press

Published 2019 by CRC Press
Taylor & Francis Group
6000 Broken Sound Parkway NW, Suite 300
Boca Raton, FL 33487-2742

First issued in hardback 2019

CRC Press is an imprint of the Taylor & Francis Group, an informa business

Visit the Taylor & Francis Web site at
http://www.taylorandfrancis.com

and the CRC Press Web site at
http://www.crcpress.com

Library of Congress Catalog Card Number: 99-64878

ISBN 13: 978-0-367-09172-9 (hbk)
ISBN 13: 978-0-7382-0233-4 (pbk)

DOI: 10.1201/9780429499210

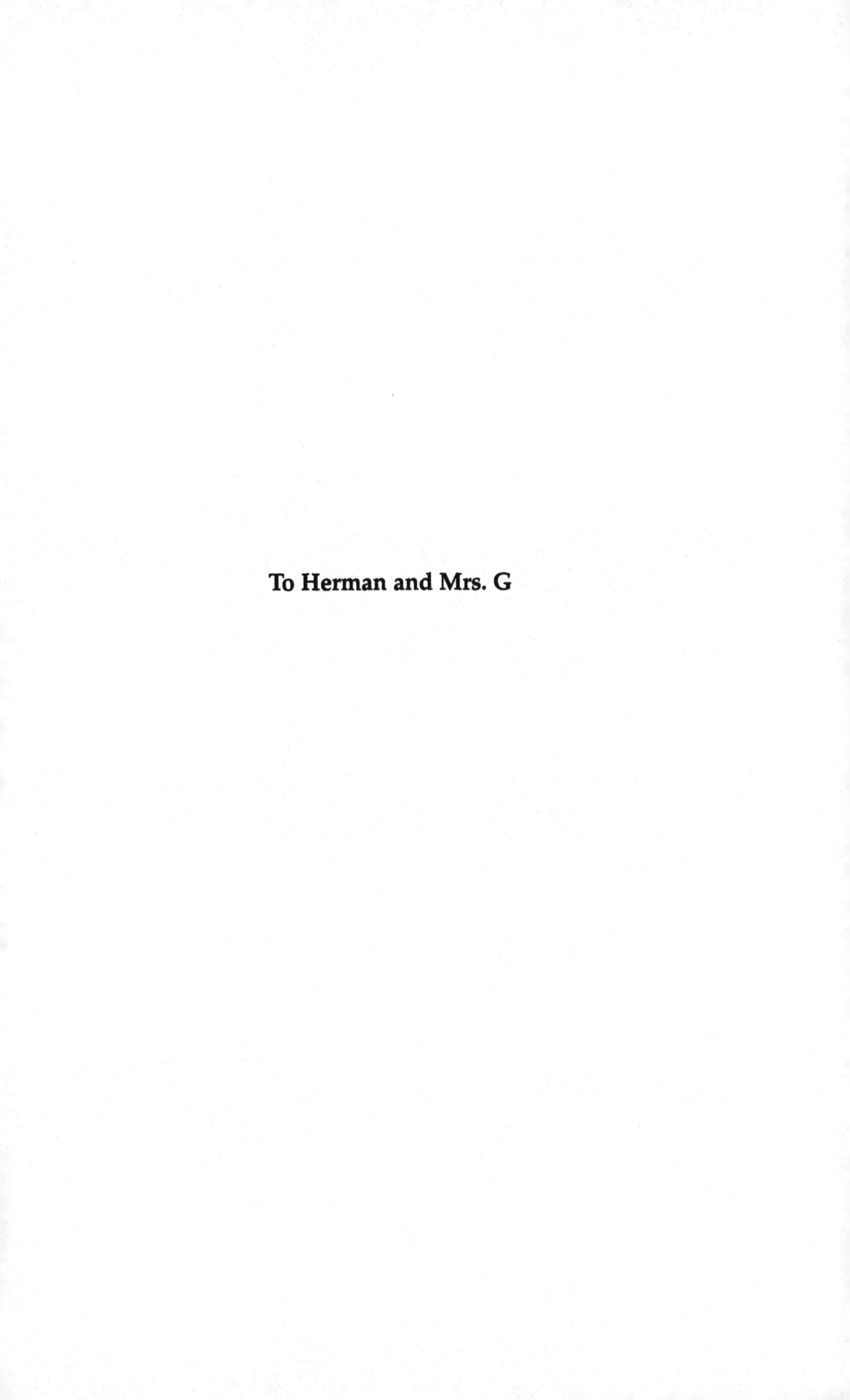

To Herman and Mrs. G

Frontiers in Physics
David Pines, Editor

Volumes of the Series published from 1961 to 1973 are not officially numbered. The parenthetical numbers shown are designed to aid librarians and bibliographers to check the completeness of their holdings.
Titles published in this series prior to 1987 appear under either the W. A. Benjamin or the Benjamin/Cummings imprint; titles published since 1986 appear under the Westview Press imprint.

1.	N. Bloembergen	Nuclear Magnetic Relaxation: A Reprint Volume, 1961
2.	G. F. Chew	S-Matrix Theory of Strong Interactions: A Lecture Note and Reprint Volume, 1961
3.	R. P. Feynman	Quantum Electrodynamics: A Lecture Note and Reprint Volume
4.	R. P. Feynman	The Theory of Fundamental Processes: A Lecture Note Volume, 1961
5.	L. Van Hove, N. M. Hugenholtz L. P. Howland	Problem in Quantum Theory of Many-Particle Systems: A Lecture Note and Reprint Volume, 1961
6.	D. Pines	The Many-Body Problem: A Lecture Note and Reprint Volume, 1961
7.	H. Frauenfelder	The Mössbauer Effect: A Review—With a Collection of Reprints, 1962
8.	L. P. Kadanoff G. Baym	Quantum Statistical Mechanics: Green's Function Methods in Equilibrium and Nonequilibrium Problems, 1962
9.	G. E. Pake	Paramagnetic Resonance: An Introductory Monograph, 1962 [cr. (42)—2nd edition]G. E. Pake
10.	P. W. Anderson	Concepts in Solids: Lectures on the Theory of Solids, 1963
11.	S. C. Frautschi	Regge Poles and S-Matrix Theory, 1963
12.	R. Hofstadter	Electron Scattering and Nuclear and Nucleon Structure: A Collection of Reprints with an Introduction, 1963
13.	A. M. Lane	Nuclear Theory: Pairing Force Correlations to Collective Motion, 1964
14.	R. Omnès	Mandelstam Theory and Regge Poles: An Introduction M. Froissart for Experimentalists, 1963
15.	E. J. Squires	Complex Angular Momenta and Particle Physics: A Lecture Note and Reprint Volume, 1963
16.	H. L. Frisch J. L. Lebowitz	The Equilibrium Theory of Classical Fluids: A Lecture Note and Reprint Volume, 1964
17.	M. Gell-Mann Y. Ne'eman	The Eightfold Way (A Review—With a Collection of Reprints), 1964
18.	M. Jacob G. F. Chew	Strong-Interaction Physics: A Lecture Note Volume, 1964
19.	P. Nozières	Theory of Interacting Fermi Systems, 1964
20.	J. R. Schrieffer	Theory of Superconductivity, 1964 (revised 3rd printing, 1983)
21.	N. Bloembergen	Nonlinear Optics: A Lecture Note and Reprint Volume, 1965

Frontiers in Physics

22.	R. Brout	Phase Transitions, 1965
23.	I. M. Khalatnikov	An Introduction to the Theory of Superfluidity, 1965
24.	P. G. deGennes	Superconductivity of Metals and Alloys, 1966
25.	W. A. Harrison	Pseudopotentials in the Theory of Metals, 1966
26.	V. Barger D. Cline	Phenomenological Theories of High Energy Scattering: An Experimental Evaluation, 1967
27.	P. Choquàrd	The Anharmonic Crystal, 1967
28.	T. Loucks	Augmented Plane Wave Method: A Guide to Performing.Electronic Structure Calculations—A Lecture Note and Reprint Volume, 1967
29.	Y. Ne'eman	Algebraic Theory of Particle Physics: Hadron Dynamics In Terms of Unitary Spin Current, 1967
30.	S. L. Adler R. F. Dashen	Current Algebras and Applications to Particle Physics, 1968
31.	A. B. Migdal	Nuclear Theory: The Quasiparticle Method, 1968
32.	J. J. J. Kokkede	The Quark Model, 1969
33.	A. B. Migdal	Approximation Methods in Quantum Mechanics, 1969
34.	R. Z. Sagdeev	Nonlinear Plasma Theory, 1969
35.	J. Schwinger	Quantum Kinematics and Dynamics, 1970
36.	R. P. Feynman	Statistical Mechanics: A Set of Lectures, 1972
37.	R. P. Feynman	Photon-Hadron Interactions, 1972
38.	E. R. Caianiello	Combinatorics and Renormalization in Quantum Field Theory, 1973
39.	G. B. Field H. Arp J. N. Bahcall	The Redshift Controversy, 1973
40.	D. Horn F. Zachariasen	Hadron Physics at Very High Energies, 1973
41.	S. Ichimaru	Basic Principles of Plasma Physics: A Statistical Approach, 1973 (2nd printing, with revisions, 1980)
42.	G. E. Pake T. L. Estle	The Physical Principles of Electron Paramagnetic Resonance, 2nd Edition, completely revised, enlarged, and reset, 1973 [cf. (9)—1st edition

Volumes published from 1974 onward are being numbered as an integral part of the bibliography.

43.	C. Davidson	Theory of Nonneutral Plasmas, 1974
44.	S. Doniach E. H. Sondheimer	Green's Functions for Solid State Physicists, 1974
45.	P. H. Frampton	Dual Resonance Models, 1974
46.	S. K. Ma	Modern Theory of Critical Phenomena, 1976
47.	D. Forster	Hydrodynamic Fluctuation, Broken Symmetry, and Correlation Functions, 1975
48.	A. B. Migdal	Qualitative Methods in Quantum Theory, 1977
49.	S. W. Lovesey	Condensed Matter Physics: Dynamic Correlations, 1980
50.	L. D. Faddeev A. A. Slavnov	Gauge Fields: Introduction to Quantum Theory, 1980
51.	P. Ramond	Field Theory: A Modern Primer, 1981 [cf. 74—2nd ed.]
52.	R. A. Broglia A. Winther	Heavy Ion Reactions: Lecture Notes Vol. I, Elastic and Inelastic Reactions, 1981
53.	R. A. Broglia A. Winther	Heavy Ion Reactions: Lecture Notes Vol. II, 1990

Frontiers in Physics

54.	H. Georgi	Lie Algebras in Particle Physics: From Isospin to Unified Theories, 1982
55.	P. W. Anderson	Basic Notions of Condensed Matter Physics, 1983
56.	C. Quigg	Gauge Theories of the Strong, Weak, and Electromagnetic Interactions, 1983
57.	S. I. Pekar	Crystal Optics and Additional Light Waves, 1983
58.	S. J. Gates M. T. Grisaru M. Rocek W. Siegel	Superspace or One Thousand and One Lessons in Supersymmetry, 1983
59.	R. N. Cahn	Semi-Simple Lie Algebras and Their Representations, 1984
60.	G. G. Ross	Grand Unified Theories, 1984
61.	S. W. Lovesey	Condensed Matter Physics: Dynamic Correlations, 2nd Edition, 1986.
62.	P. H. Frampton	Gauge Field Theories, 1986
63.	J. I. Katz	High Energy Astrophysics, 1987
64.	T. J. Ferbel	Experimental Techniques in High Energy Physics, 1987
65.	T. Appelquist A. Chodos P. G. O. Freund	Modern Kaluza-Klein Theories, 1987
66.	G. Parisi	Statistical Field Theory, 1988
67.	R. C. Richardson E. N. Smith	Techniques in Low-Temperature Condensed Matter Physics, 1988
68.	J. W. Negele H. Orland	Quantum Many-Particle Systems, 1987
69.	E. W. Kolb M. S. Turner	The Early Universe, 1990
70.	E. W. Kolb M. S. Turner	The Early Universe: Reprints, 1988
71.	V. Barger R. J. N. Phillips	Collider Physics, 1987
72.	T. Tajima	Computational Plasma Physics, 1989
73.	W. Kruer	The Physics of Laser Plasma Interactions, 1988
74.	P. Ramond	Field Theory: A Modern Primer, 2nd edition, 1989 [cf. 51—1st edition]
75.	B. F. Hatfield	Quantum Field Theory of Point Particles and Strings, 1989
76.	P. Sokolsky	Introduction to Ultrahigh Energy Cosmic Ray Physics, 1989
77.	R. Field	Applications of Perturbative QCD, 1989
80.	J. F. Gunion H. E. Haber G. Kane S. Dawson	The Higgs Hunter's Guide, 1990
81.	R. C. Davidson	Physics of Nonneutral Plasmas, 1990
82.	E. Fradkin	Field Theories of Condensed Matter Systems, 1991
83.	L. D. Faddeev A. A. Slavnov	Gauge Fields, 1990
84.	R. Broglia A. Winther	Heavy Ion Reactions, Parts I and II, 1990
85.	N. Goldenfeld	Lectures on Phase Transitions and the Renormalization Group, 1992
86.	R. D. Hazeltine J. D. Meiss	Plasma Confinement, 1992
87.	S. Ichimaru	Statistical Plasma Physics, Volume I: Basic Principles,

Frontiers in Physics

		1992
88.	S. Ichimaru	Statistical Plasma Physics, Volume II: Condensed Plasmas, 1994
89.	G. Grüner	Density Waves in Solids, 1994
90.	S. Safran	Statistical Thermodynamics of Surfaces, Interfaces, and Membranes, 1994
91.	B. d'Espagnat	Veiled Reality: An Analysis of Present Day Quantum Mechanical Concepts, 1994
92.	J. Bahcall R. Davis, Jr. P. Parker A. Smirnov R. Ulrich	Solar Neutrinos: The First Thirty Years
93.	R. Feynman F. Morinigo W. Wagner	Feynman Lectures on Gravitation
94.	M. Peskin D. Schroeder	An Introduction to Quantum Field Theory
95.	R. Feynman	Feynman Lectures on Computation
96.	M. Brack R. Bhaduri	Semiclassical Physics
97.	D. Cline	Weak Neutral Currents
98.	T. Tajima K. Shibata	Plasma Astrophysics
99.	J. Rammer	Quantum Transport Theory
100.	R. Hazeltine F. Waelbroeck	The Frameworkof Plasma Physics
101.	P. Ramond	Journeys Beyond the Standard Moel
102.	Nutku Saclioglu Turgut	Conformal Field Theory: New Non-Perturbative Methods in String and Feild Theory
103.	P. Philips	Advanced Solid State Physics

Preface to the Revised Edition

Lie Algebras in Particle Physics has been a very successful book. I have long resisted the temptation to produce a revised edition. I do so finally, because I find that there is so much new material that should be included, and so many things that I would like to say slightly differently. On the other hand, one of the good things about the first edition was that it did not do too much. The material could be dealt with in a one semester course by students with good preparation in quantum mechanics. In an attempt to preserve this advantage while including new material, I have flagged some sections that can be left out in a first reading. The titles of these sections begin with an asterisk, as do the problems that refer to them.

I may be prejudiced, but I think that this material is wonderful fun to teach, and to learn. I use this as a text for what is formally a graduate class, but it is taken successfully by many advanced undergrads at Harvard. The important prerequisite is a good background in quantum mechanics and linear algebra.

It has been over five years since I first began to revise this material and typeset it in LaTeX. Between then and now, many many students have used the evolving manuscript as a text. I am grateful to many of them for suggestions of many kinds, from typos to grammar to pedagogy.

As always, I am enormously grateful to my family for putting up with me for all this time. I am also grateful for their help with my inspirational epilogue.

Howard Georgi
Cambridge, MA
May, 1999

Contents

Why Group Theory?

Group theory is the study of symmetry. It is an incredible labor saving device. It allows us to say interesting, sometimes very detailed things about physical systems even when we don't understand exactly what the systems are! When I was a teenager, I read an essay by Sir Arthur Stanley Eddington on the Theory of Groups and a quotation from it has stuck with me for over 30 years:[1]

> We need a super-mathematics in which the operations are as unknown as the quantities they operate on, and a super-mathematician who does not know what he is doing when he performs these operations. Such a super-mathematics is the Theory of Groups.

In this book, I will try to convince you that Eddington had things a little bit wrong, as least as far as physics is concerned. A lot of what physicists use to extract information from symmetry is not the groups themselves, but group representations. You will see exactly what this means in more detail as you read on. What I hope you will take away from this book is enough about the theory of groups and Lie algebras and their representations to use group representations as labor-saving tools, particularly in the study of quantum mechanics.

The basic approach will be to alternate between mathematics and physics, and to approach each problem from several different angles. I hope that you will learn that by using several techniques at once, you can work problems more efficiently, and also understand each of the techniques more deeply.

---

[1]in **The World of Mathematics**, Ed. by James R. Newman, Simon & Schuster, New York, 1956.

 DOI: 10.1201/9780429499210-1

Chapter 1

Finite Groups

We will begin with an introduction to finite group theory. This is not intended to be a self-contained treatment of this enormous and beautiful subject. We will concentrate on a few simple facts that are useful in understanding the compact Lie algebras. We will introduce a lot of definitions, sometimes proving things, but often relying on the reader to prove them.

1.1 Groups and representations

A Group, G, is a set with a rule for assigning to every (ordered) pair of elements, a third element, satisfying:

(1.A.1) If $f, g \in G$ then $h = fg \in G$.

(1.A.2) For $f, g, h \in G$, $f(gh) = (fg)h$.

(1.A.3) There is an identity element, e, such that for all $f \in G$, $ef = fe = f$.

(1.A.4) Every element $f \in G$ has an inverse, f^{-1}, such that $ff^{-1} = f^{-1}f = e$.

Thus a group is a multiplication table specifying $g_1 g_2 \ \forall g_1, g_2 \in G$. If the group elements are discrete, we can write the multiplication table in the form

$\backslash$	e	g_1	g_2	$\cdots$
e	e	g_1	g_2	$\cdots$
g_1	g_1	$g_1 g_1$	$g_1 g_2$	$\cdots$
g_2	g_2	$g_2 g_1$	$g_2 g_2$	$\cdots$
$\vdots$	$\vdots$	$\vdots$	$\vdots$	$\ddots$

(1.1)

DOI: 10.1201/9780429499210-2

A Representation of G is a mapping, D of the elements of G onto a set of linear operators with the following properties:

1.B.1 $D(e) = 1$, where 1 is the identity operator in the space on which the linear operators act.

1.B.2 $D(g_1)D(g_2) = D(g_1g_2)$, in other words the group multiplication law is mapped onto the natural multiplication in the linear space on which the linear operators act.

1.2 Example - Z_3

A group is **finite** if it has a finite number of elements. Otherwise it is **infinite**. The number of elements in a finite group G is called the **order** of G. Here is a finite group of order 3.

\	e	a	b
e	e	a	b
a	a	b	e
b	b	e	a

(1.2)

This is Z_3, the cyclic group of order 3. Notice that every row and column of the multiplication table contains each element of the group exactly once. This must be the case because the inverse exists.

An Abelian group in one in which the multiplication law is commutative

$$g_1g_2 = g_2g_1 \,. \tag{1.3}$$

Evidently, Z_3 is Abelian.

The following is a representation of Z_3

$$D(e) = 1\,, \quad D(a) = e^{2\pi i/3}\,, \quad D(b) = e^{4\pi i/3} \tag{1.4}$$

The **dimension** of a representation is the dimension of the space on which it acts — the representation (1.4) is 1 dimensional.

1.3 The regular representation

Here's another representation of Z_3

$$D(e) = \begin{pmatrix} 1 & 0 & 0 \\ 0 & 1 & 0 \\ 0 & 0 & 1 \end{pmatrix}, \quad D(a) = \begin{pmatrix} 0 & 0 & 1 \\ 1 & 0 & 0 \\ 0 & 1 & 0 \end{pmatrix}$$
$$D(b) = \begin{pmatrix} 0 & 1 & 0 \\ 0 & 0 & 1 \\ 1 & 0 & 0 \end{pmatrix} \tag{1.5}$$

This representation was constructed directly from the multiplication table by the following trick. Take the group elements themselves to form an orthonormal basis for a vector space, $|e\rangle$, $|a\rangle$, and $|b\rangle$. Now define

$$D(g_1)|g_2\rangle = |g_1 g_2\rangle \tag{1.6}$$

The reader should show that this is a representation. It is called the **regular** representation. Evidently, the dimension of the regular representation is the order of the group. The matrices of (1.5) are then constructed as follows.

$$|e_1\rangle \equiv |e\rangle\,, \quad |e_2\rangle \equiv |a\rangle\,, \quad |e_3\rangle \equiv |b\rangle \tag{1.7}$$

$$[D(g)]_{ij} = \langle e_i|D(g)|e_j\rangle \tag{1.8}$$

The matrices are the **matrix elements** of the linear operators. (1.8) is a simple, but very general and very important way of going back and forth from operators to matrices. This works for any representation, not just the regular representation. We will use it constantly. The basic idea here is just the insertion of a complete set of intermediate states. The matrix corresponding to a product of operators is the matrix product of the matrices corresponding to the operators —

$$\begin{aligned} [D(g_1 g_2)]_{ij} &= [D(g_1)D(g_2)]_{ij} \\ &= \langle e_i|D(g_1)D(g_2)|e_j\rangle \\ &= \sum_k \langle e_i|D(g_1)|e_k\rangle\,\langle e_k|D(g_2)|e_j\rangle \\ &= \sum_k [D(g_1)]_{ik}[D(g_2)]_{kj} \end{aligned} \tag{1.9}$$

Note that the construction of the regular representation is completely general for any finite group. For any finite group, we can define a vector space in which the basis vectors are labeled by the group elements. Then (1.6) defines the regular representation. We will see the regular representation of various groups in this chapter.

1.4 Irreducible representations

What makes the idea of group representations so powerful is the fact that they live in linear spaces. And the wonderful thing about linear spaces is we are free to choose to represent the states in a more convenient way by making a linear transformation. As long as the transformation is invertible, the new states are just as good as the old. Such a transformation on the states produces a **similarity transformation** on the linear operators, so that we can always make a new representation of the form

$$D(g) \to D'(g) = S^{-1}D(g)S \tag{1.10}$$

Because of the form of the similarity transformation, the new set of operators has the same multiplication rules as the old one, so D' is a representation if D is. D' and D are said to be **equivalent** representations because they differ just by a trivial choice of basis.

Unitary operators (O such that $O^\dagger = O^{-1}$) are particularly important. A representation is unitary if all the $D(g)$s are unitary. Both the representations we have discussed so far are unitary. It will turn out that all representations of finite groups are equivalent to unitary representations (we'll prove this later - it is easy and neat).

A representation is **reducible** if it has an **invariant subspace**, which means that the action of any $D(g)$ on any vector in the subspace is still in the subspace. In terms of a projection operator P onto the subspace this condition can be written as

$$PD(g)P = D(g)P \ \forall g \in G \tag{1.11}$$

For example, the regular representation of Z_3 (1.5) has an invariant subspace projected on by

$$P = \frac{1}{3}\begin{pmatrix} 1 & 1 & 1 \\ 1 & 1 & 1 \\ 1 & 1 & 1 \end{pmatrix} \tag{1.12}$$

because $D(g)P = P \ \forall g$. The restriction of the representation to the invariant subspace is itself a representation. In this case, it is the **trivial representation for which** $D(g) = 1$ (the trivial representation, $D(g) = 1$, is always a representation — every group has one).

A representation is **irreducible** if it is not reducible.

A representation is **completely reducible** if it is equivalent to a representation whose matrix elements have the following form:

$$\begin{pmatrix} D_1(g) & 0 & \cdots \\ 0 & D_2(g) & \cdots \\ \vdots & \vdots & \ddots \end{pmatrix} \tag{1.13}$$

where $D_j(g)$ is irreducible $\forall j$. This is called **block diagonal form**.

A representation in block diagonal form is said to be the **direct sum** of the subrepresentations, $D_j(g)$,

$$D_1 \oplus D_2 \oplus \cdots \tag{1.14}$$

In transforming a representation to block diagonal form, we are decomposing the original representation into a direct sum of its irreducible components. Thus another way of defining complete reducibility is to say that **a completely reducible representation can be decomposed into a direct sum of irreducible representations.** This is an important idea. We will use it often.

We will show later that any representation of a finite group is completely reducible. For example, for (1.5), take

$$S = \frac{1}{3}\begin{pmatrix} 1 & 1 & 1 \\ 1 & \omega^2 & \omega \\ 1 & \omega & \omega^2 \end{pmatrix} \tag{1.15}$$

where

$$\omega = e^{2\pi i/3} \tag{1.16}$$

then

$$D'(e) = \begin{pmatrix} 1 & 0 & 0 \\ 0 & 1 & 0 \\ 0 & 0 & 1 \end{pmatrix} \qquad D'(a) = \begin{pmatrix} 1 & 0 & 0 \\ 0 & \omega & 0 \\ 0 & 0 & \omega^2 \end{pmatrix}$$
$$D'(b) = \begin{pmatrix} 1 & 0 & 0 \\ 0 & \omega^2 & 0 \\ 0 & 0 & \omega \end{pmatrix} \tag{1.17}$$

1.5 Transformation groups

There is a natural multiplication law for transformations of a physical system. If g_1 and g_2 are two transformations, $g_1 g_2$ means first do g_2 and then do g_1.

Note that it is purely convention whether we define our composition law to be right to left, as we have done, or left to right. Either gives a perfectly consistent definition of a transformation group.

If this transformation is a symmetry of a quantum mechanical system, then the transformation takes the Hilbert space into an equivalent one. Then for each group element g, there is a unitary operator $D(g)$ that maps the Hilbert space into an equivalent one. These unitary operators form a representation of the transformation group because the transformed quantum states represent the transformed physical system. Thus for any set of symmetries, there is a representation of the symmetry group on the Hilbert space — we say that the Hilbert space transforms according to some representation of the group. Furthermore, because the transformed states have the same energy as the originals, $D(g)$ commutes with the Hamiltonian, $[D(g), H] = 0$. As we will see in more detail later, this means that we can always choose the energy eigenstates to transform like irreducible representations of the group. It is useful to think about this in a simple example.

1.6 Application: parity in quantum mechanics

Parity is the operation of reflection in a mirror. Reflecting twice gets you back to where you started. If p is a group element representing the parity reflection, this means that $p^2 = e$. Thus this is a transformation that together with the identity transformation (that is, doing nothing) forms a very simple group, with the following multiplication law:

\	e	p
e	e	p
p	p	e

(1.18)

This group is called Z_2. For this group there are only two irreducible representations, the trivial one in which $D(p) = 1$ and one in which $D(e) = 1$ and $D(p) = -1$. Any representation is completely reducible. In particular, that means that the Hilbert space of any parity invariant system can be decomposed into states that behave like irreducible representations, that is on which $D(p)$ is either 1 or -1. Furthermore, because $D(p)$ commutes with the Hamiltonian, $D(p)$ and H can be simultaneously diagonalized. That is we can assign each energy eigenstate a definite value of $D(p)$. The energy eigenstates on which $D(p) = 1$ are said to transform according to the trivial representation. Those on which $D(p) = -1$ transform according to the other representation. This should be familiar from nonrelativistic quantum mechanics in one dimension. There you know that a particle in a potential that is symmetric about $x = 0$ has energy eigenfunctions that are either symmetric under $x \to -x$ (corresponding to the trivial representation), or antisymmetric (the representation with $D(p) = -1$).

1.7 Example: S_3

The **permutation group** (or **symmetric group**) on 3 objects, called S_3 where

$$\begin{array}{ccc} a_1 = (1,2,3) & & a_2 = (3,2,1) \\ a_3 = (1,2) & a_4 = (2,3) & a_5 = (3,1) \end{array} \tag{1.19}$$

The notation means that a_1 is a cyclic permutation of the things in positions 1, 2 and 3; a_2 is the inverse, anticyclic permutation; a_3 interchanges the objects in positions 1 and 2; and so on. The multiplication law is then determined by the transformation rule that $g_1 g_2$ means first do g_2 and then do g_1. It is

\	e	a_1	a_2	a_3	a_4	a_5
e	e	a_1	a_2	a_3	a_4	a_5
a_1	a_1	a_2	e	a_5	a_3	a_4
a_2	a_2	e	a_1	a_4	a_5	a_3
a_3	a_3	a_4	a_5	e	a_1	a_2
a_4	a_4	a_5	a_3	a_2	e	a_1
a_5	a_5	a_3	a_4	a_1	a_2	e

(1.20)

We could equally well define it to mean first do g_1 and then do g_2. These two rules define different multiplication tables, but they are related to one another by simple relabeling of the elements, so they give the same group. There is another possibility of confusion here between whether we are permuting the objects in positions 1, 2 and 3, or simply treating 1, 2 and 3 as names for the three objects. Again these two give different multiplication tables, but only up to trivial renamings. The first is a little more physical, so we will use that. The permutation group is an another example of a **transformation group** on a physical system.

S_3 is **non-Abelian** because the group multiplication law is not commutative. We will see that it is the lack of commutativity that makes group theory so interesting.

Here is a unitary irreducible representation of S_3

$$\begin{gathered} D(e)=\begin{pmatrix} 1 & 0 \\ 0 & 1 \end{pmatrix}, \quad D(a_1)=\begin{pmatrix} -\frac{1}{2} & -\frac{\sqrt{3}}{2} \\ \frac{\sqrt{3}}{2} & -\frac{1}{2} \end{pmatrix}, \\ D(a_2)=\begin{pmatrix} -\frac{1}{2} & \frac{\sqrt{3}}{2} \\ -\frac{\sqrt{3}}{2} & -\frac{1}{2} \end{pmatrix}, \quad D(a_3)=\begin{pmatrix} -1 & 0 \\ 0 & 1 \end{pmatrix}, \\ D(a_4)=\begin{pmatrix} \frac{1}{2} & \frac{\sqrt{3}}{2} \\ \frac{\sqrt{3}}{2} & -\frac{1}{2} \end{pmatrix}, \quad D(a_5)=\begin{pmatrix} \frac{1}{2} & -\frac{\sqrt{3}}{2} \\ -\frac{\sqrt{3}}{2} & -\frac{1}{2} \end{pmatrix} \end{gathered} \tag{1.21}$$

The interesting thing is that the irreducible unitary representation is more than 1 dimensional. It is necessary that at least some of the representations of a non-Abelian group must be matrices rather than numbers. Only matrices can reproduce the non-Abelian multiplication law. Not all the operators in the representation can be diagonalized simultaneously. It is this that is responsible for a lot of the power of the theory of group representations.

1.8 Example: addition of integers

The integers form an infinite group under addition.

$$xy = x + y \tag{1.22}$$

This is rather unimaginatively called the additive group of the integers. Since this group is infinite, we can't write down the multiplication table, but the rule above specifies it completely.

Here is a representation:

$$D(x)=\begin{pmatrix} 1 & x \\ 0 & 1 \end{pmatrix} \tag{1.23}$$

This representation is reducible, but you can show that it is not completely reducible and it is not equivalent to a unitary representation. It is reducible because

$$D(x)P = P \tag{1.24}$$

where

$$P=\begin{pmatrix} 1 & 0 \\ 0 & 0 \end{pmatrix} \tag{1.25}$$

However,

$$D(x)(I-P) \neq (I-P) \tag{1.26}$$

so it is not completely reducible.

The additive group of the integers is infinite, because, obviously, there are an infinite number of integers. For a finite group, all reducible representations are completely reducible, because all representations are equivalent to unitary representations.

1.9 Useful theorems

Theorem 1.1 *Every representation of a finite group is equivalent to a unitary representation.*

Proof: Suppose $D(g)$ is a representation of a finite group G. Construct the operator

$$S = \sum_{g\in G} D(g)^\dagger D(g) \tag{1.27}$$

S is hermitian and positive semidefinite. Thus it can be diagonalized and its eigenvalues are non-negative:

$$S = U^{-1} d U \tag{1.28}$$

where d is diagonal

$$d = \begin{pmatrix} d_1 & 0 & \cdots \\ 0 & d_2 & \cdots \\ \vdots & \vdots & \ddots \end{pmatrix} \tag{1.29}$$

where $d_j \geq 0\ \forall j$. Because of the group property, all of the d_js are actually positive. Proof — suppose one of the d_js is zero. Then there is a vector λ such that $S\lambda = 0$. But then

$$\lambda^\dagger S \lambda = 0 = \sum_{g\in G} ||D(g)\lambda||^2 . \tag{1.30}$$

Thus $D(g)\lambda$ must vanish for all g, which is impossible, since $D(e) = 1$. Therefore, we can construct a square-root of S that is hermitian and invertible

$$X = S^{1/2} \equiv U^{-1} \begin{pmatrix} \sqrt{d_1} & 0 & \cdots \\ 0 & \sqrt{d_2} & \cdots \\ \vdots & \vdots & \ddots \end{pmatrix} U \tag{1.31}$$

X is invertible, because none of the d_js are zero. We can now define

$$D'(g) = X\, D(g)\, X^{-1} \tag{1.32}$$

Now, somewhat amazingly, this representation is unitary!

$$D'(g)^\dagger D'(g) = X^{-1} D(g)^\dagger S D(g) X^{-1} \tag{1.33}$$

but

$$\begin{aligned} D(g)^\dagger S D(g) &= D(g)^\dagger \left(\sum_{h\in G} D(h)^\dagger D(h) \right) D(g) \\ &= \sum_{h\in G} D(hg)^\dagger D(hg) \\ &= \sum_{h\in G} D(h)^\dagger D(h) = S = X^2 \end{aligned} \tag{1.34}$$

where the last line follows because hg runs over all elements of G when h does. QED.

We saw in the representation (1.23) of the additive group of the integers an example of a reducible but not completely reducible representation. The way it works is that there is a P that projects onto an invariant subspace, but $(1-P)$ does not. This is impossible for a unitary representation, and thus representations of finite groups are always completely reducible. Let's prove it.

Theorem 1.2 *Every representation of a finite group is completely reducible.*

Proof: By the previous theorem, it is sufficient to consider unitary representations. If the representation is irreducible, we are finished because it is already in block diagonal form. If it is reducible, then $\exists$ a projector P such that $PD(g)P = D(g)P \; \forall g \in G$. This is the condition that P be an invariant subspace. Taking the adjoint gives $PD(g)^\dagger P = PD(g)^\dagger \; \forall g \in G$. But because $D(g)$ is unitary, $D(g)^\dagger = D(g)^{-1} = D(g^{-1})$ and thus since g^{-1} runs over all G when g does, $PD(g)P = PD(g) \; \forall g \in G$. But this implies that $(1-P)D(g)(1-P) = D(g)(1-P) \; \forall g \in G$ and thus $1-P$ projects onto an invariant subspace. Thus we can keep going by induction and eventually completely reduce the representation.

1.10 Subgroups

A group H whose elements are all elements of a group G is called a **subgroup** of G. The identity, and the group G are **trivial subgroups** of G. But many groups have **nontrivial subgroups** (which just means some subgroup other than G or e) as well. For example, the permutation group, S_3, has a Z_3 subgroup formed by the elements $\{e, a_1, a_2\}$.

We can use a subgroup to divide up the elements of the group into subsets called cosets. A **right-coset of the subgroup H in the group G** is a set of elements formed by the action of the elements of H on the left on a given element of G, that is all elements of the form Hg for some fixed g. You can define left-cosets as well.

For example, $\{a_3, a_4, a_5\}$ is a coset of Z_3 in S_3 in (1.20) above. The number of elements in each coset is the order of H. Every element of G must belong to one and only one coset. Thus for finite groups, the order of a subgroup H must be a factor of order of G. It is also sometimes useful to think about the **coset-space, G/H** defined by regarding each coset as a single element of the space.

A subgroup H of G is called an **invariant** or **normal subgroup** if for every $g \in G$

$$gH = Hg \tag{1.35}$$

which is (we hope) an obvious short-hand for the following: for every $g \in G$ and $h_1 \in H$ there exists an $h_2 \in H$ such that $h_1 g = g h_2$, or $g h_2 g^{-1} = h_1$. The trivial subgroups e and G are invariant for any group. It is less obvious but also true of the subgroup Z_3 of S_3 in (1.20) (you can see this by direct computation or notice that the elements of Z_3 are those permutations that involve an even number of interchanges). However, the set $\{e, a_4\}$ is a subgroup of G which is not invariant. $a_5\{e, a_4\} = \{a_5, a_1\}$ while $\{e, a_4\}a_5 = \{a_5, a_2\}$.

If H is invariant, then we can regard the coset space as a group. The multiplication law in G gives the natural multiplication law on the cosets, Hg:

$$(Hg_1)(Hg_2) = (Hg_1 H g_1^{-1})(g_1 g_2) \tag{1.36}$$

But if H is invariant $Hg_1 H g_1^{-1} = H$, so the product of elements in two cosets is in the coset represented by the product of the elements. In this case, the coset space, G/H, is called the **factor group of G by H**.

What is the factor group S_3/Z_3? The answer is Z_2.

The **center** of a group G is the set of all elements of G that commute with all elements of G. The center is always an Abelian, invariant subgroup of G. However, it may be trivial, consisting only of the identity, or of the whole group.

There is one other concept, related to the idea of an invariant subgroup, that will be useful. Notice that the condition for a subgroup to be invariant can be rewritten as

$$gHg^{-1} = H \ \forall g \in G \tag{1.37}$$

This suggests that we consider sets rather than subgroups satisfying same condition.

$$g^{-1}S\,g = S\;\forall g \in G \tag{1.38}$$

Such sets are called conjugacy classes. We will see later that there is a one-to-one correspondence between them and irreducible representations. A subgroup that is a union of conjugacy classes is invariant.

Example —

The conjugacy classes of S_3 are $\{e\}$, $\{a_1, a_2\}$ and $\{a_3, a_4, a_5\}$.

The mapping

$$G \to gGg^{-1} \tag{1.39}$$

for a fixed g is also interesting. It is called an **inner automorphism**. An **isomorphism** is a one-to-one mapping of one group onto another that preserves the multiplication law. An **automorphism** is a one-to-one mapping of a group onto itself that preserves the multiplication law. It is easy to see that (1.39) is an automorphism. Because $g^{-1}g_1g\,g^{-1}g_2g = g^{-1}g_1g_2g$, it preserves the multiplication law. Since $g^{-1}g_1g = g^{-1}g_2g \Rightarrow g_1 = g_2$, it is one to one. An automorphism of the form (1.39) where g is a group element is called an **inner automorphism**). An **outer automorphism** is one that cannot be written as $g^{-1}Gg$ for any group element g.

1.11 Schur's lemma

Theorem 1.3 *If $D_1(g)A = AD_2(g)$ $\forall g \in G$ where D_1 and D_2 are inequivalent, irreducible representations, then $A = 0$.*

Proof: This is part of **Schur's lemma**. First suppose that there is a vector $|\mu\rangle$ such that $A|\mu\rangle = 0$. Then there is a non-zero projector, P, onto the subspace that annihilates A on the right. But this subspace is invariant with respect to the representation D_2, because

$$AD_2(g)P = D_1(g)AP = 0\;\forall g \in G \tag{1.40}$$

But because D_2 is irreducible, P must project onto the whole space, and A must vanish. If A annihilates one state, it must annihilate them all. A similar argument shows that A vanishes if there is a $\langle\nu|$ which annihilates A. If no vector annihilates A on either side, then it must be an invertible square matrix. It must be square, because, for example, if the number of rows were larger than the number of columns, then the rows could not be a complete set of states, and there would be a vector that annihilates A on the right. A square matrix is invertible unless its determinant vanishes. But if the determinant vanishes, then the set of homogeneous linear equations

$$A|\mu\rangle = 0 \tag{1.41}$$

has a nontrivial solution, which again means that there is a vector that annihilates A. But if A is square and invertible, then

$$A^{-1}D_1(g)A = D_2(g) \ \forall g \in G \tag{1.42}$$

so D_1 and D_2 are equivalent, contrary to assumption. QED.

The more important half of Schur's lemma applies to the situation where D_1 and D_2 above are equivalent representations. In this case, we might as well take $D_1 = D_2 = D$, because we can do so by a simple change of basis. The other half of Schur's lemma is the following.

Theorem 1.4 *If $D(g)A = AD(g) \ \forall g \in G$ where D is a finite dimensional irreducible representation, then $A \propto I$.*

In words, if a matrix commutes with all the elements of a finite dimensional irreducible representation, it is proportional to the identity.

Proof: Note that here the restriction to a finite dimensional representation is important. We use the fact that any finite dimensional matrix has at least one eigenvalue, because the characteristic equation $\det(A - \lambda I) = 0$ has at least one root, and then we can solve the homogeneous linear equations for the components of the eigenvector $|\mu\rangle$. But then $D(g)(A - \lambda I) = (A - \lambda I)D(g) \ \forall g \in G$ and $(A - \lambda I)|\mu\rangle = 0$. Thus the same argument we used in the proof of the previous theorem implies $(A - \lambda I) = 0$. QED.

A consequence of Schur's lemma is that the form of the basis states of an irreducible representation are essentially unique. We can rewrite theorem 1.4 as the statement

$$A^{-1}D(g)A = D(g) \ \forall g \in G \Rightarrow A \propto I \tag{1.43}$$

for any irreducible representation D. This means once the form of D is fixed, there is no further freedom to make nontrivial similarity transformations on the states. The only unitary transformation you can make is to multiply all the states by the same phase factor.

In quantum mechanics, Schur's lemma has very strong consequences for the matrix elements of any operator, O, corresponding to an observable that is invariant under the symmetry transformations. This is because the matrix elements $\langle a, j, x|O|b, k, y\rangle$ behave like the A operator in (1.40). To see this, let's consider the complete reduction of the Hilbert space in more detail. The symmetry group gets mapped into a unitary representation

$$g \to D(g)\ \forall g \in G \tag{1.44}$$

where D is the (in general very reducible) unitary representation of G that acts on the entire Hilbert space of the quantum mechanical system. But if the representation is completely reducible, we know that we can choose a basis in which D has block diagonal form with each block corresponding to some unitary irreducible representation of G. We can write the orthonormal basis states as

$$|a, j, x\rangle \tag{1.45}$$

satisfying

$$\langle a, j, x \mid b, k, y\rangle = \delta_{ab}\, \delta_{jk}\, \delta_{xy} \tag{1.46}$$

where a labels the irreducible representation, $j = 1$ to n_a labels the state within the representation, and x represents whatever other physical parameters there are.

Implicit in this treatment is an important assumption that we will almost always make without talking about it. We assume that have chosen a basis in which all occurences of each irreducible representation a, is described by the same set of unitary representation matrices, $D_a(g)$. In other words, for each irreducible representation, we choose a canonical form, and use it exclusively

In this special basis, the matrix elements of $D(g)$ are

$$\langle a, j, x|\, D(g)\, |b, k, y\rangle = \delta_{ab}\, \delta_{xy}\, [D_a(g)]_{jk} \tag{1.47}$$

This is just a rewriting of (1.13) with explicit indices rather than as a matrix. We can now check that our treatment makes sense by writing the representation D in this basis by inserting a complete set of intermediate states on both sides:

$$I = \sum_{a,j,x} |a, j, x\rangle\langle a, j, x| \tag{1.48}$$

Then we can write

$$\begin{aligned} D(g) &= \sum_{a,j,x} |a, j, x\rangle\langle a, j, x|\, D(g) \sum_{b,k,y} |b, k, y\rangle\langle b, k, y| \\ &= \sum_{\substack{a,j,x \\ b,k,y}} |a, j, x\rangle\, \delta_{ab}\, \delta_{xy}\, [D_a(g)]_{jk}\, \langle b, k, y| \\ &= \sum_{a,j,k,x} |a, j, x\rangle\ [D_a(g)]_{jk}\, \langle a, k, x| \end{aligned} \tag{1.49}$$

This is another way of writing a representation that is in block diagonal form. Note that if a particular irreducible representation appears only once in D, then we don't actually need the x variable to label its states. But typically, in the full quantum mechanical Hilbert space, each irreducible representation will appear many times, and then the physical x variable distinguish states that have the same symmetry properties, but different physics. The important fact, however, is that the dependence on the physics in (1.47) is rather trivial — only that the states are orthonormal — all the group theory is independent of x and y.

Under the symmetry transformation, since the states transform like

$$|\mu\rangle \to D(g)\,|\mu\rangle \qquad \langle\mu| \to \langle\mu|\,D(g)^\dagger \tag{1.50}$$

operators transform like

$$O \to D(g)\,O\,D(g)^\dagger \tag{1.51}$$

in order that all matrix element remain unchanged. Thus an invariant observable satisfies

$$O \to D(g)\,O\,D(g)^\dagger = O \tag{1.52}$$

which implies that O commutes with $D(g)$

$$[O, D(g)] = 0\;\forall g \in G\,. \tag{1.53}$$

Then we can constrain the matrix element

$$\langle a,j,x|O|b,k,y\rangle \tag{1.54}$$

by arguing as follows:

$$\begin{aligned} 0 &= \langle a,j,x|[O,D(g)]|b,k,y\rangle \\ &= \sum_{k'}\langle a,j,x|O|b,k',y\rangle\langle b,k',y|D(g)|b,k,y\rangle \\ &-\sum_{j'}\langle a,j,x|D(g)|a,j',x\rangle\langle a,j',x|O|b,k,y\rangle \end{aligned} \tag{1.55}$$

Now we use (1.47), which exhibits the fact that the matrix elements of $D(g)$ have only trivial dependence on the physics, to write

$$\begin{aligned} 0 &= \langle a,j,x|[O,D(g)]|b,k,y\rangle \\ &= \sum_{k'}\langle a,j,x|O|b,k',y\rangle[D_b(g)]_{k'k} \\ &-\sum_{j'}[D_a(g)]_{jj'}\langle a,j',x|O|b,k,y\rangle \end{aligned} \tag{1.56}$$

Thus the matrix element (1.54) satisfies the hypotheses of Schur's lemma. It must vanish if $a \neq b$. It must be proportional to the identity (in indices, that is δ_{jk}) for $a = b$. However, the symmetry doesn't tell us anything about the dependence on the physical parameters, x and y. Thus we can write

$$\langle a, j, x|O|b, k, y\rangle = f_a(x, y)\, \delta_{ab}\, \delta_{jk} \tag{1.57}$$

The importance of this is that the physics is all contained in the function $f_a(x, y)$ — all the dependence on the group theory labels is completely fixed by the symmetry. As we will see, this can be very powerful. This is a simple example of the Wigner-Eckart theorem, which we will discuss in much more generality later.

1.12 * Orthogonality relations

The same kind of summation over the group elements that we used in the proof of theorem 1.1, can be used together with Schur's lemma to show some more remarkable properties of the irreducible representations. Consider the following linear operator (written as a "dyadic")

$$A^{ab}_{j\ell} \equiv \sum_{g\in G} D_a(g^{-1})|a, j\rangle\langle b, \ell|D_b(g) \tag{1.58}$$

where D_a and D_b are finite dimensional irreducible representations of G. Now look at

$$D_a(g_1)A^{ab}_{j\ell} = \sum_{g\in G} D_a(g_1)D_a(g^{-1})|a, j\rangle\langle b, \ell|D_b(g) \tag{1.59}$$

$$= \sum_{g\in G} D_a(g_1 g^{-1})|a, j\rangle\langle b, \ell|D_b(g) \tag{1.60}$$

$$= \sum_{g\in G} D_a((g g_1^{-1})^{-1})|a, j\rangle\langle b, \ell|D_b(g) \tag{1.61}$$

Now let $g' = g g_1^{-1}$

$$= \sum_{g'\in G} D_a(g'^{-1})|a, j\rangle\langle b, \ell|D_b(g' g_1) \tag{1.62}$$

$$= \sum_{g'\in G} D_a(g'^{-1})|a, j\rangle\langle b, \ell|D_b(g')D_b(g_1) = A^{ab}_{j\ell}D_b(g_1) \tag{1.63}$$

Now Schur's lemma (theorems 1.3 and 1.4) implies $A^{ab}_{j\ell} = 0$ if D_a and D_b are different, and further that if they are the same (remember that we have chosen a canonical form for each representation so equivalent representations are written in exactly the same way) $A^{ab}_{j\ell} \propto I$. Thus we can write

$$A^{ab}_{j\ell} \equiv \sum_{g \in G} D_a(g^{-1})|a, j\rangle\langle b, \ell| D_b(g) = \delta_{ab}\, \lambda^a_{j\ell}\, I \tag{1.64}$$

To compute $\lambda^a_{j\ell}$, compute the trace of $A^{ab}_{j\ell}$ (in the Hilbert space, not the indices) in two different ways. We can write

$$\text{Tr}\, A^{ab}_{j\ell} = \delta_{ab}\ \text{Tr}\left(\lambda^a_{j\ell} I\right) = \delta_{ab}\, \lambda^a_{j\ell}\ \text{Tr}\, I = \delta_{ab}\, \lambda^a_{j\ell}\, n_a \tag{1.65}$$

where n_a is the dimension of D_a. But we can also use the cyclic property of the trace and the fact that $A^{ab}_{jk} \propto \delta_{ab}$ to write

$$\text{Tr}\, A^{ab}_{j\ell} = \delta_{ab} \sum_{g \in G} \langle a, \ell| D_a(g) D_a(g^{-1})|a, j\rangle = N\, \delta_{ab}\, \delta_{j\ell} \tag{1.66}$$

where N is the order of the group. Thus $\lambda^a_{j\ell} = N\delta_{j\ell}/n_a$ and we have shown

$$\sum_{g \in G} D_a(g^{-1})|a, j\rangle\langle b, \ell| D_b(g) = \frac{N}{n_a}\, \delta_{ab}\, \delta_{j\ell} I \tag{1.67}$$

Taking the matrix elements of these relations yields orthogonality relations for the matrix elements of irreducible representations.

$$\sum_{g \in G} \frac{n_a}{N} [D_a(g^{-1})]_{kj} [D_b(g)]_{\ell m} = \delta_{ab} \delta_{j\ell} \delta_{km} \tag{1.68}$$

For unitary irreducible representations, we can write

$$\sum_{g \in G} \frac{n_a}{N} [D_a(g)]^*_{jk} [D_b(g)]_{\ell m} = \delta_{ab} \delta_{j\ell} \delta_{km} \tag{1.69}$$

so that with proper normalization, the matrix elements of the inequivalent unitary irreducible representations

$$\sqrt{\frac{n_a}{N}} [D_a(g)]_{jk} \tag{1.70}$$

are orthonormal functions of the group elements, g. Because the matrix elements are orthonormal, they must be linearly independent. We can also show that they are a complete set of functions of g, in the sense that an arbitrary function of g can be expanded in them. An arbitrary function of g can be written in terms of a bra vector in the space on which the regular representation acts:

$$F(g) = \langle F|g\rangle = \langle F|D_R(g)|e\rangle \tag{1.71}$$

where

$$\langle F| = \sum_{g'\in G} F(g')\langle g'| \tag{1.72}$$

and D_R is the regular representation. Thus an arbitrary $F(g)$ can be written as a linear combination of the matrix elements of the regular representation.

$$F(g) = \sum_{g'\in G} F(g')\langle g'|D_R(g)|e\rangle = \sum_{g'\in G} F(g')[D_R(g)]_{g'e} \tag{1.73}$$

But since D_R is completely reducible, this can be rewritten as a linear combination of the matrix elements of the irreducible representations. Note that while this shows that the matrix elements of the inequivalent irreducible representations are complete, it doesn't tell us how to actually find what they are. The orthogonality relations are the same. They are useful only once we actually know explicitly what the representation look like. Putting these results together, we have proved

Theorem 1.5 *The matrix elements of the unitary, irreducible representations of G are a complete orthonormal set for the vector space of the regular representation, or alternatively, for functions of* $g \in G$.

An immediate corollary is a result that is rather amazing:

$$N = \sum_i n_i^2 \tag{1.74}$$

— the order of the group N is the sum of the squares of the dimensions of the irreducible representations n_i just because this is the number of components of the matrix elements of the irreducible representations. You can check that this works for all the examples we have seen.

Example: Fourier series — cyclic group Z_N with elements a_j for $j =$ 0 to $N-1$ (with $a_0 = e$)

$$a_j a_k = a_{(j+k) \bmod N} \tag{1.75}$$

The irreducible representations of Z_N are

$$D_n(a_j) = e^{2\pi i n j/N} \tag{1.76}$$

all 1-dimensional.[1] Thus (1.69) gives

$$\frac{1}{N}\sum_{j=0}^{N-1} e^{-2\pi i n' j/N} e^{2\pi i n j/N} = \delta_{n'n} \tag{1.77}$$

which is the fundamental relation for Fourier series.

1.13 Characters

The characters $\chi_D(g)$ of a group representation D are the traces of the linear operators of the representation or their matrix elements:

$$\chi_D(g) \equiv \operatorname{Tr} D(g) = \sum_i [D(g)]_{ii} \tag{1.78}$$

The advantage of the characters is that because of the cyclic property of the trace $\operatorname{Tr}(AB) = \operatorname{Tr}(BA)$, they are unchanged by similarity transformations, thus all equivalent representations have the same characters. The characters are also different for each inequivalent irreducible representation, D_a — in fact, they are orthonormal up to an overall factor of N — to see this just sum (1.69) over $j = k$ and $\ell = m$

$$\sum_{\substack{g\in G \\ j=k \\ \ell=m}} \frac{1}{N}[D_a(g)]_{jk}^*[D_b(g)]_{\ell m} = \sum_{\substack{j=k \\ \ell=m}} \frac{1}{n_a}\delta_{ab}\delta_{j\ell}\delta_{km} = \delta_{ab}$$

or

$$\frac{1}{N}\sum_{g\in G} \chi_{D_a}(g)^* \chi_{D_b}(g) = \delta_{ab} \tag{1.79}$$

Since the characters of different irreducible representations are orthogonal, they are different.

The characters are constant on conjugacy classes because

$$\operatorname{Tr} D(g^{-1}g_1 g) = \operatorname{Tr}(D(g^{-1})\, D(g_1)\, D(g)) = \operatorname{Tr} D(g_1) \tag{1.80}$$

It is less obvious, but also true that the characters are a complete basis for functions that are constant on the conjugacy classes and we can see this by explicit calculation. Suppose that $F(g_1)$ is such a function. We already know

[1] We will prove below that Abelian finite groups have only 1-dimensional irreducible representations.

that $F(g_1)$ can be expanded in terms of the matrix elements of the irreducible representations —

$$F(g_1) = \sum_{a,j,k} c^a_{jk} [D_a(g_1)]_{jk} \tag{1.81}$$

but since F is constant on conjugacy classes, we can write it as

$$F(g_1) = \frac{1}{N} \sum_{g \in G} F(g^{-1} g_1 g) = \frac{1}{N} \sum_{a,j,k} c^a_{jk} [D_a(g^{-1} g_1 g)]_{jk} \tag{1.82}$$

and thus

$$F(g_1) = \frac{1}{N} \sum_{\substack{a,j,k \\ g,\ell,m}} c^a_{jk} [D_a(g^{-1})]_{j\ell} [D_a(g_1)]_{\ell m} [D_a(g)]_{mk} \tag{1.83}$$

But now we can do the sum over g explicitly using the orthogonality relation, (1.68).

$$F(g_1) = \sum_{\substack{a,j,k \\ \ell,m}} \frac{1}{n_a} c^a_{jk} [D_a(g_1)]_{\ell m} \delta_{jk} \delta_{\ell m} \tag{1.84}$$

or

$$F(g_1) = \sum_{a,j,\ell} \frac{1}{n_a} c^a_{jj} [D_a(g_1)]_{\ell\ell} = \sum_{a,j} \frac{1}{n_a} c^a_{jj} \chi_a(g_1) \tag{1.85}$$

This was straightforward to get from the orthogonality relation, but it has an important consequence. The characters, $\chi_a(g)$, of the independent irreducible representations form a complete, orthonormal basis set for the functions that are constant on conjugacy classes. Thus the number of irreducible representations is equal to the number of conjugacy classes. We will use this frequently.

This also implies that there is an orthogonality condition for a sum over representations. To see this, label the conjugacy classes by an integer α, and let k_α be the number of elements in the conjugacy class. Then define the matrix V with matrix elements

$$V_{\alpha a} = \sqrt{\frac{k_\alpha}{N}}\, \chi_{D_a}(g_\alpha) \tag{1.86}$$

where g_α is the conjugacy class α. Then the orthogonality relation (1.79) can be written as $V^\dagger V = 1$. But V is a square matrix, so it is unitary, and thus we also have $V V^\dagger = 1$, or

$$\sum_a \chi_{D_a}(g_\alpha)^* \chi_{D_a}(g_\beta) = \frac{N}{k_\alpha} \delta_{\alpha\beta} \tag{1.87}$$

Consequences: Let D be any representation (not necessarily irreducible). In its completely reduced form, it will contain each of the irreducible representations some integer number of times, m_a. We can compute m_a simply by using the orthogonality relation for the characters (1.79)

$$\frac{1}{N}\sum_{g\in G}\chi_{D_a}(g)^*\chi_D(g) = m_a^D \tag{1.88}$$

The point is that D is a direct sum

$$\sum_a \overbrace{D_a \oplus \cdots \oplus D_a}^{m_a^D \text{ times}} \tag{1.89}$$

For example, consider the regular representation. It's characters are

$$\chi_R(e) = N \quad \chi_R(g) = 0 \text{ for } g \neq e \tag{1.90}$$

Thus

$$m_a^R = \chi_a(e) = n_a \tag{1.91}$$

Each irreducible representation appears in the regular representation a number of times equal to its dimension. Note that this is consistent with (1.74). Note also that m_a is uniquely determined, independent of the basis.

Example: Back to S_3 once more. Let's determine the characters without thinking about the 2-dimensional representation explicitly, but knowing the conjugacy classes, $\{e\}$, $\{a_1, a_2\}$ and $\{a_3, a_4, a_5\}$. It is easiest to start with the one representation we know every group has — the trivial representation, D_0 for which $D_0(g) = 1$ for all g. This representation has characters $\chi_0(g) = 1$. Note that this is properly normalized. It follows from the condition $\sum n_a^2 = N$ that the other two representations have dimensions 1 and 2. It is almost equally easy to write down the characters for the other 1-dimensional representation. In general, when there is an invariant subgroup H of G, there are representations of G that are constant on H, forming a representation of the factor group, G/H. In this case, the factor group is Z_2, with nontrivial representation 1 for $H = \{e, a_1, a_2\}$ and -1 for $\{a_3, a_4, a_5\}$. We know that for the 2 dimensional representation, $\chi_3(e) = n_3 = 2$, thus so far the character table looks like

	e	a_1 a_2	a_3 a_4 a_5
0	1	1	1
1	1	1	-1
2	2	?	?

(1.92)

But then we can fill in the last two entries using orthogonality. We could actually have just used orthogonality without even knowing about the second representation, but using the Z_2 makes the algebra trivial.

$$\begin{array}{c||c|c|c} & e & \begin{array}{c}a_1\\a_2\end{array} & \begin{array}{c}a_3\\a_4\\a_5\end{array} \\ \hline 0 & 1 & 1 & 1 \\ \hline 1 & 1 & 1 & -1 \\ \hline 2 & 2 & -1 & 0 \end{array} \tag{1.93}$$

We can use the characters not just to find out how many irreducible representations appear in a particular reducible one, but actually to explicitly decompose the reducible representation into its irreducible components. It is easy to see that if D is an arbitrary representation, the sum

$$P_a = \frac{n_a}{N} \sum_{g \in G} \chi_{D_a}(g)^* D(g) \tag{1.94}$$

is a projection operator onto the subspace that transforms under the representation a. To see this, note that if we set $j = k$ and sum in the orthogonality relation (1.69), we find

$$\frac{n_a}{N} \sum_{g \in G} \chi_{D_a}(g)^* [D_b(g)]_{\ell m} = \delta_{ab} \delta_{\ell m} \tag{1.95}$$

Thus when D is written in block diagonal form, the sum in (1.95) gives 1 on the subspaces that transform like D_a and 0 on all the rest — thus it is the projection operator as promised. The point, however, is that (1.94) gives us the projection operator in the original basis. We did not have to know how to transform to block diagonal form. An example may help to clarify this.

Example — S_3 again

Here's a three dimensional representation of S_3

$$\begin{array}{cc} D_3(e) = \begin{pmatrix} 1 & 0 & 0 \\ 0 & 1 & 0 \\ 0 & 0 & 1 \end{pmatrix}, & D_3(a_1) = \begin{pmatrix} 0 & 0 & 1 \\ 1 & 0 & 0 \\ 0 & 1 & 0 \end{pmatrix} \\ D_3(a_2) = \begin{pmatrix} 0 & 1 & 0 \\ 0 & 0 & 1 \\ 1 & 0 & 0 \end{pmatrix}, & D_3(a_3) = \begin{pmatrix} 0 & 1 & 0 \\ 1 & 0 & 0 \\ 0 & 0 & 1 \end{pmatrix} \\ D_3(a_4) = \begin{pmatrix} 1 & 0 & 0 \\ 0 & 0 & 1 \\ 0 & 1 & 0 \end{pmatrix}, & D_3(a_5) = \begin{pmatrix} 0 & 0 & 1 \\ 0 & 1 & 0 \\ 1 & 0 & 0 \end{pmatrix} \end{array} \tag{1.96}$$

More precisely, as usual when we write down a set of matrices to represent linear operators, these are matrices which have the same **matrix elements** — that is

$$[D_3(g)]_{jk} = \langle j|D_3|k\rangle \tag{1.97}$$

One could use a different symbol to represent the operators and the matrices, but its always easy to figure out which is which from the context. The important point is that the way this acts on the states, $|j\rangle$ is by matrix multiplication **on the right**, because we can insert a complete set of intermediate states

$$D_3(g)|j\rangle = \sum_k |k\rangle\langle k|D_3(g)|j\rangle = \sum_k |k\rangle[D_3(g)]_{kj} \tag{1.98}$$

This particular representation is an important one because it is the defining representation for the group — it actually implements the permutations on the states. For example

$$\begin{aligned} D_3(a_1)|1\rangle &= \sum_k |k\rangle[D_3(a_1)]_{k1} = |2\rangle \\ D_3(a_1)|2\rangle &= \sum_k |k\rangle[D_3(a_1)]_{k2} = |3\rangle \\ D_3(a_1)|3\rangle &= \sum_k |k\rangle[D_3(a_1)]_{k3} = |1\rangle \end{aligned} \tag{1.99}$$

thus this implements the cyclic transformation (1,2,3), or $1 \to 2 \to 3 \to 1$.

Now if we construct the projection operators, we find

$$P_0 = \frac{1}{6}\left(D_3(e) + \sum_{j=1}^{5} D_3(a_j)\right) = \frac{1}{3}\begin{pmatrix} 1 & 1 & 1 \\ 1 & 1 & 1 \\ 1 & 1 & 1 \end{pmatrix} \tag{1.100}$$

$$P_1 = \frac{1}{6}\left(D_3(e) + \sum_{j=1}^{2} D_3(a_j) - \sum_{j=3}^{5} D_3(a_j)\right) = 0 \tag{1.101}$$

$$P_2 = \frac{2}{6}\left(2D_3(e) - \sum_{j=1}^{2} D_3(a_j)\right) = \frac{1}{3}\begin{pmatrix} 2 & -1 & -1 \\ -1 & 2 & -1 \\ -1 & -1 & 2 \end{pmatrix} \tag{1.102}$$

This makes good sense. P_0 projects onto the invariant combination $(|1\rangle + |2\rangle + |3\rangle)/\sqrt{3}$, which transforms trivially, while P_2 projects onto the two dimensional subspace spanned by the differences of pairs of components, $|1\rangle - |2\rangle$, etc, which transforms according to D_3.

This constructions shows that the representation D_3 decomposes into a direct sum of the irreducible representations,

$$D_3 = D_0 \oplus D_2 \tag{1.103}$$

1.14 Eigenstates

In quantum mechanics, we are often interested in the eigenstates of an invariant hermitian operator, in particular the Hamiltonian, H. We can always take these eigenstates to transform according to irreducible representations of the symmetry group. To prove this, note that we can divide up the Hilbert space into subspaces with different eigenvalues of H. Each subspace furnishes a representation of the symmetry group because $D(g)$, the group representation on the full Hilbert space, cannot change the H eigenvalue (because $[D(g), H] = 0$). But then we can completely reduce the representation in each subspace.

A related fact is that if some irreducible representation appears only once in the Hilbert space, then the states in that representation must be eigenstates of H (and any other invariant operator). This is true because $H|a, j, x\rangle$ must be in the same irreducible representation, thus

$$H\,|a, j, x\rangle = \sum_y c_y\,|a, j, y\rangle \tag{1.104}$$

and if x and y take only one value, then $|a, j, x\rangle$ is an eigenstate.

This is sufficiently important to say again in the form of a theorem:

Theorem 1.6 *If a hermitian operator, H, commutes with all the elements, $D(g)$, of a representation of the group G, then you can choose the eigenstates of H to transform according to irreducible representations of G. If an irreducible representation appears only once in the Hilbert space, every state in the irreducible representation is an eigenstate of H with the same eigenvalue.*

Notice that for Abelian groups, this procedure of choosing the H eigenstates to transform under irreducible representations is analogous to simultaneously diagonalizing H and $D(g)$. For example, for the group Z_2 associated with parity, it is the statement that we can always choose the H eigenstates to be either symmetric or antisymmetric.

In the case of parity, the linear operator representing parity is hermitian, so we know that it can be diagonalized. But in general, while we have shown that operators representing finite group elements can be chosen to be unitary, they will not be hermitian. Nevertheless, we can show that for an Abelian group that commutes with the H, the group elements can simultaneously diagonalized along with H. The reason is the following theorem:

Theorem 1.7 *All of the irreducible representations of a finite Abelian group are 1-dimensional.*

One proof of this follows from our discussion of conjugacy classes and from (1.74). For an Abelian group, conjugation does nothing, because $g\,g'\,g^{-1} = g'$ for all g and g'. Therefore, each element is in a conjugacy class all by itself. Because there is one irreducible representation for each conjugacy class, the number of irreducible representations is equal to the order of the group. Then the only way to satisfy (1.74) is to have all of the n_is equal to one. This proves the theorem, and it means that decomposing a representation of an Abelian group into its irreducible representations amounts to just diagonalizing all the representation matrices for all the group elements.

For a non-Abelian group, we cannot simultaneously diagonalize all of the $D(g)$s, but the procedure of completely reducing the representation on each subspace of constant H is the next best thing.

A classical problem which is quite analogous to the problem of diagonalizing the Hamiltonian in quantum mechanics is the problem of finding the normal modes of small oscillations of a mechanical system about a point of stable equilibrium. Here, the square of the angular frequency is the eigenvalue of the $M^{-1}K$ matrix and the normal modes are the eigenvectors of $M^{-1}K$. In the next three sections, we will work out an example.

1.15 Tensor products

We have seen that we can take reducible representations apart into direct sums of smaller representations. We can also put representations together into larger representations. Suppose that D_1 is an m dimensional representation acting on a space with basis vectors $|j\rangle$ for $j = 1$ to m and D_2 is an n dimensional representation acting on a space with basis vectors $|x\rangle$ for $x = 1$ to n. We can make an $m \times n$ dimensional space called the **tensor product space** by taking basis vectors labeled by both j and x in an ordered pair — $|j, x\rangle$. Then when j goes from 1 to m and x goes from 1 to n, the ordered pair (j, x) runs over $m \times n$ different combinations. On this large space, we can define a new representation called the **tensor product representation** $D_1 \otimes D_2$ by multiplying the two smaller representations. More precisely, the matrix elements of $D_{D_1 \otimes D_2}(g)$ are products of those of $D_1(g)$ and $D_2(g)$:

$$\langle j, x|\, D_{D_1 \otimes D_2}(g)\, |k, y\rangle \equiv \langle j|\, D_1(g)\, |k\rangle\, \langle x|\, D_2(g)\, |y\rangle \tag{1.105}$$

It is easy to see that this defines a representation of G. In general, however, it will not be an irreducible representation. One of our favorite pastimes in what follows will be to decompose reducible tensor product representations into irreducible representations.

1.16 Example of tensor products

Consider the following physics problem. Three blocks are connected by springs in a triangle as shown

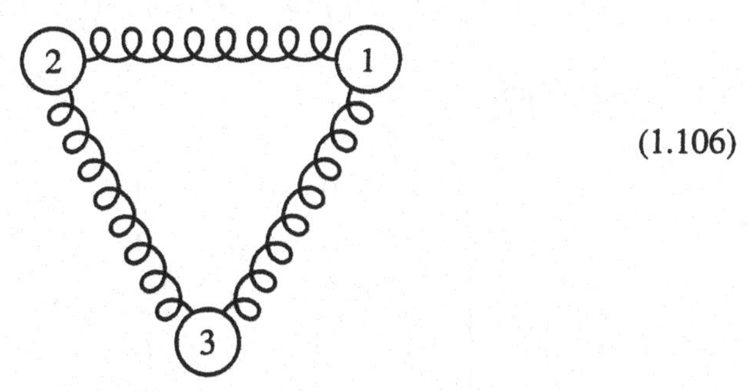

(1.106)

Suppose that these are free to slide on a frictionless surface. What can we say about the normal modes of this system. The point is that there is an S_3 symmetry of the system, and we can learn a lot about the system by using the symmetry and applying theorem 1.6. The system has 6 degrees of freedom, described by the x and y coordinates of the three blocks:

$$\begin{pmatrix} x_1 & y_1 & x_2 & y_2 & x_3 & y_3 \end{pmatrix} \tag{1.107}$$

This has the structure of a **tensor product** — the 6 dimensional space is a product of a 3 dimensional space of the blocks, and the 2 dimensional space of the x and y coordinates. We can think of these coordinates as having two indices. It is three two dimensional vectors, $\vec{r}_i$, each of the vector indices has two components. So we can write the components as $r_{j\mu}$ where j labels the mass and runs from 1 to 3 and μ labels the x or y component and runs from 1 to 2, with the connection

$$\begin{gathered} \begin{pmatrix} x_1 & y_1 & x_2 & y_2 & x_3 & y_3 \end{pmatrix} = \\ \begin{pmatrix} r_{11} & r_{12} & r_{21} & r_{22} & r_{31} & r_{32} \end{pmatrix} \end{gathered} \tag{1.108}$$

The 3 dimensional space transforms under S_3 by the representation D_3. The 2 dimensional space transforms by the representation D_2 below:

$$\begin{gathered} D_2(e) = \begin{pmatrix} 1 & 0 \\ 0 & 1 \end{pmatrix}, \quad D_2(a_1) = \begin{pmatrix} -\frac{1}{2} & -\frac{\sqrt{3}}{2} \\ \frac{\sqrt{3}}{2} & -\frac{1}{2} \end{pmatrix}, \\ D_2(a_2) = \begin{pmatrix} -\frac{1}{2} & \frac{\sqrt{3}}{2} \\ -\frac{\sqrt{3}}{2} & -\frac{1}{2} \end{pmatrix}, \; D_2(a_3) = \begin{pmatrix} -1 & 0 \\ 0 & 1 \end{pmatrix}, \\ D_2(a_4) = \begin{pmatrix} \frac{1}{2} & \frac{\sqrt{3}}{2} \\ \frac{\sqrt{3}}{2} & -\frac{1}{2} \end{pmatrix}, \; D_2(a_5) = \begin{pmatrix} \frac{1}{2} & -\frac{\sqrt{3}}{2} \\ -\frac{\sqrt{3}}{2} & -\frac{1}{2} \end{pmatrix} \end{gathered} \tag{1.109}$$

This is the same as (1.21). Then, using (1.105), the 6 dimensional representation of the coordinates is simply the product of these two representations:

$$[D_6(g)]_{j\mu k\nu} = [D_3(g)]_{jk}[D_2(g)]_{\mu\nu} \tag{1.110}$$

Thus, for example,

$$D_6(a_1) = \begin{pmatrix} 0 & 0 & 0 & 0 & -\frac{1}{2} & -\frac{\sqrt{3}}{2} \\ 0 & 0 & 0 & 0 & \frac{\sqrt{3}}{2} & -\frac{1}{2} \\ -\frac{1}{2} & -\frac{\sqrt{3}}{2} & 0 & 0 & 0 & 0 \\ \frac{\sqrt{3}}{2} & -\frac{1}{2} & 0 & 0 & 0 & 0 \\ 0 & 0 & -\frac{1}{2} & -\frac{\sqrt{3}}{2} & 0 & 0 \\ 0 & 0 & \frac{\sqrt{3}}{2} & -\frac{1}{2} & 0 & 0 \end{pmatrix} \tag{1.111}$$

This has the structure of 3 copies of $D_2(a_1)$ in place of the 1's in $D_3(a_1)$. The other generators are similar in structure.

Because the system has the S_3 symmetry, the normal modes of the system must transform under definite irreducible representations of the symmetry. Thus if we construct the projectors onto these representations, we will have gone some way towards finding the normal modes. In particular, if an irreducible representation appears only once, it must be a normal mode by theorem 1.6. If a representation appears more than once, then we need some additional information to determine the modes.

We can easily determine how many times each irreducible representation appears in D_6 by finding the characters of D_6 and using the orthogonality relations. To find the characters of D_6, we use an important general result. The character of the tensor product of two representations is the product of the characters of the factors. This follows immediately from the definition of the tensor product and the trace.

$$\chi_{D_1 \times D_2} = \chi_{D_1} \chi_{D_2} \tag{1.112}$$

So in this case,

$$\begin{aligned} \chi_6(g) &= \sum_{j\mu} [D_6(g)]_{j\mu j\mu} \\ &= [D_3(g)]_{jj}[D_2(g)]_{\mu\mu} = \chi_3(g)\chi_2(g) \end{aligned} \tag{1.113}$$

so that the product is as shown in the table below:

	e	a_1 a_2	a_3 a_4 a_5
D_3	3	0	1
D_2	2	-1	0
D_6	6	0	0

(1.114)

This is the same as the characters of the regular representation, thus this representation is equivalent to the regular representation, and contains D_0 and D_1 once and D_2 twice.

Note that (1.113) is an example of a simple but important general relation, which we might as well dignify by calling it a theorem —

Theorem 1.8 *The characters of a tensor product representation are the products of the characters of the factors.*

With these tools, we can use group theory to find the normal modes of the system.

1.17 * Finding the normal modes

The projectors onto D_0 and D_1 will be 1 dimensional.

P_0 is

$$P_0 = \frac{1}{6} \sum_{g \in G} \chi_0(g)^* D_6(g)$$

$$= \begin{pmatrix} \frac{1}{4} & \frac{\sqrt{3}}{12} & -\frac{1}{4} & \frac{\sqrt{3}}{12} & 0 & -\frac{\sqrt{3}}{6} \\ \frac{\sqrt{3}}{12} & \frac{1}{12} & -\frac{\sqrt{3}}{12} & \frac{1}{12} & 0 & -\frac{\sqrt{3}}{6} \\ -\frac{1}{4} & -\frac{\sqrt{3}}{12} & \frac{1}{4} & -\frac{\sqrt{3}}{12} & 0 & \frac{\sqrt{3}}{6} \\ \frac{\sqrt{3}}{12} & \frac{1}{12} & -\frac{\sqrt{3}}{12} & \frac{1}{12} & 0 & -\frac{\sqrt{3}}{6} \\ 0 & 0 & 0 & 0 & 0 & 0 \\ -\frac{\sqrt{3}}{6} & -\frac{1}{6} & \frac{\sqrt{3}}{6} & -\frac{1}{6} & 0 & \frac{1}{3} \end{pmatrix} \tag{1.115}$$

$$= \begin{pmatrix} \frac{1}{2} \\ \frac{\sqrt{3}}{6} \\ -\frac{1}{2} \\ \frac{\sqrt{3}}{6} \\ 0 \\ -\frac{1}{\sqrt{3}} \end{pmatrix} \begin{pmatrix} \frac{1}{2} & \frac{\sqrt{3}}{6} & -\frac{1}{2} & \frac{\sqrt{3}}{6} & 0 & -\frac{1}{\sqrt{3}} \end{pmatrix} \tag{1.116}$$

corresponding to the motion

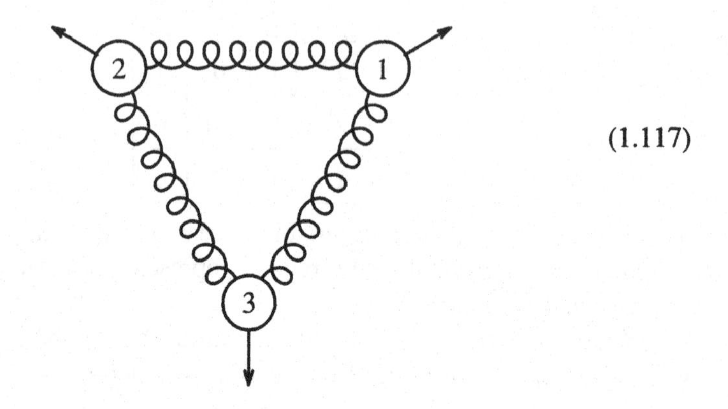

(1.117)

the so-called "breathing mode" in which the triangle grows and shrinks while retaining its shape.

P_1 is

$$P_1 = \frac{1}{6}\sum_{g\in G} \chi_1(g)^* D_6(g)$$

$$= \begin{pmatrix} \frac{1}{12} & -\frac{\sqrt{3}}{12} & \frac{1}{12} & \frac{\sqrt{3}}{12} & -\frac{1}{6} & 0 \\ -\frac{\sqrt{3}}{12} & \frac{1}{4} & -\frac{\sqrt{3}}{12} & -\frac{1}{4} & \frac{\sqrt{3}}{6} & 0 \\ \frac{1}{12} & -\frac{\sqrt{3}}{12} & \frac{1}{12} & \frac{\sqrt{3}}{12} & -\frac{1}{6} & 0 \\ \frac{\sqrt{3}}{12} & -\frac{1}{4} & \frac{\sqrt{3}}{12} & \frac{1}{4} & -\frac{\sqrt{3}}{6} & 0 \\ -\frac{1}{6} & \frac{\sqrt{3}}{6} & -\frac{1}{6} & -\frac{\sqrt{3}}{6} & \frac{1}{3} & 0 \\ 0 & 0 & 0 & 0 & 0 & 0 \end{pmatrix} \tag{1.118}$$

$$= \begin{pmatrix} -\frac{\sqrt{3}}{6} \\ \frac{1}{2} \\ -\frac{\sqrt{3}}{6} \\ -\frac{1}{2} \\ \frac{1}{\sqrt{3}} \\ 0 \end{pmatrix} \begin{pmatrix} -\frac{\sqrt{3}}{6} & \frac{1}{2} & -\frac{\sqrt{3}}{6} & -\frac{1}{2} & \frac{1}{\sqrt{3}} & 0 \end{pmatrix} \tag{1.119}$$

corresponding to the motion

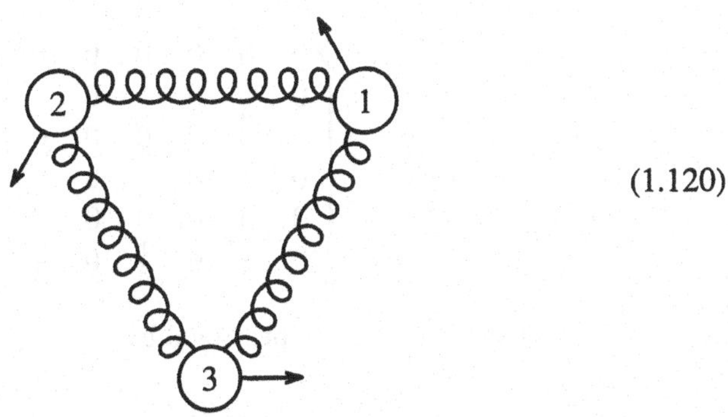

(1.120)

the mode in which the triangle rotates — this is a normal mode with zero frequency because there is no restoring force.

Notice, again, that we found these two normal modes without putting in any physics at all except the symmetry!

Finally, P_2 is

$$P_2 = \frac{2}{6} \sum_{g \in G} \chi_2(g)^* D_6(g)$$

$$= \begin{pmatrix} \frac{2}{3} & 0 & \frac{1}{6} & -\frac{\sqrt{3}}{6} & \frac{1}{6} & \frac{\sqrt{3}}{6} \\ 0 & \frac{2}{3} & \frac{\sqrt{3}}{6} & \frac{1}{6} & -\frac{\sqrt{3}}{6} & \frac{1}{6} \\ \frac{1}{6} & \frac{\sqrt{3}}{6} & \frac{2}{3} & 0 & \frac{1}{6} & -\frac{\sqrt{3}}{6} \\ -\frac{\sqrt{3}}{6} & \frac{1}{6} & 0 & \frac{2}{3} & \frac{\sqrt{3}}{6} & \frac{1}{6} \\ \frac{1}{6} & -\frac{\sqrt{3}}{6} & \frac{1}{6} & \frac{\sqrt{3}}{6} & \frac{2}{3} & 0 \\ \frac{\sqrt{3}}{6} & \frac{1}{6} & -\frac{\sqrt{3}}{6} & \frac{1}{6} & 0 & \frac{2}{3} \end{pmatrix} \tag{1.121}$$

As expected, this is a rank 4 projection operator ($\text{Tr}\, P_2 = 4$). We need some dynamical information. Fortunately, two modes are easy to get — translations of the whole triangle.

Translations in the x direction, for example, are projected by

$$T_x = \begin{pmatrix} \frac{1}{3} & 0 & \frac{1}{3} & 0 & \frac{1}{3} & 0 \\ 0 & 0 & 0 & 0 & 0 & 0 \\ \frac{1}{3} & 0 & \frac{1}{3} & 0 & \frac{1}{3} & 0 \\ 0 & 0 & 0 & 0 & 0 & 0 \\ \frac{1}{3} & 0 & \frac{1}{3} & 0 & \frac{1}{3} & 0 \\ 0 & 0 & 0 & 0 & 0 & 0 \end{pmatrix} \tag{1.122}$$

and those in the y direction by

$$T_y = \begin{pmatrix} 0 & 0 & 0 & 0 & 0 & 0 \\ 0 & \frac{1}{3} & 0 & \frac{1}{3} & 0 & \frac{1}{3} \\ 0 & 0 & 0 & 0 & 0 & 0 \\ 0 & \frac{1}{3} & 0 & \frac{1}{3} & 0 & \frac{1}{3} \\ 0 & 0 & 0 & 0 & 0 & 0 \\ 0 & \frac{1}{3} & 0 & \frac{1}{3} & 0 & \frac{1}{3} \end{pmatrix} \tag{1.123}$$

So the nontrivial modes are projected by

$$P_2 - T_x - T_y =$$

$$\begin{pmatrix} \frac{1}{3} & 0 & -\frac{1}{6} & -\frac{\sqrt{3}}{6} & -\frac{1}{6} & \frac{\sqrt{3}}{6} \\ 0 & \frac{1}{3} & \frac{\sqrt{3}}{6} & -\frac{1}{6} & -\frac{\sqrt{3}}{6} & -\frac{1}{6} \\ -\frac{1}{6} & \frac{\sqrt{3}}{6} & \frac{1}{3} & 0 & -\frac{1}{6} & -\frac{\sqrt{3}}{6} \\ -\frac{\sqrt{3}}{6} & -\frac{1}{6} & 0 & \frac{1}{3} & \frac{\sqrt{3}}{6} & -\frac{1}{6} \\ -\frac{1}{6} & -\frac{\sqrt{3}}{6} & -\frac{1}{6} & \frac{\sqrt{3}}{6} & \frac{1}{3} & 0 \\ \frac{\sqrt{3}}{6} & -\frac{1}{6} & -\frac{\sqrt{3}}{6} & -\frac{1}{6} & 0 & \frac{1}{3} \end{pmatrix} \tag{1.124}$$

To see what the corresponding modes look like, act with this on the vector $\begin{pmatrix} 0 & 0 & 0 & 0 & 0 & 1 \end{pmatrix}$ to get

$$\begin{pmatrix} \frac{\sqrt{3}}{6} & -\frac{1}{6} & -\frac{\sqrt{3}}{6} & -\frac{1}{6} & 0 & \frac{1}{3} \end{pmatrix} \tag{1.125}$$

corresponding to the motion

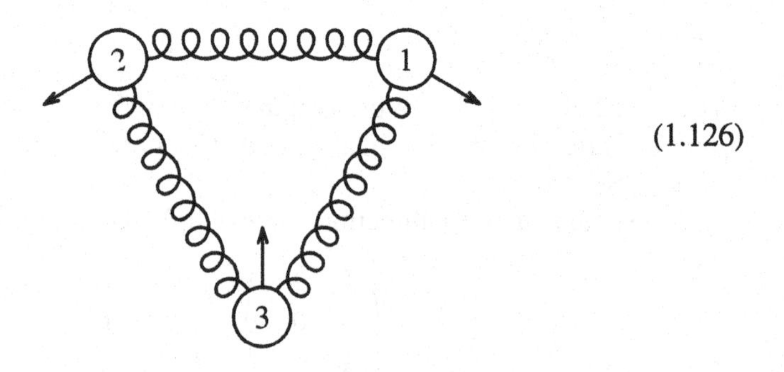

(1.126)

Then rotating by $2\pi/3$ gives a linearly independent mode.

1.18 * Symmetries of $2n$+1-gons

This is a nice simple example of a transformation group for which we can work out the characters (and actually the whole set of irreducible representations) easily. Consider a regular polygon with $2n + 1$ vertices, like the 7-gon shown below.

(1.127)

The group of symmetries of the $2n$+1-gon consists of the identity, the $2n$ rotations by $\frac{\pm 2\pi j}{2n+1}$ for $j = 1$ to n,

$$\text{rotations by } \frac{\pm 2\pi j}{2n+1} \text{ for } j = 1 \text{ to } n \tag{1.128}$$

and the $2n$+1 reflections about lines through the center and a vertex, as show below:

reflections about lines through center and vertex (1.129)

(1.130)

Thus the order of the group of symmetries is $N = 2 \times (2n + 1)$.

There are $n + 2$ conjugacy classes:

1 — the identity, e;

2 — the $2n$+1 reflections;

3 to n+2 — the rotations by $\frac{\pm 2\pi j}{2n+1}$ for $j = 1$ to n — each value of j is a separate conjugacy class.

The way this works is that the reflections are all in the same conjugacy class because by conjugating with rotations, you can get from any one reflection to any other. The rotations are unchanged by conjugation by rotations, but a conjugation by a reflection changes the sign of the rotation, so there is a ± pair in each conjugacy class.

Furthermore, the n conjugacy classes of rotations are equivalent under cyclic permutations and relabeling of the vertices, as shown below:

(1.131)

(1.132)

The characters look like

e	r	$j=1$	$j=2$	$\cdots$	$j=n$
1	1	1	1	$\cdots$	1
1	-1	1	1	$\cdots$	1
2	0	$2\cos\frac{2\pi m}{2n+1}$	$2\cos\frac{4\pi m}{2n+1}$	$\cdots$	$2\cos\frac{2n\pi m}{2n+1}$

(1.133)

In the last line, the different values of m give the characters of the n *different* 2-dimensional representations.

1.19 Permutation group on n objects

Any element of the permutation group on n objects, called S_n, can be written in term of cycles, where a cycle is a cyclic permutation of a subset. We will use a notation that makes use of this, where each cycle is written as a set of numbers in parentheses, indicating the set of things that are cyclicly permuted. For example:

(1) means $x_1 \to x_1$

(1372) means $x_1 \to x_3 \to x_7 \to x_2 \to x_1$

Each element of S_n involves each integer from 1 to n in exactly one cycle.

Examples:

The identity element looks like e =**(1)(2)**$\cdots$**(n)** — n 1-cycles — there is only one of these.

An interchange of two elements looks like **(12)(3)···(n)** — a 2-cycle and $n-2$ 1-cycles — there are $n(n-1)/2$ of these — $(j_1j_2)(j_3)\cdots(j_n)$.

An arbitrary element has k_j j-cycles, where

$$\sum_{j=1}^{n} j\,k_j = n \tag{1.134}$$

For example, the permutation **(123)(456)(78)(9)** has two 3-cycles, 1 2-cycle and a 1-cycle, so $k_1 = k_2 = 1$ and $k_3 = 2$.

There is an simple (but reducible) n dimensional representation of S_n called the **defining representation** where the "objects" being permuted are just the basis vectors of an n dimensional vector space,

$$|1\rangle\,,\ |2\rangle\,,\cdots|n\rangle \tag{1.135}$$

If the permutation takes x_j to x_k, the corresponding representation operator D takes $|j\rangle$ to $|k\rangle$, so that

$$D\,|j\rangle = |k\rangle \tag{1.136}$$

and thus

$$\langle\ell|\,D\,|j\rangle = \delta_{k\ell} \tag{1.137}$$

Each matrix in the representation has a single 1 in each row and column.

1.20 Conjugacy classes

The conjugacy classes are just the cycle structure, that is they can be labeled by the integers k_j. For example, all interchanges are in the same conjugacy class — it is enough to check that the inner automorphism gg_1g^{-1} doesn't change the cycle structure of g_1 when g is an interchange, because we can build up any permutation from interchanges. Let us see how this works in some examples. In particular, we will see that conjugating an arbitrary permutation by the interchange (12)(3)··· just interchanges 1 and 2 without changing the cycle structure

Examples — **(12)(3)(4)·(1)(23)(4)·(12)(3)(4)** (note that an interchange is its own inverse)

$$\begin{array}{cl} 1234 & \\ \downarrow & (12)(3)(4) \\ 2134 & \\ \downarrow & \mathbf{(1)(23)(4)} \\ 2314 & \\ \downarrow & (12)(3)(4) \\ 3214 & \\ & = \\ 1234 & \\ \downarrow & \mathbf{(2)(13)(4)} \\ 3214 & \end{array} \tag{1.138}$$

(12)(3)(4)·(1)(234)·(12)(3)(4)

$$\begin{array}{cl} 1234 & \\ \downarrow & (12)(3)(4) \\ 2134 & \\ \downarrow & \mathbf{(1)(234)} \\ 2341 & \\ \downarrow & (12)(3)(4) \\ 3241 & \\ & = \\ 1234 & \\ \downarrow & \mathbf{(2)(134)} \\ 3241 & \end{array} \tag{1.139}$$

If 1 and 2 are in different cycles, they just get interchanged by conjugation by (12), as promised.

The same thing happens when 1 and 2 are in the same cycle. For example

$$\begin{array}{cl} 1234 & \\ \downarrow & (12)(3)(4) \\ 2134 & \\ \downarrow & \mathbf{(123)(4)} \\ 1324 & \\ \downarrow & (12)(3)(4) \\ 3124 & \\ & = \\ 1234 & \\ \downarrow & \mathbf{(213)(4)} \\ 3124 & \end{array} \tag{1.140}$$

Again, in the same cycle this time, 1 and 2 just get interchanged.

Another way of seeing this is to notice that the conjugation is analogous to a similarity transformation. In fact, in the defining, n dimensional representation of (1.135) the conjugation by the interchange (12) is just a change of basis that switches $|1\rangle \leftrightarrow |2\rangle$. Then it is clear that conjugation does not change the cycle structure, but simply interchanges what the permutation does to 1 and 2. Since we can put interchanges together to form an arbitrary permutation, and since by repeated conjugations by interchanges, we can get from any ordering of the integers in the given cycle structure to any other, **the conjugacy classes must consist of all possible permutations with a particular cycle structure.**

Now let us count the number of group elements in each conjugacy class. Suppose a conjugacy class consists of permutations of the form of k_1 1-cycles, k_2 2-cycles, etc, satisfying (1.134). The number of different permutations in the conjugacy class is

$$\frac{n!}{\prod_j j^{k_j} k_j!} \tag{1.141}$$

because each permutation of number 1 to n gives a permutation in the class, but cyclic order doesn't matter within a cycle

$$(123) \text{ is the same as } (231) \tag{1.142}$$

and order doesn't matter at all between cycles of the same length

$$(12)(34) \text{ is the same as } (34)(12) \tag{1.143}$$

1.21 Young tableaux

It is useful to represent each j-cycle by a column of boxes of length j, top-justified and arranged in order of decreasing j as you go to the right. The total number of boxes is n. Here is an example:

(1.144)

is four 1-cycles in S_4 — that is the identity element — always a conjugacy class all by itself. Here's another:

(1.145)

is a 4-cycle, a 3-cycle and a 1-cycle in S_8. These collections of boxes are called Young tableaux. Each different tableaux represents a different conjugacy class, and therefore the tableaux are in one-to-one correspondence with the irreducible representations.

1.22 Example — our old friend S_3

The conjugacy classes are

(1.146)

with numbers of elements

$$\frac{3!}{3!} = 1 \qquad \frac{3!}{2} = 3 \qquad \frac{3!}{3} = 2 \tag{1.147}$$

1.23 Another example — S_4

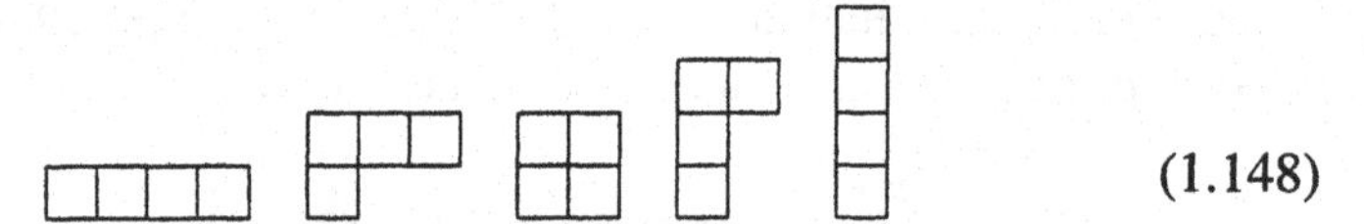

(1.148)

with numbers of elements

$$\frac{4!}{4!} = 1 \quad \frac{4!}{4} = 6 \quad \frac{4!}{8} = 3 \quad \frac{4!}{3} = 8 \quad \frac{4!}{4} = 6 \tag{1.149}$$

The characters of S_4 look like this (with the conjugacy classes which label the columns in the same order as in (1.148)):

conjugacy classes

1	1	1	1	1
3	1	−1	0	−1
2	0	2	−1	0
3	−1	−1	0	1
1	−1	1	1	−1

(1.150)

The first row represents the trivial representation.

1.24 * Young tableaux and representations of S_n

We have seen that a Young tableau with n boxes is associated with an irreducible representation of S_n. We can actually use the tableau to explicitly construct the irreducible representation by identifying an appropriate subspace of the regular representation of S_n.

To see what the irreducible representation is, we begin by putting the integers from 1 to n in the boxes of the tableau in all possible ways. There are $n!$ ways to do this. We then identify each assignment of integers 1 to n to the boxes with a state in the regular representation of S_n by defining a standard ordering, say from left to right and then top down (like reading words on a page) to translate from integers in the boxes to a state associated with a particular permutation. So for example

$$\begin{array}{|c|c|c|c|}\hline 6&5&3&2\\\hline 1&7\\\cline{1-2} 4\\\cline{1-1}\end{array} \to |6532174\rangle \tag{1.151}$$

where $|6532174\rangle$ is the state corresponding to the permutation

$$1234567 \to 6532174 \tag{1.152}$$

Now each of the $n!$ assignment of boxes to the tableau describes one of the $n!$ states of the regular representation.

Next, for a particular tableau, symmetrize the corresponding state in the numbers in each row, and antisymmetrize in the numbers in each column. For example

$$\begin{array}{|c|c|}\hline 1&2\\\hline\end{array} \to |12\rangle + |21\rangle \tag{1.153}$$

and

$$\begin{array}{|c|c|}\hline 1&2\\\hline 3\\\cline{1-1}\end{array} \to |123\rangle + |213\rangle - |321\rangle - |231\rangle \tag{1.154}$$

Now the set of states constructed in this ways spans some subspace of the regular representation. We can construct the states explicitly, and we know how permutations act on these states. That the subspace constructed in this way is a representation of S_n, because a permutation just corresponds to starting with a different assignment of numbers to the tableau, so acting with the permutation on any state in the subspace gives another state in the subspace. In fact, this representation is irreducible, and is the irreducible representation we say is associated with the Young tableau.

Consider the example of S_3. The tableau

$$\begin{array}{|c|c|c|}\hline \ & \ & \ \\\hline\end{array} \tag{1.155}$$

gives completely symmetrized states, and so is associated with a one dimensional subspace that transforms under the trivial representation. The tableau

$$\begin{array}{|c|}\hline \ \\\hline \ \\\hline \ \\\hline\end{array} \tag{1.156}$$

gives completely antisymmetrized states, and so, again is associated with a one dimensional subspace, this time transforming under the representation in which interchanges are represented by -1. Finally

$$\begin{array}{|c|c|}\hline \; & \; \\ \hline \; \\ \cline{1-1}\end{array} \tag{1.157}$$

gives the following states:

$$\begin{array}{|c|c|}\hline 1 & 2 \\ \hline 3 \\ \cline{1-1}\end{array} \to |123\rangle + |213\rangle - |321\rangle - |231\rangle \tag{1.158}$$

$$\begin{array}{|c|c|}\hline 3 & 2 \\ \hline 1 \\ \cline{1-1}\end{array} \to |321\rangle + |231\rangle - |123\rangle - |213\rangle \tag{1.159}$$

$$\begin{array}{|c|c|}\hline 2 & 3 \\ \hline 1 \\ \cline{1-1}\end{array} \to |231\rangle + |321\rangle - |132\rangle - |312\rangle \tag{1.160}$$

$$\begin{array}{|c|c|}\hline 1 & 3 \\ \hline 2 \\ \cline{1-1}\end{array} \to |132\rangle + |312\rangle - |231\rangle - |321\rangle \tag{1.161}$$

$$\begin{array}{|c|c|}\hline 3 & 1 \\ \hline 2 \\ \cline{1-1}\end{array} \to |312\rangle + |132\rangle - |213\rangle - |123\rangle \tag{1.162}$$

$$\begin{array}{|c|c|}\hline 2 & 1 \\ \hline 3 \\ \cline{1-1}\end{array} \to |213\rangle + |123\rangle - |312\rangle - |132\rangle \tag{1.163}$$

Note that interchanging two numbers in the same column of a tableau just changes the sign of the state. This is generally true. Furthermore, you can see explicitly that the sum of three states related by cyclic permutations vanishes. Thus the subspace is two dimensional and transforms under the two dimensional irreducible representation of S_3.

It turns out that the dimension of the representation constructed in this way is

$$\frac{n!}{H} \tag{1.164}$$

where the quantity H is the "hooks" factor for the Young tableau, computed as follows. A hook is a line passing vertically up through the bottom of some column of boxes, making a right hand turn in some box and passing out through the row of boxes. There is one hook for each box. Call the number of boxes the hook passes through h. Then H is the product of the hs for all hooks. We will come back to hooks when we discuss the application of Young tableaux to the representations of $SU(N)$ in chapter XIII.

This procedure for constructing the irreducible representations of S_n is entirely mechanical (if somewhat tedious) and can be used to construct all the representations of S_n from the Young tableaux with n boxes.

We could say much more about finite groups and their representations, but our primary subject is continuous groups, so we will leave finite groups for now. We will see, however, that the representations of the permutation groups play an important role in the representations of continuous groups. So we will come back to S_n now and again.

Problems

1.A. Find the multiplication table for a group with three elements and prove that it is unique.

1.B. Find all essentially different possible multiplication tables for groups with four elements (which cannot be related by renaming elements).

1.C. Show that the representation (1.135) of the permutation group is reducible.

1.D. Suppose that D_1 and D_2 are equivalent, irreducible representations of a finite group G, such that

$$D_2(g) = S\, D_1(g)\, S^{-1} \quad \forall g \in G$$

What can you say about an operator A that satisfies

$$A\, D_1(g) = D_2(g)\, A \quad \forall g \in G?$$

1.E. Find the group of all the discrete rotations that leave a regular tetrahedron invariant by labeling the four vertices and considering the rotations as permutations on the four vertices. This defines a four dimensional representation of a group. Find the conjugacy classes and the characters of the irreducible representations of this group.

***1.F.** Analyze the normal modes of the system of four blocks sliding on a frictionless plane, connected by springs as shown below:

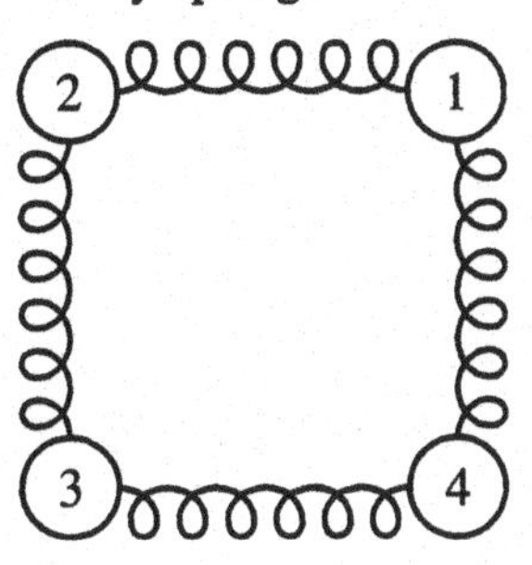

just as we did for the triangle, but using the 8-element symmetry group of the square. Assume that the springs are rigidly attached to the masses (rather than pivoted, for example), so that the square has some rigidity.

Chapter 2

Lie Groups

Suppose our group elements $g \in G$ depend smoothly on a set of continuous parameters —

$$g(\alpha) \tag{2.1}$$

What we mean by smooth is that there is some notion of closeness on the group such that if two elements are "close together" in the space of the group elements, the parameters that describe them are also close together.

2.1 Generators

Since the identity is an important element in the group, it is useful to parameterize the elements (at least those close to the identity element) in such a way that $\alpha = 0$ corresponds to the identity element. Thus we assume that in some neighborhood of the identity, the group elements can be described by a function of N real parameters, α_a for $a = 1$ to N, such that

$$g(\alpha)|_{\alpha=0} = e \tag{2.2}$$

Then if we find a representation of the group, the linear operators of the representation will be parameterized the same way, and

$$D(\alpha)|_{\alpha=0} = 1 \tag{2.3}$$

Then in some neighborhood of the identity element, we can Taylor expand $D(\alpha)$, and if we are close enough, just keep the first term:

$$D(d\alpha) = 1 + i d\alpha_a X_a + \cdots \tag{2.4}$$

 DOI: 10.1201/9780429499210-3

where we have called the parameter $d\alpha$ to remind you that it is infinitesimal. In (2.4), a sum over repeated indices is understood (the "Einstein summation convention") and

$$X_a \equiv -i \left. \frac{\partial}{\partial \alpha_a} D(\alpha) \right|_{\alpha=0} \tag{2.5}$$

The X_a for $a = 1$ to N are called the **generators of the group**. If the parameterization is parsimonious (that is — all the parameters are actually needed to distinguish different group elements), the X_a will be independent. The i is included in the definition (2.5) so that if the representation is unitary, the X_a will be hermitian operators.

Sophus Lie showed how the generators can actually be defined in the abstract group without mentioning representations at all. As a result of his work, groups of this kind are called **Lie groups**. I am not going to talk about them this way because I am more interested in representations than in groups, but it is a beautiful theoretical construction that you may want to look up if you haven't seen it.

As we go away from the identity, there is enormous freedom to parameterize the group elements in different ways, but we may as well choose our parameterization so that the group multiplication law and thus the multiplication law for the representation operators in the Hilbert space looks nice. In particular, we can go away from the identity in some fixed direction by simply raising an infinitesimal group element

$$D(d\alpha) = 1 + id\alpha_a X_a \tag{2.6}$$

to some large power. Because of the group property, this always gives another group element. This suggests defining the representation of the group elements for finite α as

$$D(\alpha) = \lim_{k\to\infty} (1 + i\alpha_a X_a/k)^k = e^{i\alpha_a X_a} \tag{2.7}$$

In the limit, this must go to the representation of a group element because $1 + i\alpha_a X_a/k$ becomes the representation of a group element in (2.4) as k becomes large. This defines a particular parameterization of the representations (sometimes called the **exponential parameterization**), and thus of the group multiplication law itself. In particular, this means that we can write the group elements (at least in some neighborhood of e) in terms of the generators. That's nice, because unlike the group elements, the generators form a vector space. They can be added together and multiplied by real numbers. In fact, we will often use the term generator to refer to any element in the real linear space spanned by the X_as.

2.2 Lie algebras

Now in any particular direction, the group multiplication law is uncomplicated. There is a one parameter family of group elements of the form

$$U(\lambda) = e^{i\lambda\alpha_a X_a} \tag{2.8}$$

and the group multiplication law is simply

$$U(\lambda_1)U(\lambda_2) = U(\lambda_1 + \lambda_2) \tag{2.9}$$

However, if we multiply group elements generated by two different linear combinations of generators, things are not so easy. In general,

$$e^{i\alpha_a X_a} e^{i\beta_b X_b} \neq e^{i(\alpha_a+\beta_a)X_a} \tag{2.10}$$

On the other hand, because the exponentials form a representation of the group (at least if we are close to the identity), it must be true that the product is some exponential of a generator,

$$e^{i\alpha_a X_a} e^{i\beta_b X_b} = e^{i\delta_a X_a} \tag{2.11}$$

for some δ. And because everything is smooth, we can find δ_a by expanding both sides and equating appropriate powers of α and β. When we do this, something interesting happens. We find that it only works if the generators form an **algebra** under commutation (or a commutator algebra). To see this, let's actually do it to leading nontrivial order. We can write

$$i\delta_a X_a = \ln\left(1 + e^{i\alpha_a X_a} e^{i\beta_b X_b} - 1\right) \tag{2.12}$$

I will now expand this, keeping terms up to second order in the parameters α and β, using the Taylor expansion of $\ln(1+K)$ where

$$\begin{aligned}
K &= e^{i\alpha_a X_a} e^{i\beta_b X_b} - 1 \\
&= (1 + i\alpha_a X_a - \frac{1}{2}(\alpha_a X_a)^2 + \cdots) \\
&\quad (1 + i\beta_b X_b - \frac{1}{2}(\beta_b X_b)^2 + \cdots) - 1 \\
&= i\alpha_a X_a + i\beta_a X_a - \alpha_a X_a \beta_b X_b \\
&\quad -\frac{1}{2}(\alpha_a X_a)^2 - \frac{1}{2}(\beta_a X_a)^2 + \cdots
\end{aligned} \tag{2.13}$$

This gives

$$\begin{aligned} i\delta_a X_a &= K - \frac{1}{2}K^2 + \cdots \\ &= i\alpha_a X_a + i\beta_a X_a - \alpha_a X_a \beta_b X_b \\ &-\frac{1}{2}(\alpha_a X_a)^2 - \frac{1}{2}(\beta_a X_a)^2 \\ &+\frac{1}{2}(\alpha_a X_a + \beta_a X_a)^2 + \cdots \end{aligned} \tag{2.14}$$

Now here is the point. The higher order terms in (2.14) are trying to cancel. If the Xs were numbers, they would cancel, because the product of the exponentials is the exponential of the sum of the exponents. They fail to cancel only because the Xs are linear operators, and don't commute with one another. Thus the extra terms beyond $i\alpha_a X_a + i\beta_a X_a$ in (2.14) are proportional to the commutator. Sure enough, explicit calculation in (2.14) gives

$$\begin{aligned} i\delta_a X_a &= K - \frac{1}{2}K^2 + \cdots \\ &= i\alpha_a X_a + i\beta_a X_a \\ &-\frac{1}{2}\left[\alpha_a X_a, \beta_b X_b\right] + \cdots \end{aligned} \tag{2.15}$$

We obtained (2.15) using only the group property and smoothness, which allowed us to use the Taylor expansion. From (2.15) we can calculate δ_a, again in an expansion in α and β. We conclude that

$$[\alpha_a X_a, \beta_b X_b] = -2i(\delta_c - \alpha_c - \beta_c)X_c + \cdots \equiv i\gamma_c X_c \tag{2.16}$$

where the i is put in to make γ real and the $\cdots$ represent terms that have more than two factors of α or β. Since (2.16) must be true for all α and β, we must have

$$\gamma_c = \alpha_a \beta_b f_{abc} \tag{2.17}$$

for some constants f_{abc}, thus

$$[X_a, X_b] = i\, f_{abc} X_c\,. \tag{2.18}$$

where

$$f_{abc} = -f_{bac} \tag{2.19}$$

because $[A, B] = -[B, A]$. Note that we can now write

$$\delta_a = \alpha_a + \beta_a - \frac{1}{2}\gamma_a + \cdots \tag{2.20}$$

so that if γ and the higher terms vanish, we would restore the equality in (2.10).

(2.18) is what is meant by the statement that the generators form an algebra under commutation. We have just shown that this follows from the group properties for Lie groups, because the Lie group elements depend smoothly on the parameters. The commutator in the algebra plays a role similar to the multiplication law for the group.

Now you might worry that if we keep expanding (2.12) beyond second order, we would need additional conditions to make sure that the group multiplication law is maintained. **The remarkable thing is that we don't. The commutator relation (2.18) is enough.** In fact, if you know the constants, f_{abc}, you can reconstruct δ as accurately as you like for any α and β in some finite neighborhood of the origin! Thus the f_{abc} are tremendously important — they summarize virtually the entire group multiplication law. The f_{abc} are called the **structure constants** of the group. They can be computed in any nontrivial representation, that is unless the X_a vanish.

The commutator relation (2.18) is called the **Lie algebra** of the group. The Lie algebra is completely determined by the structure constants. Each group representation gives a representation of the algebra in an obvious way, and the structure constants are the same for all representations because they are fixed just by the group multiplication law and smoothness. Equivalence, reducibility and irreducibility can be transferred from the group to the algebra with no change.

Note that if there is any unitary representation of the algebra, then the f_{abc}s are real, because if we take the adjoint of the commutator relation for hermitian Xs, we get

$$\begin{aligned} [X_a, X_b]^\dagger &= -i\, f_{abc}^* X_c \\ = [X_b, X_a] = i\, f_{bac} X_c &= -i\, f_{abc} X_c \end{aligned} \tag{2.21}$$

Since we are interested in groups which have unitary representations, we will just assume that the f_{abc} are real.

2.3 The Jacobi identity

The matrix generators also satisfy the following identity:

$$[X_a, [X_b, X_c]] + \text{cyclic permutations} = 0\,. \tag{2.22}$$

called the **Jacobi identity**, which you can check by just expanding out the commutators.[1]

The Jacobi identity can be written in a different way that is sometimes easier to use and is also instructive:

$$[X_a, [X_b, X_c]] = [[X_a, X_b], X_c] + [X_b, [X_a, X_c]] \,. \tag{2.23}$$

This is a generalization of the product rule for commutation:

$$[X_a, X_b X_c] = [X_a, X_b] X_c + X_b [X_a, X_c] \,. \tag{2.24}$$

The Jacobi identity is rather trivial for the Lie algebras with only finite dimensional representations that we will study in this book. But it is worth noting that in Lie's more general treatment, it makes sense in situations in which the product of generators is not even well defined.

2.4 The adjoint representation

The structure constants themselves generate a representation of the algebra called the **adjoint representation**. If we use the algebra(2.18), we can compute

$$\begin{aligned} &[X_a, [X_b, X_c]] \\ &= i\, f_{bcd}\, [X_a, X_d] \\ &= -f_{bcd} f_{ade} X_e \end{aligned} \tag{2.25}$$

so (because the X_a are independent), 2.22) implies

$$f_{bcd} f_{ade} + f_{abd} f_{cde} + f_{cad} f_{bde} = 0 \,. \tag{2.26}$$

Defining a set of matrices T_a

$$[T_a]_{bc} \equiv -i f_{abc} \tag{2.27}$$

then (2.26) can be rewritten as

$$[T_a, T_b] = i\, f_{abc} T_c \tag{2.28}$$

Thus the structure constants themselves furnish a representation of the algebra. This is called the **adjoint representation**. The **dimension** of a representation is the dimension of the linear space on which it acts (just as for a

[1]The Jacobi identity is really more subtle than this. We could have proved it directly in the abstract group, where the generators are not linear operators on a Hilbert space. Then the algebra involves a "Lie product" which is not necessarily a commutator, but nevertheless satisfies the Jacobi identity.

finite group). The dimension of the adjoint representation is just the number of independent generators, which is the number of real parameters required to describe a group element. Note that since the f_{abc}s are real, the generators of the adjoint representation are pure imaginary.

We would like to have a convenient scalar product on the linear space of the generators in the adjoint representation, (2.27), to turn it into a vector space. A good one is the trace in the adjoint representation

$$\mathrm{Tr}(T_a T_b) \tag{2.29}$$

This is a real symmetric matrix. We will next show that we can put it into a very simple canonical form. We can change its form by making a linear transformation on the X_a, which in turn, induces a linear transformation on the structure constants. Suppose

$$X_a \to X'_a = L_{ab} X_b \tag{2.30}$$

then

$$\begin{aligned} [X'_a, X'_b] &= i\, L_{ad} L_{be} f_{dec} X_c \\ &= i\, L_{ad} L_{be} f_{deg} L^{-1}_{gh} L_{hc} X_c \\ &= i\, L_{ad} L_{be} f_{deg} L^{-1}_{gc} X'_c \end{aligned} \tag{2.31}$$

so[2]

$$f_{abc} \to f'_{abc} = L_{ad} L_{be} f_{deg} L^{-1}_{gc} \tag{2.32}$$

If we then define a new T_as with the transformed fs,

$$[T_a]_{bc} \to [T'_a]_{bc} = L_{ad} L_{be} [T_d]_{eg} L^{-1}_{gc} \tag{2.33}$$

or

$$[T_a] \to [T'_a] = L_{ad} L [T_d] L^{-1} \tag{2.34}$$

In other words, a linear transformation on the X_as induces a linear transformation on the T_as which involves both a similarity transformation and the same linear transformation on the a index that labels the generator. But in the trace the similarity transformation doesn't matter, so

$$\mathrm{Tr}(T_a T_b) \to \mathrm{Tr}(T'_a T'_b) = L_{ac} L_{bd}\, \mathrm{Tr}(T_c T_d) \tag{2.35}$$

---

[2]Because of the L^{-1} in (2.32), it would be make sense to treat the third index in f_{abc} differently, and write it as an upper index — f^c_{ab}. We will not bother to do this because we are going to move very quickly to a restricted set of groups and basis sets in which $\mathrm{Tr}(T_a T_b) \propto \delta_{ab}$. Then only orthogonal transformation on the X_as are allowed, $L^{-1} = L^T$, so that all three indices are treated in the same way.

Thus we can diagonalize the trace by choosing an appropriate L (here we only need an orthogonal matrix). Suppose we have done this (and dropped the primes), so that

$$\mathrm{Tr}(T_a T_b) = k^a \delta_{ab} \text{ no sum} \tag{2.36}$$

We still have the freedom to rescale the generators (by making a diagonal L transformation), so for example, we could choose all the non-zero k^as to have absolute value 1. But, we cannot change the sign of the k^as (because L appears squared in the transformation (2.35)).

For now, **we will assume that the k^as are positive. This defines the class of algebras that we study in this book.** They are called the **compact Lie algebras**. We will come back briefly below to algebras in which some are zero.[3] And we will take

$$\mathrm{Tr}(T_a T_b) = \lambda\, \delta_{ab} \tag{2.37}$$

for some convenient positive λ. In this basis, the structure constants are completely antisymmetric, because we can write

$$f_{abc} = -i\, \lambda^{-1}\, \mathrm{Tr}([T_a, T_b]\, T_c) \tag{2.38}$$

which is completely antisymmetric because of the cyclic property of the trace.

$$\begin{aligned} \mathrm{Tr}([T_a, T_b]\, T_c) &= (T_a T_b T_c - T_b T_a T_c) \\ &= (T_b T_c T_a - T_c T_b T_a) = \mathrm{Tr}([T_b, T_c]\, T_a) \end{aligned} \tag{2.39}$$

which implies

$$f_{abc} = f_{bca}\,. \tag{2.40}$$

Taken together, (2.19) and (2.40) imply the complete antisymmetry of f_{abc}

$$\begin{aligned} f_{abc} &= f_{bca} = f_{cab} \\ &= -f_{bac} = -f_{acb} = -f_{cba}\,. \end{aligned} \tag{2.41}$$

In this basis, the adjoint representation is unitary, because the T_a are imaginary and antisymmetric, and therefore hermitian.

---

[3]Algebras in which some of the k_as are negative have no nontrivial finite dimensional unitary representations. This does not mean that they are not interesting (the Lorentz group is one such), but we will not discuss them.

2.5 Simple algebras and groups

An **invariant subalgebra** is some set of generators which goes into itself under commutation with any element of the algebra. That is, if X is any generator in the invariant subalgebra and Y is any generator in the whole algebra, $[Y, X]$ is a generator in the invariant subalgebra. When exponentiated, an invariant subalgebra generates an invariant subgroup. To see this note that

$$h = e^{iX}, \quad g = e^{iY} \tag{2.42}$$

$$g^{-1} h\, g = e^{iX'} \tag{2.43}$$

where

$$X' = e^{-iY} X e^{iY} = X - i\,[Y, X] - \frac{1}{2}\,[Y, [Y, X]] + \cdots . \tag{2.44}$$

Note that the easy way to see this is to consider

$$X'(\epsilon) = e^{-i\epsilon Y} X e^{i\epsilon Y} \tag{2.45}$$

then Taylor expand in ϵ and set $\epsilon = 1$. Each derivative brings another commutator. Evidently, each of the terms in X' is in the subalgebra, and thus $e^{iX'}$ is in the subgroup, which is therefore invariant.

The whole algebra and 0 are trivial invariant subalgebras. An algebra which has no nontrivial invariant subalgebra is called **simple**. A simple algebra generates a **simple group**.

The adjoint representation of a simple Lie algebra satisfying (2.37) is irreducible. To see this, assume the contrary. Then there is an invariant subspace in the adjoint representation. But the states of the adjoint representation correspond to generators, so this means that we can find a basis in which the invariant subspace is spanned by some subset of the generators, T_r for $r = 1$ to K. Call the rest of the generators T_x for $x = K + 1$ to N. Then because the rs span an invariant subspace, we must have

$$[T_a]_{xr} = -i\, f_{axr} = 0 \tag{2.46}$$

for all a, x and r. Because of the complete antisymmetry of the structure constants, this means that all components of f that have two rs and one x or two xs and one r vanish. But that means that the nonzero structures constants involve either three rs or three xs, and thus the algebra falls apart into two nontrivial invariant subalgebras, and is not simple. Thus the adjoint representation of a simple Lie algebra satisfying (2.37) is irreducible.

We will often find it useful to discuss special Abelian invariant subalgebras consisting of a single generator which commutes with all the generators of the group (or of some subgroup we are interested in). We will call such an algebra a $U(1)$ factor of the group. $U(1)$ is the group of phase transformations. $U(1)$ factors do not appear in the structure constants at all. These Abelian invariant subalgebras correspond to directions in the space of generators for which $k_a = 0$ in (2.36). If X_a is a $U(1)$ generator, $f_{abc} = 0$ for all b and c. That also means that the corresponding k^a is zero, so the trace scalar product does not give a norm on the space. The structure constants do not tell us anything about the $U(1)$ subalgebras.

Algebras without Abelian invariant subalgebras are called **semisimple**. They are built, as we will see, by putting simple algebras together. In these algebras, every generator has a non-zero commutator with some other generator. Because of the cyclic property of the structure constants, (2.38), this also implies that every generator is a linear combination of commutators of generators. In such a case, the structure constants carry a great deal of information. We will use them to determine the entire structure of the algebra and its representations. From here on, unless explicitly stated, we will discuss semisimple algebras, and we will deal with representations by unitary operators.

2.6 States and operators

The generators of a representation (like the elements of the representations they generate) can be thought of as either linear operators or matrices, just as we saw when we were discussing representations of finite groups —

$$X_a|i\rangle = |j\rangle\langle j|X_a|i\rangle = |j\rangle[X_a]_{ji} \tag{2.47}$$

with the sum on j understood. As in (1.98), the states form row vectors and the matrix representing a linear operator acts on the right.

In the Hilbert space on which the representation acts, the group elements can be thought of as transformations on the states. The group element $e^{i\alpha_a X_a}$ maps or **transforms** the kets as follows:

$$|i\rangle \to |i'\rangle = e^{i\alpha_a X_a}|i\rangle\,. \tag{2.48}$$

Taking the adjoint shows that the corresponding bras transform as

$$\langle i| \to \langle i'| = \langle i|e^{-i\alpha_a X_a}\,. \tag{2.49}$$

The ket obtained by acting on $|i\rangle$ with an operator O is a sum of kets, and therefore must also transform as in (2.48).

$$\begin{aligned} O|i\rangle &\to e^{i\alpha_a X_a} O|i\rangle \\ &= e^{i\alpha_a X_a} O e^{-i\alpha_a X_a} e^{i\alpha_a X_a} |i\rangle = O'|i'\rangle . \end{aligned} \tag{2.50}$$

This implies that any operator O transforms as follows:

$$O \to O' = e^{i\alpha_a X_a} O e^{-i\alpha_a X_a} . \tag{2.51}$$

The transformation leaves all matrix elements invariant.

The action of the algebra on these objects is related to the change in the state of operator under an infinitesimal transformation.

$$-i\delta|i\rangle = -i\left((1 + i\alpha_a X_a)|i\rangle - |i\rangle\right) = \alpha_a X_a |i\rangle \tag{2.52}$$

$$-i\delta\langle i| = -\langle i|\alpha_a X_a \tag{2.53}$$

$$-i\delta O = [\alpha_a X_a, O] . \tag{2.54}$$

Thus, corresponding to the action of the generator X_a on a ket

$$X_a|i\rangle \tag{2.55}$$

is $-X_a$ acting on a bra[4]

$$-\langle i|X_a \tag{2.56}$$

and the commutator of X_a with an operator

$$[X_a, O] . \tag{2.57}$$

Then the invariance of a matrix element $\langle i|O|i\rangle$ is expressed by the fact,

$$\langle i|O\left(X_a|i\rangle\right) + \langle i|\left[X_a, O\right]|i\rangle - \left(\langle i|X_a\right) O|i\rangle = 0 . \tag{2.58}$$

2.7 Fun with exponentials

Consider the exponential

$$e^{i\alpha_a X_a} \tag{2.59}$$

[4]The argument above can be summarized by saying that the minus signs in (2.56) and in the commutator in (2.57) come ultimately from the unitarity of the transformation, (2.48).

where X_a is a representation matrix. We can always define the exponential as a power series,

$$e^{i\alpha_a X_a} = \sum_{n=0}^{\infty} \frac{(i\alpha_a X_a)^n}{n!} \tag{2.60}$$

However, it is useful to develop some rules for dealing with these things without expanding, like our simple rules for exponentials of commuting numbers. We have already seen that the multiplication law is not as simple as just adding the exponents. You might guess that the calculus is also more complicated. In particular,

$$\frac{\partial}{\partial \alpha_b} e^{i\alpha_a X_a} \neq i X_b \, e^{i\alpha_a X_a} \tag{2.61}$$

However, it is true that

$$\frac{d}{ds} e^{is\alpha_a X_a} = i\alpha_b X_b \, e^{is\alpha_a X_a} = i e^{is\alpha_a X_a} \, \alpha_b X_b \tag{2.62}$$

because $\alpha_a X_a$ commutes with itself. This is very important, because you can often use it to derive other useful results. It is also true that

$$\frac{\partial}{\partial \alpha_b} \, e^{i\alpha_a X_a} \bigg|_{\alpha=0} = i X_b \tag{2.63}$$

because this can be shown directly from the expansion. It is occasionally useful to have a general expression for the derivative. Besides, it is a beautiful formula, so I will write it down and tell you how to derive it. The formula is

$$\frac{\partial}{\partial \alpha_b} e^{i\alpha_a X_a} = \int_0^1 ds \, e^{is\alpha_a X_a} \, (i X_b) \, e^{i(1-s)\alpha_c X_c} \tag{2.64}$$

I love this relation because it is so nontrivial, yet so easy to remember. The integral just expresses the fact that the derivative may act anywhere "inside" the exponential, so the result is the average of all the places where the derivative can act. One way of deriving this is to define the exponential as a limit as in (2.7).

$$e^{i\alpha_a X_a} = \lim_{k\to\infty} (1 + i\alpha_a X_a / k)^k \tag{2.65}$$

and differentiate both sides — the result (2.64) is then just an exercise in defining an integral as a limit of a sum. Another way of doing it is to expand both sides and use the famous integral

$$\int_0^1 ds \, s^m \, (1-s)^n = \frac{m! \, n!}{(m+n+1)!} \tag{2.66}$$

We will see other properties of exponentials of matrices as we go along.

Problems

2.A. Find all components of the matrix $e^{i\alpha A}$ where

$$A = \begin{pmatrix} 0 & 0 & 1 \\ 0 & 0 & 0 \\ 1 & 0 & 0 \end{pmatrix}$$

2.B. If $[A, B] = B$, calculate

$$e^{i\alpha A} \, B \, e^{-i\alpha A}$$

2.C. Carry out the expansion of δ_c in (2.11) and (2.12) to third order in α and β (one order beyond what is discussed in the text).

Chapter 3

SU(2)

The $SU(2)$ algebra is familiar.[1]

$$[J_j, J_k] = i\epsilon_{jk\ell}J_\ell \tag{3.1}$$

This is the simplest of the compact Lie algebras because ϵ_{ijk} for $i, j, k =$ 1 to 3 is the simplest possible completely antisymmetric object with three indices. (3.1) is equivalent (in units in which $\hbar = 1$) to the angular momentum algebra that you studied in quantum mechanics. In fact we will only do two things differently here. One is to label the generators by 1, 2 and 3 instead of x, y and z. This is obviously a great step forward. More important is the fact that we will not make any use of the operator $J_a J_a$. Initially, this will make the analysis slightly more complicated, but it will start us on a path that generalizes beautifully to all the other compact Lie algebras.

3.1 J_3 eigenstates

Our ultimate goal is to completely reduce the Hilbert space of the world to block diagonal form. To start the process, let us think about some finite space, of dimension N, and assume that it transforms under some irreducible representation of the algebra. Then we can see what the form of the algebra tells us about the representation. Clearly, we want to diagonalize as many of the elements of the algebra as we can. In this case, since nothing commutes with anything else, we can only diagonalize one element, which we may as well take to be J_3. When we have done that, we pick out the states with the highest value of J_3 (we can always do that because we have assumed that the space

[1]We will see below why the name $SU(2)$ is appropriate.

DOI: 10.1201/9780429499210-4

is finite dimensional). Call the highest value of J_3 j. Then we have a set of states

$$J_3|j,\alpha\rangle = j|j,\alpha\rangle \tag{3.2}$$

where α is another label, only necessary if there is more than one state of highest J_3 (of course, you know that we really don't need α because the highest state is unique, but we haven't shown that yet, so we will keep it). We can also always choose the states so that

$$\langle j,\alpha|j,\beta\rangle = \delta_{\alpha\beta} \tag{3.3}$$

3.2 Raising and lowering operators

Now, just as in introductory quantum mechanics, we define raising and lowering operators,

$$J^{\pm} = (J_1 \pm iJ_2)/\sqrt{2} \tag{3.4}$$

satisfying

$$[J_3, J^{\pm}] = \pm J^{\pm} \tag{3.5}$$

$$[J^+, J^-] = J_3 \tag{3.6}$$

so they raise and lower the value of J_3 on the states. If

$$J_3|m\rangle = m|m\rangle \tag{3.7}$$

then

$$J_3 J^{\pm}|m\rangle = J^{\pm}J_3|m\rangle \pm J^{\pm}|m\rangle = (m \pm 1)\, J^{\pm}|m\rangle \tag{3.8}$$

The key idea is that we can use the raising and lowering operators to construct the irreducible representations and to completely reduce reducible representations. This idea is very simple for $SU(2)$, but it is very useful to see how it works in this simple case before we generalize it to an arbitrary compact Lie algebra.

There is no state with J_3=j+1 because we have assumed that j is the highest value of J_3. Thus it must be that

$$J^+|j,\alpha\rangle = 0 \;\forall\; \alpha \tag{3.9}$$

because any non-zero states would have J_3=j+1. The states obtained by acting with the lowering operator have J_3=$j-1$, so it makes sense to define

$$J^-|j,\alpha\rangle \equiv N_j(\alpha)|j-1,\alpha\rangle \tag{3.10}$$

where $N_j(\alpha)$ is a normalization factor. But we easily see that states with different α are orthogonal, because

$$\begin{aligned}
&N_j(\beta)^* N_j(\alpha)\langle j-1,\beta|j-1,\alpha\rangle \\
&= \langle j,\beta|J^+J^-|j,\alpha\rangle \\
&= \langle j,\beta|\,[J^+,J^-]\,|j,\alpha\rangle \\
&= \langle j,\beta|J_3|j,\alpha\rangle \\
&= j\langle j,\beta|j,\alpha\rangle = j\;\delta_{\alpha\beta}
\end{aligned} \tag{3.11}$$

Thus we can choose the states $|j-1,\alpha\rangle$ to be orthonormal by choosing

$$N_j(\alpha) = \sqrt{j} \equiv N_j \tag{3.12}$$

Then in addition to (3.10), we have

$$\begin{aligned}
J^+|j-1,\alpha\rangle &= \frac{1}{N_j}J^+J^-|j,\alpha\rangle \\
&= \frac{1}{N_j}\,[J^+,J^-]\,|j,\alpha\rangle \\
&= \frac{j}{N_j}|j,\alpha\rangle = N_j|j,\alpha\rangle
\end{aligned} \tag{3.13}$$

The point is that because of the algebra, we can define the states so that the raising and lowering operators act without changing α. That is why the parameter α is eventually going to go away. Now an analogous argument shows that there are orthonormal states $|j-2,\alpha\rangle$ satisfying

$$\begin{aligned}
J^-|j-1,\alpha\rangle &= N_{j-1}|j-2,\alpha\rangle \\
J^+|j-2,\alpha\rangle &= N_{j-1}|j-1,\alpha\rangle
\end{aligned} \tag{3.14}$$

Continuing the process, we find a whole tower of orthonormal states, $|j-k,\alpha\rangle$ satisfying

$$\begin{aligned}
J^-|j-k,\alpha\rangle &= N_{j-k}|j-k-1,\alpha\rangle \\
J^+|j-k-1,\alpha\rangle &= N_{j-k}|j-k,\alpha\rangle
\end{aligned} \tag{3.15}$$

The Ns can be chosen to be real, and because of the algebra, they satisfy

$$\begin{aligned}
N_{j-k}^2 &= \langle j-k,\alpha|J^+J^-|j-k,\alpha\rangle \\
&= \langle j-k,\alpha|\,[J^+,J^-]\,|j-k,\alpha\rangle \\
&\quad +\langle j-k,\alpha|J^-J^+|j-k,\alpha\rangle \\
&= N_{j-k+1}^2 + j - k
\end{aligned} \tag{3.16}$$

This is a recursion relation for the Ns which is easy to solve by starting with N_j:

$$\begin{array}{rcll} N_j^2 & & & = j \\ N_{j-1}^2 & - & N_j^2 & = j-1 \\ & \vdots & & \vdots \\ N_{j-k}^2 & - & N_{j-k+1}^2 & = j-k \\ \hline N_{j-k}^2 & = & \multicolumn{2}{l}{(k+1)j - k(k+1)/2} \\ & = & \multicolumn{2}{l}{\frac{1}{2}(k+1)(2j-k)} \end{array} \tag{3.17}$$

or setting $k = j - m$

$$N_m = \frac{1}{\sqrt{2}}\sqrt{(j+m)(j-m+1)} \tag{3.18}$$

Because the representation is finite dimensional (by assumption — we haven't proved this) there must be some maximum number of lowering operators, ℓ, that we can apply to $|j,\alpha\rangle$. We must eventually come to some $m = j - \ell$ such that applying any more lowering operators gives 0. Then ℓ is a non-negative integer specifying the number of times we can lower the states with highest J_3. Another lowering operator annihilates the state —

$$J^-|j-\ell,\alpha\rangle = 0\,. \tag{3.19}$$

But then the norm of $J^-|j-\ell,\alpha\rangle$ must vanish, which means that

$$N_{j-\ell} = \frac{1}{\sqrt{2}}\sqrt{(2j-\ell)(\ell+1)} = 0 \tag{3.20}$$

the factor $\ell + 1$ cannot vanish, thus we must have

$$\ell = 2j\,. \tag{3.21}$$

Thus

$$j = \frac{\ell}{2} \quad \text{for some integer } \ell. \tag{3.22}$$

Now we can get rid of α. It is now clear that the space breaks up into subspaces that are invariant under the algebra, one for each value of α, because the generators do not change α. Thus from our original assumption of irreducibility, there must be only one α value, so we can drop the α entirely.

Furthermore, there can be no other states, or the subspace we just constructed would be nontrivial (and invariant). Thus we have learned how the generators act on all the finite dimensional irreducible representations. In fact, though we won't prove it, there are no others — that is all representations are finite dimensional, so we know all of them.

3.3 The standard notation

We can now switch to the standard notation in which we label the states of the irreducible representations by the highest J_3 value in the representation and the J_3 value:[2]

$$|j,m\rangle \tag{3.23}$$

and the matrix elements of the generators are determined by the matrix elements of J_3 and the raising and lowering operators, $J^{\pm}$:[3]

$$\begin{aligned} \langle j,m'|J_3|j,m\rangle &= m\,\delta_{m'm} \\ \langle j,m'|J^+|j,m\rangle &= \sqrt{(j+m+1)(j-m)/2}\,\delta_{m',m+1} \\ \langle j,m'|J^-|j,m\rangle &= \sqrt{(j+m)(j-m+1)/2}\,\delta_{m',m-1} \end{aligned} \tag{3.24}$$

These matrix elements define the **spin** j representation of the $SU(2)$ algebra:

$$[J_a^j]_{k\ell} = \langle j,j+1-k|J_a|j,j+1-\ell\rangle \tag{3.25}$$

Here we have written the matrix elements in the conventional language where the rows and columns are labeled from 1 to $2j+1$. In this case, it is often convenient to label the rows and columns directly by their m values, which are just $j+1-\ell$ and $j+1-k$ above in (3.25). In this notation, (3.25) would read

$$[J_a^j]_{m'm} = \langle j,m'|J_a|j,m\rangle \tag{3.26}$$

where m and m' run from j to $-j$ in steps of -1. We will use these interchangeably — choosing whichever is most convenient for the problem at hand.

[2]Well, not completely standard — in some books, including the first edition of this one, the j and m are written in the other order.

[3]The $\sqrt{2}$ factors are the result of our definition of the raising and lowering operators and are absent in some other treatments.

For example, for 1/2, this gives the spin 1/2 representation

$$
\begin{aligned}
J_1^{1/2} &= \frac{1}{2}\begin{pmatrix} 0 & 1 \\ 1 & 0 \end{pmatrix} = \frac{1}{2}\sigma_1 \\
J_2^{1/2} &= \frac{1}{2}\begin{pmatrix} 0 & -i \\ i & 0 \end{pmatrix} = \frac{1}{2}\sigma_2 \\
J_3^{1/2} &= \frac{1}{2}\begin{pmatrix} 1 & 0 \\ 0 & -1 \end{pmatrix} = \frac{1}{2}\sigma_3
\end{aligned}
\tag{3.27}
$$

where the σs are the Pauli matrices.

$$
\sigma_1 = \begin{pmatrix} 0 & 1 \\ 1 & 0 \end{pmatrix} \quad \sigma_2 = \begin{pmatrix} 0 & -i \\ i & 0 \end{pmatrix} \quad \sigma_3 = \begin{pmatrix} 1 & 0 \\ 0 & -1 \end{pmatrix}
\tag{3.28}
$$

satisfying

$$
\sigma_a \sigma_b = \delta_{ab} + i\epsilon_{abc}\sigma_c
\tag{3.29}
$$

The spin 1/2 representation is the simplest representation of $SU(2)$. It is called the "defining" representation of $SU(2)$, and is responsible for the name SU, which is an acronym for "Special Unitary". Exponentiating the generators of the spin 1/2 representation to get the representation of finite group elements gives matrices of the form

$$
e^{i\vec{\alpha}\cdot\vec{\sigma}/2}
\tag{3.30}
$$

which are the most general 2×2 unitary matrices with determinant 1. The "special", in Special Unitary means that the determinant is 1, rather than an arbitrary complex number of absolute value 1.

All the other irreducible representations can be constructed similarly. For example, the spin 1 representation looks like

$$
\begin{aligned}
J_1^1 &= \frac{1}{\sqrt{2}}\begin{pmatrix} 0 & 1 & 0 \\ 1 & 0 & 1 \\ 0 & 1 & 0 \end{pmatrix} \\
J_2^1 &= \frac{1}{\sqrt{2}}\begin{pmatrix} 0 & -i & 0 \\ i & 0 & -i \\ 0 & i & 0 \end{pmatrix} \\
J_3^1 &= \begin{pmatrix} 1 & 0 & 0 \\ 0 & 0 & 0 \\ 0 & 0 & -1 \end{pmatrix}
\end{aligned}
\tag{3.31}
$$

while the spin 3/2 representation is

$$J_1^{3/2} = \begin{pmatrix} 0 & \sqrt{\frac{3}{2}} & 0 & 0 \\ \sqrt{\frac{3}{2}} & 0 & 2 & 0 \\ 0 & 2 & 0 & \sqrt{\frac{3}{2}} \\ 0 & 0 & \sqrt{\frac{3}{2}} & 0 \end{pmatrix}$$

$$J_2^{3/2} = \begin{pmatrix} 0 & -\sqrt{\frac{3}{2}}i & 0 & 0 \\ \sqrt{\frac{3}{2}}i & 0 & -2i & 0 \\ 0 & 2i & 0 & -\sqrt{\frac{3}{2}}i \\ 0 & 0 & \sqrt{\frac{3}{2}}i & 0 \end{pmatrix} \tag{3.32}$$

$$J_3^{3/2} = \begin{pmatrix} \frac{3}{2} & 0 & 0 & 0 \\ 0 & \frac{1}{2} & 0 & 0 \\ 0 & 0 & -\frac{1}{2} & 0 \\ 0 & 0 & 0 & -\frac{3}{2} \end{pmatrix}$$

The construction of the irreducible representations above generalizes to any compact Lie algebra, as we will see. The J_3 values are called **weights**, and the analysis we have just done is called the **highest weight construction** because it starts with the unique highest weight of the representation. Note that the same construction provides a systematic procedure for bringing an arbitrary finite dimensional representation into block diagonal form. The procedure is as follows:

1. Diagonalize J_3.

2. Find the states with the highest J_3 value, j.

3. For each such state, explicitly construct the states of the irreducible spin j representation by applying the lowering operator to the states with highest J_3.

4. Now set aside the subspace spanned by these representations, which is now in canonical form, and concentrate on the subspace orthogonal to it.

5. Take these remaining states, go to step 2 and start again with the states with next highest J_3 value.

(3.33)

The end result will be the construction of a basis for the Hilbert space of the form

$$|j, m, \alpha\rangle \tag{3.34}$$

where m and j refer to the J_3 value and the representation as usual (as in (3.23) and α refers to all the other observables that can be diagonalized to characterize the state. These satisfy

$$\langle j', m', \alpha' | j, m, \alpha\rangle = \delta_{m'm}\, \delta_{j'j}\, \delta_{\alpha'\alpha} \tag{3.35}$$

The Kronecker δs are automatic consequences of our construction. They are also required by Schur's lemma, because the matrix elements satisfy

$$\begin{aligned} &\langle j', m', \alpha' | J_a | j, m, \alpha\rangle \\ &= [J_a^{j'}]_{m'm''} \langle j', m'', \alpha' | j, m, \alpha\rangle \\ &= \langle j', m', \alpha' | j, m'', \alpha\rangle [J_a^j]_{m''m} \end{aligned} \tag{3.36}$$

because we can insert a complete set of intermediate states on either side of J_a. Thus $\langle j', m', \alpha' | j, m, \alpha\rangle$ commutes with all the elements of an irreducible representation, and is either 0 if $j \neq j'$ or proportional to the identity, $\delta_{m'm}$ if $j = j'$.

3.4 Tensor products

You have probably all used the highest weight scheme, possibly without knowing it, to do what in introductory quantum mechanics is called **addition of angular momentum**. This occurs when we form a **tensor product** of two sets of states which transform under the group.[4] This happens, in turn, whenever a system responds to the group transformation in more than one way. The classic example of this is a particle that carries both spin and orbital angular momentum. In this case, the system can be described in a space that you can think of as built of a product of two different kinds of kets.

$$|i, x\rangle \equiv |i\rangle\, |x\rangle \tag{3.37}$$

where the first states, $|i\rangle$ transforms under representation D_1 of the group and the second, $|x\rangle$, under D_2. Then the product, called the tensor product,

[4] We saw an example of this in the normal modes of the triangle in our discussion of finite groups.

transforms as follows:

$$
\begin{aligned}
D(g)\,|i,x\rangle &= |j,y\rangle\,[D_{1\otimes 2}(g)]_{jyix} \\
&= |j\rangle\,|y\rangle\,[D_1(g)]_{ji}\,[D_2(g)]_{yx} \\
&= (|j\rangle\,[D_1(g)]_{ji})\;(|y\rangle\,[D_2(g)]_{yx})
\end{aligned}
\tag{3.38}
$$

In other words, the two kets are just transforming independently under their own representations. If we look at this near the identity, for infinitesimal α_a,

$$
\begin{aligned}
&(1+i\alpha_a J_a)\,|i,x\rangle \\
&= |j,y\rangle\langle j,y|\,(1+i\alpha_a J_a)\,|i,x\rangle \\
&= |j,y\rangle\,\left(\delta_{ji}\delta_{yx} + i\alpha_a[J_a^{1\otimes 2}(g)]_{jyix}\right) \\
&= |j,y\rangle\,\left(\delta_{ji} + i\alpha_a[J_a^1]_{ji}\right)\left(\delta_{yx} + i\alpha_a[J_a^2]_{yx}\right)
\end{aligned}
\tag{3.39}
$$

Thus identifying first powers of α_a

$$
[J_a^{1\otimes 2}(g)]_{jyix} = [J_a^1]_{ji}\delta_{yx} + \delta_{ji}[J_a^2]_{yx} \tag{3.40}
$$

When we multiply the representations, the generators add, in the sense shown in (3.40). This is what happens with addition of angular momenta. We will often write (3.40) simply as

$$
J_a^{1\otimes 2} = J_a^1 + J_a^2 \tag{3.41}
$$

leaving you to figure out from the context where the indices go, and ignoring the δ-functions which, after all, are just identity operators on the appropriate space. In fact, you can think of this in terms of the action of the generators as follows:

$$
J_a\Big(|j\rangle|x\rangle\Big) = \Big(J_a|j\rangle\Big)|x\rangle + |j\rangle\Big(J_a|x\rangle\Big) \tag{3.42}
$$

3.5 J_3 values add

This is particularly simple for the generator J_3 because we work in a basis in which J_3 is diagonal. Thus **the J_3 values of tensor product states are just the sums of the J_3 values of the factors:**

$$
J_3\Big(|j_1,m_1\rangle|j_2,m_2\rangle\Big) = (m_1+m_2)\,\Big(|j_1,m_1\rangle|j_2,m_2\rangle\Big) \tag{3.43}
$$

This is what we would expect, classically, for addition of angular momentum, of course. But in quantum mechanics, we can only make it work for one component. We can, however, use this in the highest weight construction, (3.33).

Consider, for example, the tensor product of a spin 1/2 and spin 1 representation. The highest weight procedure (3.33) is what you would use to decompose the product space into irreducible representations. Let's do it explicitly. There is a unique highest weight state,

$$|3/2, 3/2\rangle = |1/2, 1/2\rangle|1, 1\rangle \tag{3.44}$$

We can now construct the rest of the spin 3/2 states by applying lowering operators to both sides. For example using (3.42)

$$\begin{gathered} J^-|3/2, 3/2\rangle = J^-\Big(|1/2, 1/2\rangle|1, 1\rangle\Big) \\ = \sqrt{\frac{3}{2}}|3/2, 1/2\rangle = \sqrt{\frac{1}{2}}|1/2, -1/2\rangle|1, 1\rangle + |1/2, 1/2\rangle|1, 0\rangle \end{gathered} \tag{3.45}$$

or

$$|3/2, 1/2\rangle = \sqrt{\frac{1}{3}}|1/2, -1/2\rangle|1, 1\rangle + \sqrt{\frac{2}{3}}|1/2, 1/2\rangle|1, 0\rangle \tag{3.46}$$

Continuing the process gives

$$\begin{gathered} |3/2, -1/2\rangle = \sqrt{\frac{2}{3}}|1/2, -1/2\rangle|1, 0\rangle + \sqrt{\frac{1}{3}}|1/2, 1/2\rangle|1, -1\rangle \\ |3/2, -3/2\rangle = |1/2, -1/2\rangle|1, -1\rangle \end{gathered} \tag{3.47}$$

Then the remaining states are orthogonal to these —

$$\sqrt{\frac{2}{3}}|1/2, -1/2\rangle|1, 1\rangle - \sqrt{\frac{1}{3}}|1/2, 1/2\rangle|1, 0\rangle \tag{3.48}$$

and

$$\sqrt{\frac{1}{3}}|1/2, -1/2\rangle|1, 0\rangle - \sqrt{\frac{2}{3}}|1/2, 1/2\rangle|1, -1\rangle \tag{3.49}$$

applying the highest weight scheme to this reduced space gives

$$\begin{gathered} |1/2, 1/2\rangle = \sqrt{\frac{2}{3}}|1/2, -1/2\rangle|1, 1\rangle - \sqrt{\frac{1}{3}}|1/2, 1/2\rangle|1, 0\rangle \\ |1/2, -1/2\rangle = \sqrt{\frac{1}{3}}|1/2, -1/2\rangle|1, 0\rangle - \sqrt{\frac{2}{3}}|1/2, 1/2\rangle|1, -1\rangle \end{gathered} \tag{3.50}$$

In this case, we have used up all the states, so the process terminates. Note that the signs of the spin 1/2 states were not determined when we found the states orthogonal to the spin 3/2 states, but that the **relative** sign is fixed because the $J_3 = \pm 1/2$ states are related by the raising and lowering operators.

Problems

3.A. Use the highest weight decomposition, (3.33), to show that

$$\{j\} \otimes \{s\} = \sum_{\oplus \ell=|s-j|}^{s+j} \{\ell\}$$

where the $\oplus$ in the summation just means that the sum is a direct sum, and $\{k\}$ denotes the spin k representation of $SU(2)$. To do this problem, you do not need to construct the precise linear combinations of states that appear in each irreducible representation, but you must at least show how the counting of states goes at each stage of the highest weight decomposition.

3.B. Calculate

$$e^{i\vec{r}\cdot\vec{\sigma}}$$

where $\vec{\sigma}$ are the Pauli matrices. Hint: write $\vec{r} = |\vec{r}|\ \hat{r}$.

3.C. Show explicitly that the spin 1 representation obtained by the highest weight procedure with $j = 1$ is equivalent to the adjoint representation, with $f_{abc} = \epsilon_{abc}$ by finding the similarity transformation that implements the equivalence.

3.D. Suppose that $[\sigma_a]_{ij}$ and $[\eta_a]_{xy}$ are Pauli matrices in two different two dimensional spaces. In the four dimensional tensor product space, define the basis

$$|1\rangle = |i=1\rangle|x=1\rangle \qquad |2\rangle = |i=1\rangle|x=2\rangle$$

$$|3\rangle = |i=2\rangle|x=1\rangle \qquad |4\rangle = |i=2\rangle|x=2\rangle$$

Write out the matrix elements of $\sigma_2 \otimes \eta_1$ in this basis.

3.E. We will often abbreviate the tensor product notation by leaving out the indices and the identity matrices. This makes for a very compact notation, but you must keep your wits about you to stay in the right space. In the example of problem 3.D, we could write:

$$[\sigma_a]_{ij}[\eta_b]_{xy} \quad \text{as} \quad \sigma_a\eta_b$$

$$[\sigma_a]_{ij}\delta_{xy} \quad \text{as} \quad \sigma_a$$

$$\delta_{ij}[\eta_b]_{xy} \quad \text{as} \quad \eta_b$$

$$\delta_{ij}\delta_{xy} \quad \text{as} \quad 1$$

So for example, $(\sigma_1)(\sigma_2\eta_1) = i\sigma_3\eta_1$ and $(\sigma_1\eta_2)(\sigma_1\eta_3) = i\eta_1$.

To get some practice with this notation, calculate

(a) $[\sigma_a, \sigma_b\eta_c]$,

(b) $\mathrm{Tr}\left(\sigma_a\left\{\eta_b, \sigma_c\eta_d\right\}\right)$,

(c) $[\sigma_1\eta_1, \sigma_2\eta_2]$.

where σ_a and η_a are independent sets of Pauli matrices and $\{A, B\} = AB + BA$ is the "anticommutator."

Chapter 4

Tensor Operators

A **tensor operator** is a set of operators that transforms under commutation with the generators of some Lie algebra like an irreducible representation of the algebra. In this chapter, we will define and discuss tensor operators for the $SU(2)$ algebra discussed in chapter 3. A tensor operator transforming under the spin-s representation of $SU(2)$ consists of a set of operators, O^s_ℓ for $\ell = 1$ to $2s+1$ (or $-s$ to s), such that

$$[J_a, O^s_\ell] = O^s_m \, [J^s_a]_{m\ell} \,. \tag{4.1}$$

It is true, though we have not proved it, that every irreducible representation is finite dimensional and equivalent to one of the representations that we found with the highest weight construction. We can always choose all tensor operators for $SU(2)$ to have this form.

4.1 Orbital angular momentum

Here is an example — a particle in a spherically symmetric potential. If the particle has no spin, then J_a is the orbital angular momentum operator,

$$J_a = L_a = \epsilon_{abc} \, r_b \, p_c \tag{4.2}$$

The position vector is related to a tensor operator because it transforms under the adjoint representation

$$\begin{aligned} [J_a, r_b] &= \epsilon_{acd} \, [r_c \, p_d, r_b] = -i \, \epsilon_{acd} \, r_c \, \delta_{bd} \\ &= -i \, \epsilon_{acb} \, r_c = r_c \, [J^{\text{adj}}_a]_{cb} \end{aligned} \tag{4.3}$$

DOI: 10.1201/9780429499210-5

where $J^{\rm adj}$ is the adjoint representation, and we know from problem 3.C that this representation is equivalent to the standard spin 1 representation from the highest weight procedure.

4.2 Using tensor operators

Note that the transformation of the position operator in (4.3) does not have quite the right form, because the representation matrices $J_a^{\rm adj}$ are not the standard form. The first step in using tensor operators is to choose the operator basis so that the conventional spin s representation appears in the commutation relation (4.1). This is not absolutely necessary, but it makes things easier, as we will see. We will discuss this process in general, and then see how it works for r_a.

Suppose that we are given a set of operators, Ω_x for $x = 1$ to $2s+1$ that transforms according a representation D that is equivalent to the spin-s representation of $SU(2)$:

$$[J_a, \Omega_x] = \Omega_y \, [J_a^D]_{yx} \tag{4.4}$$

Since by assumption, D is equivalent to the spin-s representation, we can find a matrix S such that

$$S \, J_a^D \, S^{-1} = J_a^s \tag{4.5}$$

or in terms of matrix elements

$$[S]_{\ell x} \, [J_a^D]_{xy} \, [S^{-1}]_{y\ell'} = [J_a^s]_{\ell\ell'} \tag{4.6}$$

Then we define a new set of operators

$$O_\ell^s = \Omega_y \, [S^{-1}]_{y\ell} \quad \text{for } \ell = -s \text{ to } s \tag{4.7}$$

Now O_ℓ^s satisfies

$$\begin{aligned}
&[J_a, O_\ell^s] \\
&= [J_a, \Omega_y] \, [S^{-1}]_{y\ell} \\
&= \Omega_z [J_a^D]_{zy} \, [S^{-1}]_{y\ell} \\
&= \Omega_z [S^{-1}]_{z\ell'} \, [S]_{\ell' z'} \, [J_a^D]_{z'y} \, [S^{-1}]_{y\ell} \\
&= O_{\ell'}^s \, [J_a^s]_{\ell'\ell}
\end{aligned} \tag{4.8}$$

which is what we want. Notice that (4.8) is particularly simple for J_3, because in our standard basis in which the indices ℓ label the J_3 value, J_3^1, (or J_3^s for any s) is a diagonal matrix

$$[J_3^s]_{\ell'\ell} = \ell\, \delta_{\ell\ell'} \quad \text{for } \ell, \ell' = -s \text{ to } s\,. \tag{4.9}$$

Thus

$$[J_3, O_\ell^s] = O_{\ell'}^s\, [J_3^s]_{\ell'\ell} = \ell\, O_\ell^s\,. \tag{4.10}$$

In practice, it is usually not necessary to find the matrix S explicitly. If we can find any linear combination of the Ω_x which has a definite value of J_3 (that means that it is proportional to its commutator with J_3), we can take that to be a component of O^s, and then build up all the other O^s components by applying raising and lowering operators.

For the position operator it is easiest to start by finding the operator r_0. Since $[J_3, r_3] = 0$, we know that r_3 has $J_3 = 0$ and therefore that $r_3 \propto r_0$. Thus we can take

$$r_0 = r_3 \tag{4.11}$$

Then the commutation relations for the spin 1 raising and lowering operators give the rest

$$\begin{aligned} [J^\pm, r_0] &= r_{\pm 1} \\ &= \mp(r_1 \pm i\, r_2)/\sqrt{2} \end{aligned} \tag{4.12}$$

4.3 The Wigner-Eckart theorem

The interesting thing about tensor operators is how the product $O_\ell^s\,|j, m, \alpha\rangle$ transforms.

$$\begin{aligned} & J_a\, O_\ell^s\,|j, m, \alpha\rangle \\ &= [J_a, O_\ell^s]\,|j, m, \alpha\rangle + O_\ell^s\, J_a\,|j, m, \alpha\rangle \\ &= O_{\ell'}^s\,|j, m, \alpha\rangle\, [J_a^s]_{\ell'\ell} + O_\ell^s\,|j, m', \alpha\rangle\, [J_a^j]_{m'm} \end{aligned} \tag{4.13}$$

This is the transformation law for a tensor product of spin s and spin j, $s \otimes j$. Because we are using the standard basis for the states and operators in which J_3 is diagonal, this is particularly simple for the generator J_3, for which (4.13) becomes

$$J_3\, O_\ell^s\,|j, m, \alpha\rangle = (\ell + m)\, O_\ell^s\,|j, m, \alpha\rangle \tag{4.14}$$

The J_3 value of the product of a tensor operator with a state is just the sum of the J_3 values of the operator and the state.

The remarkable thing about this is that the product of the tensor operator and the ket behaves under the algebra just like the tensor product of two kets. Thus we can decompose it into irreducible representations in exactly the same way, using the highest weight procedure. That is, we note that $O^s_s\,|j,j,\alpha\rangle$ with $J_3 = j + s$ is the highest weight state. We can lower it to construct the rest of the spin $j + s$ representation. Then we can find the linear combination of $J_3 = j + s - 1$ states that is the highest weight of the spin $j + s - 1$ representation, and lower it to get the entire representation, and so on. In this way, we find explicit representations for the states of the irreducible components of the tensor product in terms of linear combinations of the $O^s_\ell\,|j,m,\alpha\rangle$. You probably know, and have shown explicitly in problem 3.A, that in this decomposition, each representation from $j + s$ to $|j - s|$ appears exactly once. We can write the result of the highest weight analysis as follows:

$$\sum_\ell O^s_\ell\,|j,M-\ell,\alpha\rangle\,\langle s,j,\ell,M-\ell\mid J,M\rangle = k_J\,|J,M\rangle \tag{4.15}$$

Here $|J,M\rangle$ is a normalized state that transforms like the $J_3 = M$ component of the spin J representation and k_J is an unknown constant for each J (but does not depend on M). The coefficients $\langle s,j,\ell,M-\ell\mid J,M\rangle$ are determined by the highest weight construction, and can be evaluated from the tensor product of kets, where all the normalizations are known and the constants k_J are equal to 1:

$$\sum_\ell |s,\ell\rangle\,|j,M-\ell\rangle\,\langle s,j,\ell,M-\ell\mid J,M\rangle = |J,M\rangle \tag{4.16}$$

One way to prove[1] that the coefficients can be taken to be the same in (4.15) and (4.16) is to notice that in both cases, $J^+\,|J,J\rangle$ must vanish and that this condition determines the coefficients $\langle s,j,\ell,J-\ell\mid J,J\rangle$ up to a multiplicative constant. Since the transformation properties of $O^s_\ell\,|j,m\rangle$ and $|s,\ell\rangle\,|j,m\rangle$ are identical, the coefficients must be proportional. The only difference is the factor of k_J in (4.15).

We can invert (4.15) and express the original product states as linear combinations of the states with definite total spin J.

$$O^s_\ell\,|j,m,\alpha\rangle = \sum_{J=|j-s|}^{j+s} \langle J,\ell+m\mid s,j,\ell,m\rangle\,k_J\,|J,\ell+m\rangle \tag{4.17}$$

[1]This is probably obvious, but as we will emphasize below, the operators are different because we do not have a scalar product for them.

The coefficients $\langle J, M \mid s, j, \ell, M-\ell\rangle$ are thus entirely determined by the algebra, up to some choices of the phases of the states. Once we have a convention for fixing these phases, we can make tables of these coefficients once and for all, and be done with it. The notation $\langle J, \ell+m|s, j, \ell, m\rangle$ just means the coefficient of $|J, \ell+m\rangle$ in the product $|s, \ell\rangle\,|j, m\rangle$. These are called **Clebsch-Gordan coefficients.**

The Clebsch-Gordan coefficients are all group theory. The physics comes in when we reexpress the $|J, \ell+m\rangle$ in terms of the Hilbert space basis states $|J, \ell+m, \beta\rangle$ —

$$k_J\,|J, \ell+m\rangle = \sum_\beta k_{\alpha\beta}\,|J, \ell+m, \beta\rangle \tag{4.18}$$

We have absorbed the unknown coefficients k_J into the equally unknown coefficients $k_{\alpha\beta}$. These depend on α, j, O^s and s, because the original products do, and on β and J, of course. But they do not depend at all on ℓ or m. We only need to know the coefficients for one value of $\ell+m$. The $k_{\alpha\beta}$ are called **reduced matrix elements** and denoted

$$k_{\alpha\beta} = \langle J, \beta|\, O^s\,|j, \alpha\rangle \tag{4.19}$$

Putting all this together, we get the Wigner-Eckart theorem for matrix elements of tensor operators:

$$\begin{aligned}&\langle J, m', \beta|\, O^s_\ell\,|j, m, \alpha\rangle\\ &= \delta_{m',\ell+m}\,\langle J, \ell+m|s, j, \ell, m\rangle \cdot \langle J, \beta|\, O^s\,|j, \alpha\rangle\end{aligned} \tag{4.20}$$

If we know any non-zero matrix element of a tensor operator between states of some given J, β and j, α, we can compute all the others using the algebra. This sounds pretty amazing, but all that is really going on is that we can use the raising and lowering operators to go up and down within representations using pure group theory. Thus by clever use of the raising and lowering operators, we can compute any matrix element from another. The Wigner-Eckart theorem just expresses this formally.

4.4 Example

Suppose

$$\langle 1/2, 1/2, \alpha|\, r_3\,|1/2, 1/2, \beta\rangle = A \tag{4.21}$$

Find

$$\langle 1/2, 1/2, \alpha|\, r_1\,|1/2, -1/2, \beta\rangle = ? \tag{4.22}$$

First, since $r_0 = r_3$,

$$\langle 1/2, 1/2, \alpha | \, r_0 \, | 1/2, 1/2, \beta \rangle = A \tag{4.23}$$

Then we know from (4.12) that

$$r_1 = \frac{1}{\sqrt{2}}(-r_{+1} + r_{-1}) \tag{4.24}$$

Thus

$$\begin{aligned} &\langle 1/2, 1/2, \alpha | \, r_1 \, | 1/2, -1/2, \beta \rangle \\ &= \langle 1/2, 1/2, \alpha | \, \frac{1}{\sqrt{2}}(-r_{+1} + r_{-1}) \, | 1/2, -1/2, \beta \rangle \\ &= -\frac{1}{\sqrt{2}} \langle 1/2, 1/2, \alpha | \, r_{+1} \, | 1/2, -1/2, \beta \rangle \end{aligned} \tag{4.25}$$

Now we could plug this into the formula, and you could find the Clebsch-Gordan coefficients in a table. But I'll be honest with you. I can never remember what the definitions in the formula are long enough to use it. Instead, I try to understand what the formula means, and I suggest that you do the same. We could also just use what we have already done, decomposing $1/2 \otimes 1$ into irreducible representations. For example, we know from the highest weight construction that

$$|3/2, 3/2\rangle \equiv r_{+1} \, |1/2, 1/2, \beta\rangle \tag{4.26}$$

is a 3/2,3/2 state because it is the highest weight state that we can get as a product of an r_ℓ operator acting on an $|1/2, m\rangle$ state. Then we can get the corresponding $|3/2, 1/2\rangle$ state in the same representation by acting with the lowering operator J^-

$$\begin{aligned} |3/2, 1/2\rangle &= \sqrt{\frac{2}{3}} \, J^- \, |3/2, 3/2\rangle \\ &= \sqrt{\frac{2}{3}} \, r_0 \, |1/2, 1/2, \beta\rangle + \sqrt{\frac{1}{3}} \, r_{+1} \, |1/2, -1/2, \beta\rangle \end{aligned} \tag{4.27}$$

But we know that this spin-3/2 state has zero matrix element with any spin-1/2 state, and thus

$$\begin{aligned} 0 &= \langle 1/2, 1/2, \alpha | \, 3/2, 1/2 \rangle \\ &= \sqrt{\frac{2}{3}} \langle 1/2, 1/2, \alpha | \, r_0 \, | 1/2, 1/2, \beta \rangle \\ &+ \sqrt{\frac{1}{3}} \langle 1/2, 1/2, \alpha | \, r_{+1} \, | 1/2, -1/2, \beta \rangle \end{aligned} \tag{4.28}$$

so

$$\begin{aligned}&\langle 1/2,1/2,\alpha|\, r_{+1}\,|1/2,-1/2,\beta\rangle\\&=-\sqrt{2}\langle 1/2,1/2,\alpha|\, r_0\,|1/2,1/2,\beta\rangle\\&=-\sqrt{2}\,A\end{aligned}\tag{4.29}$$

so

$$\langle 1/2,1/2,\alpha|\, r_1\,|1/2,-1/2,\beta\rangle = A \tag{4.30}$$

Although we did not need it here, we can also conclude that

$$|1/2,1/2\rangle = \sqrt{\frac{1}{3}}\, r_0\,|1/2,1/2,\alpha\rangle - \sqrt{\frac{2}{3}}\, r_{+1}\,|1/2,-1/2,\alpha\rangle \tag{4.31}$$

is a 1/2,1/2 state. This statement is actually a little subtle, and shows the power of the algebra. When we did this analysis for the tensor product of j=1 and j=1/2 states, we used the fact that the $|1/2,1/2\rangle$ must be orthogonal to the $|3/2,1/2\rangle$ states to find the form of the $|1/2,1/2\rangle$ state. We cannot do this here, because we do not know from the symmetry alone how to determine the norms of the states

$$r_\ell\,|1/2,m\rangle \tag{4.32}$$

However, we know from the analysis with the states and the fact that the transformation of these objects is analogous that

$$J^+\,|1/2,1/2\rangle = 0 \tag{4.33}$$

Thus it is a 1/2,1/2 state because it is the highest weight state in the representation. We will return to this issue later.

There are several ways of approaching such questions. Here is another way. Consider the matrix elements

$$\langle 1/2,m,\alpha|\, r_a\,|1/2,m',\beta\rangle \tag{4.34}$$

The Wigner-Eckart theorem implies that these matrix elements are all proportional to a single parameter, the $k_{\alpha\beta}$. Furthermore, this result is a consequence of the algebra alone. Any operator that has the same commutation relations with J_a will have matrix elements proportional to r_a. But J_a itself has the same commutation relations. Thus the matrix elements of r_a are proportional to those of J_a. This is only helpful if the matrix elements of J_a are not zero (if they are all zero, the Wigner-Eckart theorem is trivially satisfied). In this case, they are not (at least if $\alpha=\beta$)

$$\langle 1/2,m,\alpha|\, J_a\,|1/2,m',\beta\rangle = \delta_{\alpha\beta}\frac{1}{2}[\sigma_a]_{mm'} \tag{4.35}$$

Thus

$$\langle 1/2, m, \alpha | \, r_a \, | 1/2, m', \beta \rangle \propto [\sigma_a]_{mm'} \tag{4.36}$$

This gives the same result.

4.5 * Making tensor operators

If often happens that you come upon a set of operators which transforms under commutation with the generators like a reducible representation of the algebra

$$[J_a, \Omega_x] = \Omega_y [J_a^D]_{yx} \tag{4.37}$$

where D is reducible. In this case, some work is required to turn these into tensor operators, but the work is essentially just the familiar highest weight construction again. The first step is to make linear combinations of the Ω_x operators that have definite J_3 values

$$[J_3, O_{m,\alpha}] = m \, O_{m,\alpha} \tag{4.38}$$

This is always possible because D can be decomposed into irreducible representations that have this property. Then we can apply the highest weight procedure and conclude that the **operators**, with the highest weight, $O_{j,\alpha}$ are components of a tensor operator with spin j, one for each α. If there are any operators with weight $j-1/2$, $O_{j-1/2,\beta}$, they will be components of tensor operators with spin $j-1/2$. However, things can get subtle at the next level. To find the tensor operators with spin $j-1$, you must find linear combinations of the operators with weight $j-1$ which have vanishing commutator with J^+ — then they correspond to the highest weights of the spin $j-1$ reps

$$[J^+, O_{j-1,\gamma}] \tag{4.39}$$

The point is, if you get the operators in a random basis, you have nothing like a scalar product, so you cannot simply find the operators that are "orthogonal" to the ones you have already assigned to representations. I hope that an example will make this clearer. Consider seven operators, $a_{\pm 1}$, $b_{\pm 1}$ and a_0, b_0 and c_0, with the following commutation relations with the generators:

$$\begin{gathered} [J_3, a_{+1}] = a_{+1} \qquad [J_3, b_{+1}] = b_{+1} \\ [J_3, a_0] = [J_3, b_0] = [J_3, c_0] = 0 \\ [J_3, a_{-1}] = -a_{-1} \qquad [J_3, b_{-1}] = -b_{-1} \end{gathered} \tag{4.40}$$

$$
\begin{gathered}
[J^+, a_{+1}] = [J^+, b_{+1}] = 0 \\
[J^+, a_0] = a_{+1} \qquad [J^+, b_0] = b_{+1} \qquad [J^+, c_0] = a_{+1} - b_{+1} \\
[J^+, a_{-1}] = c_0 \qquad [J^+, b_{-1}] = \frac{1}{2}(a_0 + b_0 - 3c_0)
\end{gathered}
\tag{4.41}
$$

$$
\begin{gathered}
[J^-, a_{+1}] = \frac{1}{2}(a_0 + b_0 + c_0) \qquad [J^-, b_{+1}] = \frac{1}{2}(a_0 + b_0 - c_0) \\
[J^-, a_0] = 2a_{-1} + b_{-1} \qquad [J^-, b_0] = a_{-1} + b_{-1} \qquad [J^-, c_0] = a_{-1} \\
[J^-, a_{-1}] = [J^-, b_{-1}] = 0
\end{gathered}
\tag{4.42}
$$

To construct the tensor operators, we start with the highest weight states, and define

$$
A_{+1} = a_{+1} \qquad B_{+1} = b_{+1} \tag{4.43}
$$

Then we construct the rest of the components by applying the lowering operators

$$
A_0 = \frac{1}{2}(a_0 + b_0 + c_0) \qquad B_0 = \frac{1}{2}(a_0 + b_0 - c_0) \tag{4.44}
$$

and

$$
A_{-1} = 2a_{-1} + b_{-1} \qquad B_{-1} = a_{-1} + b_{-1} \tag{4.45}
$$

You can check that the raising operators now just move us back up within the representations.

Now there is one operator left, so it must be a spin 0 representation. **But which one is it?** It must be the linear combination that has vanishing commutator with $J^{\pm}$ — therefore it is

$$
C_0 = a_0 - b_0 - c_0 \tag{4.46}
$$

Let me emphasize again that we went through this analysis explicitly to show the differences between dealing with states and dealing with tensor operators. Had this been a set of seven states transforming similarly under the algebra, we could have constructed the singlet state by simply finding the linear combination of $J_3 = 0$ states orthogonal to the $J_3 = 0$ states in the triplets. Here we do not have this crutch, but we can still find the singlet operator directly from the commutation relations. We could do the same thing for states, of course, but it is usually easier for states to use the nice properties of the scalar product.

4.6 Products of operators

One of the reasons that tensor operators are important is that a product of two tensor operators, $O^{s_1}_{m_1}$ and $O^{s_2}_{m_2}$ in the spin s_1 and spin s_2 representations, transforms under the tensor product representation, $s_1 \otimes s_2$ because

$$\begin{aligned} &[J_a, O^{s_1}_{m_1}\, O^{s_2}_{m_2}] \\ &= [J_a, O^{s_1}_{m_1}]\; O^{s_2}_{m_2} + O^{s_1}_{m_1}\; [J_a, O^{s_2}_{m_2}] \\ &= O^{s_1}_{m_1'}\, O^{s_2}_{m_2} [X^{s_1}_a]_{m_1' m_1} + O^{s_1}_{m_1}\, O^{s_2}_{m_2'} [X^{s_2}_a]_{m_2' m_2} \end{aligned} \tag{4.47}$$

Thus the product can be decomposed into tensor operators using the highest weight procedure.

Note that as usual, things are particularly simple for the generator J_3. (4.47) implies

$$[J_3, O^{s_1}_{m_1}\, O^{s_2}_{m_2}] = (m_1 + m_2)\, O^{s_1}_{m_1}\, O^{s_2}_{m_2} \tag{4.48}$$

The J_3 value of the product of two tensor operators is just the sum of the J_3 values of the two operators in the product.

Problems

4.A. Consider an operator O_x, for $x = 1$ to 2, transforming according to the spin 1/2 representation as follows:

$$[J_a, O_x] = O_y\, [\sigma_a]_{yx}/2$$

where σ_a are the Pauli matrices. Given

$$\langle 3/2, -1/2, \alpha|\, O_1\, |1, -1, \beta\rangle = A$$

find

$$\langle 3/2, -3/2, \alpha|\, O_2\, |1, -1, \beta\rangle$$

4.B. The operator $(r_{+1})^2$ satisfies

$$\left[L^+, (r_{+1})^2\right] = 0$$

It is therefore the O_{+2} component of a spin 2 tensor operator. Construct the other components, O_m. Note that the product of tensor operators transforms like the tensor product of their representations. What is the connection of this with the spherical harmonics, $Y_{l,m}(\theta,\phi)$? Hint: let $r_1 = \sin\theta\cos\phi$, $r_2 = \sin\theta\sin\phi$, and $r_3 = \cos\theta$. Can you generalize this construction to arbitrary ℓ and explain what is going on?

4.C. Find

$$e^{i\alpha_a X_a^1}$$

where the X_a^1 are given by (3.31) **Hint:** There is a trick that makes this one easy. Write

$$\alpha_a X_a^1 = \alpha\,\hat{\alpha}_a X_a^1$$

where

$$\alpha = \sqrt{\alpha_a\alpha_a}\,, \qquad \hat{\alpha}_a\hat{\alpha}_a = 1$$

You know that $\hat{\alpha}_a X_a^1$ has eigenvalues ± 1 and 0, just like X_3^1 (because all directions are equivalent). Thus $(\hat{\alpha}_a X_a^1)^2$ is a projection operator and

$$(\hat{\alpha}_a X_a^1)^3 = (\hat{\alpha}_a X_a^1)$$

You should be able to use this to manipulate the expansion of the exponential and get an explicit expression for $e^{i\alpha_a X_a^1}$.

Chapter 5

Isospin

The idea of isospin arose in nuclear physics in the early thirties. Heisenberg introduced a notation in which the proton and neutron were treated as two components of a **nucleon** doublet

$$N \equiv \begin{pmatrix} p \\ n \end{pmatrix} \tag{5.1}$$

He did this originally because he was trying to think about the forces between nucleons in nuclei, and it was mathematically convenient to write things in this notation. In fact, his first ideas about this were totally wrong — he really didn't have the right idea about the relation between the proton and the neutron. He was thinking of the neutron as a sort of tightly bound state of proton and electron, and imagined that forces between nucleons could arise by exchange of electrons. In this way you could get a force between proton and neutron by letting the electron shuttle back and forth — in analogy with an H_2^+ ion, and a force between neutron and neutron — an analogy with a neutral H_2 molecule. But no force between proton and proton.

5.1 Charge independence

It was soon realized that the model was crazy, and the force had to be **charge independent** — the same between pp, pn and nn to account for the pattern of nuclei that were observed. But while his model was crazy, he had put the p and n together in a doublet, and he had used the Pauli matrices to describe their interactions. Various people soon realized that charge independence would be automatic if there were really a conserved "spin" that acted on the doublet of p and n just as ordinary spin acts on the two J_3 components of a spin-1/2 representation. Some people called this "isobaric spin",

DOI: 10.1201/9780429499210-6

which made sense, because isobars are nuclei with the same total number of **baryons**,[1] protons plus neutrons, and thus the transformations could move from one isobar to another. Unfortunately, Wigner called it **isotopic spin** and that name stuck. This name makes no sense at all because the isotopes have the same number of protons and different numbers of neutrons, so eventually, the "topic" got dropped, and it is now called **isospin**.

5.2 Creation operators

Isospin really gets interesting in particle physics, where particles are routinely created and destroyed. The natural language for describing this dynamics is based on creation and annihilation operators (and this language is very useful for nuclear physics, as we will see). For example, for the nucleon doublet in (5.1), we can write

$$\begin{aligned} |p,\alpha\rangle &= a^\dagger_{N,\frac{1}{2},\alpha}|0\rangle \\ |n,\alpha\rangle &= a^\dagger_{N,-\frac{1}{2},\alpha}|0\rangle \end{aligned} \tag{5.2}$$

where the

$$a^\dagger_{N,\pm\frac{1}{2},\alpha} \tag{5.3}$$

are **creation operators** for proton $(+\frac{1}{2})$ and neutron $(-\frac{1}{2})$ respectively in the state α, and $|0\rangle$ is the **vacuum state** — the state with no particles in it. The N stands for nucleon, and it is important to give it a name because we will soon discuss creation operators for other particles as well. The creation operators are not hermitian. Their adjoints are **annihilation operators**,

$$a_{N,\pm\frac{1}{2},\alpha} \tag{5.4}$$

These operators annihilate a proton (or a neutron) if they can find one, and otherwise annihilate the state, so they satisfy

$$a_{N,\pm\frac{1}{2},\alpha}|0\rangle = 0 \tag{5.5}$$

The whole notation assumes that the symmetry that rotates proton into neutron is at least approximately correct. If the proton and the neutron were not in some sense similar, it wouldn't make any sense to talk about them being in the same state.

[1] Baryons are particles like protons and neutrons. More generally, the baryon number is one third the number of quarks. Because, as we will discuss in more detail later, the proton and the neutron are each made of three quarks, each has baryon number 1.

Because the p and n are fermions, their creation and annihilation operators satisfy anticommutation relations:

$$\begin{gathered}\{a_{N,m,\alpha}, a^\dagger_{N,m',\beta}\} = \delta_{mm'}\delta_{\alpha\beta} \\ \{a^\dagger_{N,m,\alpha}, a^\dagger_{N,m',\beta}\} = \{a_{N,m,\alpha}, a_{N,m',\beta}\} = 0\end{gathered} \tag{5.6}$$

With creation and annihilation operators, we can make **multiparticle states** by simply applying more than one creation operator to the vacuum state. For example

$$\begin{gathered}\overbrace{a^\dagger_{N,\frac{1}{2},\alpha_1} \cdots a^\dagger_{N,\frac{1}{2},\alpha_n}}^{n \text{ proton creation operators}} |0\rangle \\ \propto |n \text{ protons}; \alpha_1, \cdots, \alpha_n\rangle\end{gathered} \tag{5.7}$$

produces an n proton state, with the protons in states, α_1 through α_n. The anticommutation relation implies that the state is completely antisymmetric in the labels of the particles. This guarantees that the state vanishes if any two of the αs are the same. It means (among other things) that the Pauli exclusion principle is automatically satisfied. What is nice about the creation and annihilation operators is that we can construct states with both protons and neutrons in the same way. For example,

$$\begin{gathered}\overbrace{a^\dagger_{N,m_1,\alpha_1} \cdots a^\dagger_{N,m_n,\alpha_n}}^{n \text{ nucleon creation operators}} |0\rangle \\ \propto |n \text{ nucleon}; m_1, \alpha_1; \cdots; m_n, \alpha_n\rangle\end{gathered} \tag{5.8}$$

is an n nucleon state, with the nucleons in states described by the m variable (which tells you whether it is a proton or a neutron) and the α label, which tells you what state the nucleon is in. Now the anticommutation relation implies that the state is completely antisymmetric under exchange of the pairs of labels, m and α.

$$\begin{gathered}|n \text{ nucleon}; m_1, \alpha_1; m_2, \alpha_2 \cdots; m_n, \alpha_n\rangle \\ = -|n \text{ nucleon}; m_2, \alpha_2; m_1, \alpha_1; \cdots; m_n, \alpha_n\rangle\end{gathered} \tag{5.9}$$

If you haven't seen this before, it should bother you. It is one thing to assume that the proton creation operators anticommute, because two protons really cannot be in the same state. But why should proton and neutron creation operators anticommute? This principle is called the "generalized exclusion principle." Why should it be true? This is an important question, and we will come back to it below. For now, however, we will just see how the creation and annihilation operators behave in some examples.

5.3 Number operators

We can make operators that count the number of protons and neutrons by putting creation and annihilation operators together (the summation convention is assumed):

$$\begin{array}{ll} a^\dagger_{N,+\frac{1}{2},\alpha}\, a_{N,+\frac{1}{2},\alpha} & \text{counts protons} \\ a^\dagger_{N,-\frac{1}{2},\alpha}\, a_{N,-\frac{1}{2},\alpha} & \text{counts neutrons} \\ a^\dagger_{N,m,\alpha}\, a_{N,m,\alpha} & \text{counts nucleons} \end{array} \tag{5.10}$$

Acting on any state with N_p protons and N_n neutrons, these operators have eigenvalues N_p, N_n and $N_p + N_n$ respectively. This works because of (5.5) and the fact that for a generic pair of creation and annihilation operators

$$\left[a^\dagger a, a^\dagger\right] = a^\dagger \tag{5.11}$$

Notice that the number operators in (5.10) are summed over all the possible quantum states of the proton and neutron, labeled by α. If we did not sum over α, the operators would just count the number of protons or neutrons or both in the state α. We could get fancy and devise more restricted number operators where we sum over some α and not others, but we won't talk further about such things. The total number operators, summed over all α, will be particularly useful.

5.4 Isospin generators

For the one-particle states, we know how the generators of isospin symmetry should act, in analogy with the spin generators:

$$T_a|m,\alpha\rangle = |m',\alpha\rangle\,[J_a^{1/2}]_{m'm} = \frac{1}{2}|m',\alpha\rangle\,[\sigma_a]_{m'm} \tag{5.12}$$

Or in terms of creation operators

$$T_a\, a^\dagger_{N,m,\alpha}\,|0\rangle = a^\dagger_{N,m',\alpha}\,|0\rangle\,[J_a^{1/2}]_{m'm} = \frac{1}{2}a^\dagger_{N,m',\alpha}\,|0\rangle\,[\sigma_a]_{m'm} \tag{5.13}$$

Furthermore, the state with no particles should transform like the trivial representation —

$$T_a\,|0\rangle = 0 \tag{5.14}$$

Thus we will get the right transformation properties for the one particle states if the creation operators transform like a tensor operator in the spin 1/2 representation under isospin:

$$\left[T_a, a^\dagger_{N,m,\alpha}\right] = a^\dagger_{N,m',\alpha}\,[J^{1/2}_a]_{m'm} = \frac{1}{2}a^\dagger_{N,m',\alpha}\,[\sigma_a]_{m'm} \tag{5.15}$$

It is easy to check that the following form for T_a does the trick:

$$\begin{aligned} T_a &= a^\dagger_{N,m',\alpha}\,[J^{1/2}_a]_{m'm}\,a_{N,m,\alpha} + \cdots \\ &= \frac{1}{2}a^\dagger_{N,m',\alpha}\,[\sigma_a]_{m'm}\,a_{N,m,\alpha} + \cdots \\ &= \frac{1}{2}a^\dagger_{N,\alpha}\,\sigma_a\,a_{N,\alpha} + \cdots \end{aligned} \tag{5.16}$$

where $\cdots$ commutes with the nucleon creations and annihilation operators (and also annihilates $|0\rangle$). The last line is written in matrix form, where we think of the annihilation operators as column vectors and the creation operators as row vectors. Let us check that (5.16) has the right commutation relations with the creation operators so that (5.15) is satisfied.

$$\begin{aligned} &\left[T_a, a^\dagger_{N,m,\alpha}\right] \\ &= \left[a^\dagger_{N,m',\beta}\,[J^{1/2}_a]_{m'm''}\,a_{N,m'',\beta}, a^\dagger_{N,m,\alpha}\right] \\ &= a^\dagger_{N,m',\beta}\,[J^{1/2}_a]_{m'm''}\,\left\{a_{N,m'',\beta}, a^\dagger_{N,m,\alpha}\right\} \\ &- \left\{a^\dagger_{N,m',\beta}, a^\dagger_{N,m,\alpha}\right\}\,[J^{1/2}_a]_{m'm''}\,a_{N,m'',\beta} \\ &= a^\dagger_{N,m',\alpha}\,[J^{1/2}_a]_{m'm} \end{aligned} \tag{5.17}$$

The advantage of thinking about the generators in this way is that we now immediately see how multiparticle states transform. Since the multiparticle states are built by applying more tensor (creation) operators to the vacuum state, the multiparticle states transform like tensor products — not a surprising result, but not entirely trivial either.

5.5 Symmetry of tensor products

We pause here to discuss an important fact about the combination of spin states (either ordinary spin or isospin). We will use it in the next section to discuss the deuteron. The result is this: when the tensor product of two identical spin 1/2 representations is decomposed into irreducible representations, the spin 1 representation appears symmetrically, while the spin 0 appears antisymmetrically. To see what this means, suppose that the spin 1/2 states are

$$|1/2, \pm 1/2, \alpha\rangle \tag{5.18}$$

where α indicates whatever other parameters are required to describe the state. Now consider the highest weight state in the tensor product. This is the spin 1 combination of two identical J_3=1/2 states, and is thus symmetric in the exchange of the other labels:

$$|1,1\rangle = |1/2,1/2,\alpha\rangle|1/2,1/2,\beta\rangle = |1/2,1/2,\beta\rangle|1/2,1/2,\alpha\rangle \tag{5.19}$$

The lowering operators that produce the other states in the spin 1 representation preserve this symmetry because they act in the same way on the two spin 1/2 states.

$$\begin{aligned} |1,0\rangle &= \frac{1}{\sqrt{2}}\Big(|1/2,-1/2,\alpha\rangle|1/2,1/2,\beta\rangle \\ &\quad +|1/2,1/2,\alpha\rangle|1/2,-1/2,\beta\rangle\Big) \\ |1,-1\rangle &= |1/2,-1/2,\alpha\rangle|1/2,-1/2,\beta\rangle \end{aligned} \tag{5.20}$$

Then the orthogonal spin 0 state is antisymmetric in the exchange of α and β:

$$\begin{aligned} |0,0\rangle &= \frac{1}{\sqrt{2}}\Big(|1/2,-1/2,\alpha\rangle|1/2,1/2,\beta\rangle \\ &\quad -|1/2,1/2,\alpha\rangle|1/2,-1/2,\beta\rangle\Big) \end{aligned} \tag{5.21}$$

5.6 The deuteron

The nucleons have spin 1/2 as well as isospin 1/2, so the α in the nucleon creation operator actually contains a J_3 label, in addition to whatever other parameters are required to determine the state.

As a simple example of the transformation of a multiparticle state, consider a state of two nucleons in an s-wave — a zero angular momentum state. Then the total angular momentum of the state is simply the spin angular momentum, the sum of the two nucleon spins. Furthermore, in an s-wave state, the wave function is symmetrical in the exchange of the position variables of the two nucleons. Then because the two-particle wave function is proportional to the product of two anticommuting creation operators acting on the vacuum state, it is antisymmetric under the simultaneous exchange of the isospin and spin labels of the two nucleons — if the spin representation is symmetric, the isospin representation must be antisymmetric, and vice versa. When combined with the results of the previous section, this has physical consequences. The only allowed states are those with isospin 1 and spin 0 or with isospin 0 and spin 1. The deuteron is an isospin 0 conbination, and has spin 1, as expected.

5.7 Superselection rules

It appears, in this argument, that we have assigned some fundamental physical significance to the anticommutation of the creation operators for protons and neutrons. As I mentioned above, this seems suspect, because in fact, the proton and neutron are not identical particles. What we actually know directly from the Pauli exclusion principle is that the creation operator, $a_\alpha^\dagger$ for any state of a particle obeying Fermi-Dirac statistics satisfies

$$\left(a_\alpha^\dagger\right)^2 = 0 \tag{5.22}$$

If we have another creation operator for the same particle in another state, $a_\beta^\dagger$, we can form the combination $a_\alpha^\dagger + a_\beta^\dagger$, which when acting on the vacuum creates the particle in the state $\alpha + \beta$ (with the wrong normalization). Thus the exclusion principle also implies

$$\left(a_\alpha^\dagger + a_\beta^\dagger\right)^2 = 0 \tag{5.23}$$

and thus

$$\left\{a_\alpha^\dagger, a_\beta^\dagger\right\} = 0 \tag{5.24}$$

This argument is formally correct, but it doesn't really make much physical sense if $a_\alpha^\dagger$ and $a_\beta^\dagger$ create states of different particles, because it doesn't really make sense to superpose the states — this superposition is forbidden by a superselection rule. A superselection rule is a funny concept. It is the statement that you never need to think about superposing states with different values of an exactly conserved quantum number because those states must be orthogonal. Anything you can derive by such a superposition must also be derivable in some other way that does not involve the "forbidden" superposition. Thus as you see, the superposition is not so much forbidden as it is irrelevant. In this case, it is possible to show that one can choose the creation operators to anticommute without running into inconsistencies, but there is a much stronger argument. The anticommutation is required by the fact that the creation operators transform like tensor operators. Let's see how this implies the stated result for the two nucleon system.

Call the creation operators for the baryons $a^\dagger_{\pm\pm}$ (dropping the N for brevity) where the first sign is the sign of the third component of isospin and the second is the sign of third component of spin. Since $\left(a^\dagger_{++}\right)^2 = 0$, there is no two nucleon state with $T_3 = 1$ and $J_3 = 1$. But this means that there is no state with isospin 1 and spin 1, since the highest weight state would have to have $T_3 = 1$ and $J_3 = 1$. In terms of creation operators, for example

$$\left[T^-, \left(a^\dagger_{++}\right)^2\right] = \left\{a^\dagger_{-+}, a^\dagger_{++}\right\} = 0 \tag{5.25}$$

Similar arguments show that the operators must anticommute whenever they have one common index and the others are different.

The argument for operators that have no index in common is a little more subtle. First compute

$$\left[J^-, \left[T^-, \left(a^\dagger_{++}\right)^2\right]\right] = \left\{a^\dagger_{--}, a^\dagger_{++}\right\} + \left\{a^\dagger_{-+}, a^\dagger_{+-}\right\} = 0 \tag{5.26}$$

But the two terms in the sum must separately vanish because they are physically distinguishable. There cannot be a relation like (5.26) unless the two operators

$$\left\{a^\dagger_{--}, a^\dagger_{++}\right\} |0\rangle \tag{5.27}$$

and

$$\left\{a^\dagger_{-+}, a^\dagger_{+-}\right\} |0\rangle \tag{5.28}$$

separately vanish, because these two operators, if they did not vanish, would do physically distinguishable things — the creation of a proton with spin up and a neutron with spin down is not the same as the creation of proton with spin down and a neutron with spin up. Thus the operators (5.27) and (5.28) must separately vanish. Thus, not only does the isospin 1, spin 1 state (5.26) vanish but so also does the isospin 0, spin 0 state

$$\left\{a^\dagger_{--}, a^\dagger_{++}\right\} |0\rangle - \left\{a^\dagger_{-+}, a^\dagger_{+-}\right\} |0\rangle \tag{5.29}$$

5.8 Other particles

When isospin was introduced, the only known particles that carried it were the proton and neutron, and the nuclei built out of them. But as particle physicists explored further, at higher energies, new particles appeared that are not built out of nucleons. The first of these were the pions, three spinless bosons (that is obeying Bose-Einstein, rather than Fermi-Dirac statistics) with charges $Q = +1$, 0 and -1, and $T_3 = Q$, forming an isospin triplet.[2] The creation and annihilation operators for the pions can be written as

$$a^\dagger_{\pi,m,\alpha}\,, \quad a_{\pi,m,\alpha} \quad \text{for } m = -1 \text{ to } 1 \tag{5.30}$$

They satisfy commutation, rather than anticommutation relations

$$\begin{aligned} \left[a_{\pi,m,\alpha}, a^\dagger_{\pi,m',\beta}\right] &= \delta_{mm'}\delta_{\alpha\beta} \\ \left[a^\dagger_{\pi,m,\alpha}, a^\dagger_{\pi,m',\beta}\right] = \left[a_{\pi,m,\alpha}, a_{\pi,m',\beta}\right] &= 0 \end{aligned} \tag{5.31}$$

so that the particle states will be completely symmetric. They also commute with nucleon creation and annihilation operators.

The isospin generators look like

$$T_a = a^\dagger_{\pi,m,\alpha}\,[J^1_a]_{mm'}\,a_{\pi,m',\alpha} + \cdots \tag{5.32}$$

where as in (5.16) the $\cdots$ refers to the contributions of other particles (like nucleons). Again, then the creation operators are tensor operators.

There are many many other particles like the nucleons and the pions that participate in the strong interactions and carry isospin. The formalism of creation and annihilation operators gives us a nice way of writing the generators of isospin that acts on all these particles. The complete form of the isospin generators is

$$T_a = \sum_{\substack{\text{particles } x \\ \text{states } \alpha \\ T_3 \text{ values } m,m'}} a^\dagger_{x,m,\alpha}\,[J^{j_x}_a]_{mm'}\,a_{x,m',\alpha} \tag{5.33}$$

where $a^\dagger_{x,m,\alpha}$ and $a_{x,m',\alpha}$ are creation and annihilation operators for x-type particles satisfying commutation or anticommutation relations depending on whether they are bosons or fermions,

$$\begin{aligned} \left[a_{x,m,\alpha}, a^\dagger_{x',m',\beta}\right]_\pm &= \delta_{mm'}\delta_{\alpha\beta}\delta_{xx'} \\ \left[a^\dagger_{x,m,\alpha}, a^\dagger_{x',m',\beta}\right]_\pm = \left[a_{x,m,\alpha}, a_{x',m',\beta}\right]_\pm &= 0 \end{aligned} \tag{5.34}$$

The rule for the $\pm$ (+ for anticommutator, $-$ for commutator) is that the anticommutator is used when both x and x' are fermions, otherwise the commutator is used. The j_x in (5.33) is the isospin of the x particles.

[2] When these particles were discovered, it was not completely obvious that they were not built out of nucleons and their antiparticles. When very little was known about the strong interactions, it was possible to imagine, for example, that the π^+ was a bound state of a proton and an antineutron. This has all the right quantum numbers — even the isospin is right. It just turns out that this model of the pion is wrong. Group theory can never tell you this kind of thing. You need real dynamical information about the strong interactions.

5.9 Approximate isospin symmetry

Isospin is an approximate symmetry. What this means in general is that the Hamiltonian can be written as

$$H = H_0 + \Delta H \tag{5.35}$$

where H_0 commutes with the symmetry generators and ΔH does not, but in some sense ΔH is small compared to H_0. It is traditional to say in the case of isospin that the "strong" interactions are isospin symmetric while the weak and electromagnetic interactions are not, and so take $H_0 = H_S$ and $\Delta H = H_{EM} + H_W$ where H_S, H_{EM} and H_W are the contributions to the Hamiltonian describing the strong interactions (including the kinetic energy), the electromagnetic interactions, and the weak interactions, respectively. From our modern perspective, this division is a bit misleading for two reasons. Firstly, the division between electromagnetic and weak interactions is not so obvious because of the partial unification of the two forces. Secondly, part of the isospin violating interaction arises from the difference in mass between the u and d quarks which is actually part of the kinetic energy. It seems to be purely accidental that this effect is roughly the same size as the effect of the electromagnetic interactions. But this accident was important historically, because it made it easy to understand isospin as an approximate symmetry. There are so many such accidents in particle physics that it makes one wonder whether there is something more going on. At any rate, we will simply lump all isospin violation into ΔH. The group theory doesn't care about the dynamics anyway, as long as the symmetry structure is properly taken into account.

5.10 Perturbation theory

The way (5.35) is used is in perturbation theory. The states are classified into eigenstates of the zeroth order, isospin symmetric part of the Hamiltonian, H_0. Sometimes, just H_0 is good enough to approximate the physics of interest. If not, one must treat the effects of ΔH as perturbations. In the scattering of strongly interacting particles, for example, the weak and electromagnetic interactions can often be ignored. Thus in pion-nucleon scattering, all the different possible charge states have either isospin 1/2 or 3/2 (because $1 \otimes 1/2 = 3/2 \oplus 1/2$), so this scattering process can be described approximately by only two amplitudes.

The mathematics here is exactly the same as that which appears in the decomposition of a spin-1/2 state with an orbital angular momentum 1 into states with total angular momentum 3/2 and 1/2. The state with one pion and one nucleon can be described as a tensor product of an isospin 1/2 nucleon state with an isospin 1 pion state, just as the state with both spin and orbital angular momentum can be described as a tensor product, having both spin and angular momentum indices.

Problems

5.A. Suppose that in some process, a pair of pions is produced in a state with zero relative orbital angular momentum. What total isospin values are possible for this state?

5.B. Show that the operators defined in (5.33) have the commutation relations of isospin generators.

5.C. Δ^{++}, Δ^{+}, Δ^{0} and Δ^{-} are isospin 3/2 particles ($T_3 = 3/2$, 1/2, $-1/2$ and $-3/2$ respectively) with baryon number 1. They are produced by strong interactions in π-nucleon collisions. Compare the probability of producing Δ^{++} in $\pi^{+}P \to \Delta^{++}$ with the probability of producing Δ^{0} in $\pi^{-}P \to \Delta^{0}$.

Chapter 6

Roots and Weights

Now we are going to generalize the analysis of the representations of the $SU(2)$ algebra to an arbitrary simple Lie algebra. The idea is simple. First, we do what we always try to do in quantum mechanics — find the largest possible set of commuting hermitian observables and use their eigenvalues to label the states. In this case, our observables will be the largest set of hermitian generators we can find that commute with one another, and can therefore be simultaneously diagonalized. Their eigenvalues will be the analog of J_3. The rest of the generators will be analogous to the raising and lowering operators in $SU(2)$. We will find that every raising operator corresponds to an $SU(2)$ subgroup of the Lie algebra, and then we can use what we know about $SU(2)$ to learn about the larger algebra.

6.1 Weights

We want the largest possible set of commuting hermitian generators because we want to diagonalize as much as possible. A subset of commuting hermitian generators which is as large as possible is called a **Cartan subalgebra**. It will turn out that the Cartan subalgebra is essentially unique, in that any one we choose will give the same results.

In a particular irreducible representation, D, there will be a number of hermitian generators, H_i for $i = 1$ to m, corresponding to the elements of the Cartan Subalgebra called the **Cartan generators** satisfying

$$H_i = H_i^\dagger , \qquad \text{and} \qquad [H_i, H_j] = 0 . \tag{6.1}$$

The Cartan generators form a linear space. Thus we can choose a basis in

DOI: 10.1201/9780429499210-7

which they satisfy

$$\mathrm{Tr}\,(H_i H_j) = k_D \delta_{ij} \quad \text{for } i,j = 1 \text{ to } m \tag{6.2}$$

where k is some constant that depends on the representation and on the normalization of the generators. The integer m, the number of independent Cartan generators, is called the **rank** of the algebra.

Of course, the point is that the Cartan generators can be simultaneously diagonalized. After diagonalization of the Cartan generators, the states of the representation D can be written as $|\mu, x, D\rangle$ where

$$H_i|\mu, x, D\rangle = \mu_i|\mu, x, D\rangle \tag{6.3}$$

and x is any other label that is necessary to specify the state.

The eigenvalues μ_i are called **weights.** They are real, because they are eigenvalues of hermitian operators. The m-component vector with components μ_i is the **weight vector**. We will often use a vector notation in which

$$\alpha \cdot \mu \equiv \alpha_i \mu_i \quad \text{and} \quad \alpha^2 \equiv \alpha_i \alpha_i \tag{6.4}$$

6.2 More on the adjoint representation

The adjoint representation, defined by (2.27), is particularly important. Because the rows and columns of the matrices defined by (2.27) are labeled by the same index that labels the generators, the states of the adjoint representation correspond to the generators themselves. We will denote the state in the adjoint representation corresponding to an arbitrary generator X_a as

$$|X_a\rangle\,. \tag{6.5}$$

Linear combinations of these states correspond to linear combinations of the generators —

$$\alpha|X_a\rangle + \beta|X_b\rangle = |\alpha X_a + \beta X_b\rangle\,. \tag{6.6}$$

A convenient scalar product on this space is the following:[1]

$$\langle X_a|X_b\rangle = \lambda^{-1}\,\mathrm{Tr}\left(X_a^\dagger X_b\right)\,, \tag{6.7}$$

(λ is what we called k_D for the adjoint representation — see (2.37)). Now using (6.6) and (2.27), we can compute the action of a generator on a state, as follows:

$$\begin{aligned} X_a|X_b\rangle &= |X_c\rangle\langle X_c|X_a|X_b\rangle = |X_c\rangle\,[T_a]_{cb} = -i\,f_{acb}|X_c\rangle \\ &= i\,f_{abc}|X_c\rangle = |i\,f_{abc}X_c\rangle = |[X_a, X_b]\rangle\,. \end{aligned} \tag{6.8}$$

[1]We need the dagger because we will be led to consider complex linear combinations of the generators, analogous to the raising and lowering operators for $SU(2)$.

6.3 Roots

The **roots** are the weights of the adjoint representation. Because $[H_i, H_j] = 0$, the states corresponding to the Cartan generators have zero weight vectors

$$H_i|H_j\rangle = |[H_i, H_j]\rangle = 0 \tag{6.9}$$

Furthermore, all states in the adjoint representation with zero weight vectors correspond to Cartan generators. Because of (6.2), the Cartan states are orthonormal,

$$\langle H_i|H_j\rangle = \lambda^{-1}\,\mathrm{Tr}\,(H_i H_j) = \delta_{ij} \tag{6.10}$$

The other states of the adjoint representation, those not corresponding to the Cartan generators, have non-zero weight vectors, α, with components α_i,

$$H_i|E_\alpha\rangle = \alpha_i|E_\alpha\rangle \tag{6.11}$$

which means that the corresponding generators satisfy

$$[H_i, E_\alpha] = \alpha_i E_\alpha \tag{6.12}$$

It will turn out (and we will prove it below) that for the adjoint representation, the non-zero weights uniquely specify the corresponding states, so there is no need for another parameter (like x in (6.3) in the arbitrary representation D). Like the $SU(2)$ raising and lowering operators, the E_α are not hermitian. They cannot be hermitian because we can take the adjoint of (6.12) and get

$$\left[H_i, E_\alpha^\dagger\right] = -\alpha_i E_\alpha^\dagger \tag{6.13}$$

thus we can take

$$E_\alpha^\dagger = E_{-\alpha}\,. \tag{6.14}$$

This should remind you of the raising and lowering operators J^+ and J^- in $SU(2)$.

States corresponding to different weights must be orthogonal, because they have different eigenvalues of at least one of the Cartan generators. Thus we can choose the normalization of the states in the adjoint representation (that is, the generators) so that

$$\langle E_\alpha|E_\beta\rangle = \lambda^{-1}\,\mathrm{Tr}\left(E_\alpha^\dagger E_\beta\right) = \delta_{\alpha\beta}\;(= \prod_i \delta_{\alpha_i\beta_i})\,. \tag{6.15}$$

The weights α_i are called roots, and the special weight vector α with components α_i is a **root vector**.

6.4 Raising and lowering

— The $E_{\pm\alpha}$ are raising and lowering operators for the weights, because the state $E_{\pm\alpha}|\mu, D\rangle$ has weight $\mu \pm \alpha$ —

$$H_i E_{\pm\alpha}|\mu, D\rangle = [H_i, E_{\pm\alpha}]\,|\mu, D\rangle + E_{\pm\alpha} H_i|\mu, D\rangle = (\mu \pm \alpha)_i E_{\pm\alpha}|\mu, D\rangle\,. \tag{6.16}$$

At this point, we have no notion of positivity, so it doesn't make sense to ask which is raising and which is lowering. But we will introduce this later.

Equation (6.16) is true for any representation, but it is particularly important for the adjoint representation. To see why, consider the state $E_\alpha|E_{-\alpha}\rangle$. This has weight $\alpha - \alpha = 0$, thus it is a linear combination of states corresponding to Cartan generators. This in turn implies that $[E_\alpha, E_{-\alpha}]$ is a linear combination of Cartan generators:

$$E_\alpha|E_{-\alpha}\rangle = \beta_i\,|H_i\rangle = |\beta_i H_i\rangle = |\beta \cdot H\rangle = |[E_\alpha, E_{-\alpha}]\rangle\,. \tag{6.17}$$

But we can actually compute β —

$$\begin{aligned}
\beta_i &= \langle H_i|E_\alpha|E_{-\alpha}\rangle \\
&= \lambda^{-1}\,\mathrm{Tr}\,(H_i\,[E_\alpha, E_{-\alpha}]) && \text{this follows from (6.8)} \\
&= \lambda^{-1}\,\mathrm{Tr}\,(E_{-\alpha}\,[H_i, E_\alpha]) && \text{from the cyclic property of Tr} \\
&= \lambda^{-1}\,\alpha_i\;\mathrm{Tr}\,(E_{-\alpha}E_\alpha) && \text{from (6.12)} \\
&= \alpha_i && \text{from (6.15).}
\end{aligned} \tag{6.18}$$

Thus

$$[E_\alpha, E_{-\alpha}] = \alpha \cdot H\,. \tag{6.19}$$

This should remind you of the $SU(2)$ commutation relation $[J^+, J^-] = J_3$. It is this analogy that we will exploit to learn about the compact Lie groups and their representations.

6.5 Lots of $SU(2)$s

For each non-zero pair of root vectors, $\pm\alpha$, there is an $SU(2)$ subalgebra of the group, with generators

$$\begin{aligned}
E^\pm &\equiv |\alpha|^{-1} E_{\pm\alpha} \\
E_3 &\equiv |\alpha|^{-2} \alpha \cdot H\,.
\end{aligned} \tag{6.20}$$

To see this, note that

$$\begin{aligned}[E_3, E^{\pm}] &= |\alpha|^{-3} [\alpha \cdot H, E_{\pm\alpha}] \\ &= |\alpha|^{-3} \alpha \cdot (\pm\alpha) E_{\pm\alpha} = \pm|\alpha|^{-1} E_{\pm\alpha} = \pm E^{\pm}\end{aligned} \tag{6.21}$$

and from (6.19)

$$\begin{aligned}[E^+, E^-] &= |\alpha|^{-2} [E_\alpha, E_{-\alpha}] \\ &= |\alpha|^{-2} \alpha \cdot H = E_3 .\end{aligned} \tag{6.22}$$

We know on general grounds that the states of each irreducible representation of the full algebra can be decomposed into irreducible representations of any one of these $SU(2)$ subalgebras, and we already know everything about the irreducible representations of $SU(2)$. This puts very strong constraints on the nature of the roots. For example, we can now easily prove that the root vectors correspond to unique generators. Suppose the contrary, so that there are two generators, E_α and E'_α. We can choose linear combinations of these two so that they are orthogonal in the adjoint representation (I will use the same names for the two generators, assuming that I chose them to be orthogonal from the beginning, just to avoid useless notation) — thus we can write

$$\langle E_\alpha | E'_\alpha \rangle = \lambda^{-1} \operatorname{Tr}\left(E_\alpha^\dagger E'_\alpha\right) = \lambda^{-1} \operatorname{Tr}\left(E_{-\alpha} E'_\alpha\right) = 0 . \tag{6.23}$$

Consider the behavior of the state $|E'_\alpha\rangle$ under the action of the $SU(2)$ subalgebra (6.20). $E^-|E'_\alpha\rangle$ has zero weight vector, and thus it is a linear combination of Cartan states. But

$$\begin{aligned}\langle H_i | E^- | E'_\alpha \rangle &= \lambda^{-1} \operatorname{Tr}\left(H_i \left[E^-, E'_\alpha\right]\right) \\ &= -\lambda^{-1} \operatorname{Tr}\left(E^- \left[H_i, E'_\alpha\right]\right) \\ &= -\alpha_i \lambda^{-1} \operatorname{Tr}\left(E'_\alpha E^-\right) = 0\end{aligned} \tag{6.24}$$

for all i, and thus the coefficient of every Cartan state in $E^-|E'_\alpha\rangle$ vanishes, and therefore

$$E^- | E'_\alpha \rangle = 0 . \tag{6.25}$$

But we also have

$$E_3 | E'_\alpha \rangle = |\alpha|^{-2} \alpha \cdot H | E'_\alpha \rangle . = | E'_\alpha \rangle . \tag{6.26}$$

Equations (6.25) and (6.26) are inconsistent, because (6.25) implies that $|E'_\alpha\rangle$ is the lowest J_3 state in an $SU(2)$ representation, and (6.26) implies that it has $J_3 = 1$. But the lowest J_3 state of an $SU(2)$ representation cannot have positive J_3 — the J_3 value is always $-j$ for a non-negative half integer j. Thus E'_α cannot exist, and we have shown, as promised, that $|E_\alpha\rangle$ is uniquely specified by α — no other labels are required.

In fact, if α is a root, then no non-zero multiple of α (except $-\alpha$) is a root. To see this, note that the three states $|E_3\rangle$ and $|E^\pm\rangle$ form a spin 1 representation of the algebra (6.20), because they form the adjoint representation. Now suppose $k\alpha$ a root for $k \neq \pm 1$. Clearly, k must be a half-integer, because the E_3 value of the corresponding state must be a half integer. But if k is an integer not equal to ± 1, the state is part of a representation that contains another state with root α, which is impossible, by the argument we just gave. And if k is half an odd integer, then there is a state with root $\alpha/2$, and we can repeat the argument using the $SU(2)$ associated with that generator and get a contradiction in the same way.

6.6 Watch carefully - this is important!

More generally, for any weight μ of a representation D, the E_3 value is

$$E_3|\mu, x, D\rangle = \frac{\alpha \cdot \mu}{\alpha^2}|\mu, x, D\rangle \,. \tag{6.27}$$

Because the E_3 values must be integers or half integers,

$$\frac{2\alpha \cdot \mu}{\alpha^2} \text{ is an integer.} \tag{6.28}$$

The general state $|\mu, x, D\rangle$ can always be written as a linear combination of states transforming according to definite representations of the $SU(2)$ defined by (6.20). Suppose that the highest spin state that appears in the linear combination is j. Then there is some non-negative integer p such that

$$\left(E^+\right)^p |\mu, x, D\rangle \neq 0 \tag{6.29}$$

with weight $\mu + p\,\alpha$ is the highest E_3 state of the $SU(2)$ spin j representation, so that

$$\left(E^+\right)^{p+1} |\mu, x, D\rangle = 0 \,. \tag{6.30}$$

The E_3 value of the state (6.29) is

$$\frac{\alpha \cdot (\mu + p\,\alpha)}{\alpha^2} = \frac{\alpha \cdot \mu}{\alpha^2} + p = j \,. \tag{6.31}$$

Likewise, there is some non-negative integer q such that

$$\left(E^-\right)^q |\mu, x, D\rangle \neq 0 \tag{6.32}$$

with weight $\mu - q\,\alpha$ is the lowest E_3 state of the $SU(2)$ spin j representation, so that

$$(E^-)^{q+1}\,|\mu, x, D\rangle = 0 \tag{6.33}$$

and the E_3 value of the state (6.32) is

$$\frac{\alpha \cdot (\mu - q\,\alpha)}{\alpha^2} = \frac{\alpha \cdot \mu}{\alpha^2} - q = -j\,. \tag{6.34}$$

Adding (6.31) and (6.34) gives

$$\frac{2\alpha \cdot \mu}{\alpha^2} + p - q = 0 \tag{6.35}$$

or

$$\frac{\alpha \cdot \mu}{\alpha^2} = -\frac{1}{2}(p-q)\,. \tag{6.36}$$

We will refer to (6.36) as the "master formula". The relations (6.31), (6.34) and (6.36) are the basic relations that lead to a geometrical classification of all the compact Lie groups. They don't look like much, but when we augment them with some geometrical intuition, we can exploit them to great effect, as you will see.

Here is a simple first step. Applying (6.36) to the roots gives a particularly strong constraint, because we can apply it twice for any pair of distinct roots, α and β. Defining the $SU(2)$ algebra with E_α gives

$$\frac{\alpha \cdot \beta}{\alpha^2} = -\frac{1}{2}(p-q)\,. \tag{6.37}$$

Defining the $SU(2)$ algebra with E_β gives

$$\frac{\beta \cdot \alpha}{\beta^2} = -\frac{1}{2}(p'-q')\,. \tag{6.38}$$

Multiplying these gives a remarkable formula for the angle $\theta_{\alpha\beta}$ between the roots α and β:

$$\cos^2\theta_{\alpha\beta} = \frac{(\alpha \cdot \beta)^2}{\alpha^2\beta^2} = \frac{(p-q)(p'-q')}{4}\,. \tag{6.39}$$

What is remarkable about this is that $(p-q)(p'-q')$ must be an integer, so (because it must be non-negative) there are only four interesting possibilities (up to complements) for angles between roots!

$$
\begin{array}{cc}
(p-q)(p'-q') & \theta_{\alpha\beta} \\
0 & 90^\circ \\
1 & 60^\circ \text{ or } 120^\circ \\
2 & 45^\circ \text{ or } 135^\circ \\
3 & 30^\circ \text{ or } 150^\circ
\end{array}
\tag{6.40}
$$

The possibility $(p-q)(p'-q') = 4$, corresponds to 0° or 180° — neither is interesting. 0° is already ruled out by our theorem on uniqueness. 180° is trivial because roots always come in pairs with opposite signs, both in the same $SU(2)$ subgroup.

Problems

6.A. Show that $[E_\alpha, E_\beta]$ must be proportional to $E_{\alpha+\beta}$. What happens if $\alpha + \beta$ is not a root?

6.B. Suppose that the raising lowering operators of some Lie algebra satisfy

$$[E_\alpha, E_\beta] = N\, E_{\alpha+\beta}$$

for some nonzero N. Calculate

$$[E_\alpha, E_{-\alpha-\beta}]$$

6.C. Consider the simple Lie algebra formed by the ten matrices:

$$\sigma_a \qquad \sigma_a\tau_1 \qquad \sigma_a\tau_3 \qquad \tau_2$$

for $a = 1$ to 3 where σ_a and τ_a are Pauli matrices in orthogonal spaces (see problem 3.E). Take $H_1 = \sigma_3$ and $H_2 = \sigma_3\tau_3$ as the Cartan subalgebra. Find
(a) the weights of the four dimensional representation generated by these matrices, and
(b) the weights of the adjoint representation.
Hint: Although you have enough information to do the problem after reading this chapter, it may be easier after you have seen the example of $SU(3)$ worked out in the next chapter.

Chapter 7

$SU(3)$

After $SU(2)$ the most important algebra in particle physics is $SU(3)$. Maybe it is more important. I'm not sure. $SU(3)$ is the group of 3×3 unitary matrices with determinant 1 (again, as in (3.30), the U stands for "unitary" and the S stands for "special", which means determinant 1).

7.1 The Gell-Mann matrices

$SU(3)$ is generated by the 3×3 hermitian, traceless matrices. There are various ways of seeing that the tracelessness constraint is what gives determinant 1. If we exponentiate the hermitian generators to get unitary matrices

$$U(\alpha) = e^{i\alpha_a X_a} \tag{7.1}$$

we can compute the determinant in any basis. In particular, if we diagonalize $\alpha_a X_a$,

$$V\alpha_a X_a V^{-1} = D \tag{7.2}$$

where D is diagonal, we have

$$\det(U(\alpha)) = \det(e^{iD}) = \prod_j e^{i[D]_{jj}} = e^{i\,\mathrm{Tr}\,D} = e^{i\,\mathrm{Tr}\,\alpha_a X_a} \tag{7.3}$$

Thus if $\mathrm{Tr}\,\alpha_a X_a = 0$, the determinant is 1.

The standard basis for the hermitian 3×3 matrices in the physics literature is in terms of a generalization of the Pauli matrices, called the Gell-Mann matrices:

$$\begin{aligned}
\lambda_1 &= \begin{pmatrix} 0 & 1 & 0 \\ 1 & 0 & 0 \\ 0 & 0 & 0 \end{pmatrix} & \lambda_2 &= \begin{pmatrix} 0 & -i & 0 \\ i & 0 & 0 \\ 0 & 0 & 0 \end{pmatrix} \\
\lambda_3 &= \begin{pmatrix} 1 & 0 & 0 \\ 0 & -1 & 0 \\ 0 & 0 & 0 \end{pmatrix} & \lambda_4 &= \begin{pmatrix} 0 & 0 & 1 \\ 0 & 0 & 0 \\ 1 & 0 & 0 \end{pmatrix} \\
\lambda_5 &= \begin{pmatrix} 0 & 0 & -i \\ 0 & 0 & 0 \\ i & 0 & 0 \end{pmatrix} & \lambda_6 &= \begin{pmatrix} 0 & 0 & 0 \\ 0 & 0 & 1 \\ 0 & 1 & 0 \end{pmatrix} \\
\lambda_7 &= \begin{pmatrix} 0 & 0 & 0 \\ 0 & 0 & -i \\ 0 & i & 0 \end{pmatrix} & \lambda_8 &= \frac{1}{\sqrt{3}} \begin{pmatrix} 1 & 0 & 0 \\ 0 & 1 & 0 \\ 0 & 0 & -2 \end{pmatrix}
\end{aligned} \tag{7.4}$$

These are generalizations of the Pauli matrices in the sense that the first three Gell-Mann matrices contain the Pauli matrices acting on a subspace:

$$\lambda_a = \begin{pmatrix} \sigma_a & 0 \\ 0 & 0 \end{pmatrix} \qquad \text{for } a = 1 \text{ to } 3 \tag{7.5}$$

You can imagine that we could go on and construct 4×4 matrices that contain these 3×3 matrices in the same way, and so on.

The $SU(3)$ generators are conventionally defined by

$$T_a = \frac{1}{2}\lambda_a \tag{7.6}$$

and they satisfy

$$\text{Tr}\,(T_a T_b) = \frac{1}{2}\delta_{ab} \tag{7.7}$$

Clearly, T_a for $a = 1$ to 3 generate an $SU(2)$ subgroup of $SU(3)$. This is sometimes called the isospin subgroup, for reasons that will become apparent when we discuss $SU(3)$ as an approximate symmetry of the strong interactions — where this subgroup is in fact Heisenberg's isospin. It is convenient to put T_3 in the Cartan subalgebra. There is one generator, T_8, that commutes with T_3, so we put it in the Cartan subalgebra as well, and take

$$H_1 = T_3 \qquad H_2 = T_8 \tag{7.8}$$

7.2 Weights and roots of $SU(3)$

The weights of this representation are easy to find because T_3 and T_8 are already diagonal

$$T_3 = \begin{pmatrix} \frac{1}{2} & 0 & 0 \\ 0 & -\frac{1}{2} & 0 \\ 0 & 0 & 0 \end{pmatrix} \quad T_8 = \frac{\sqrt{3}}{6} \begin{pmatrix} 1 & 0 & 0 \\ 0 & 1 & 0 \\ 0 & 0 & -2 \end{pmatrix} \tag{7.9}$$

The eigenvectors, and associated weights are

$$\begin{gathered} \begin{pmatrix} 1 \\ 0 \\ 0 \end{pmatrix} \to (1/2, \sqrt{3}/6) \\ \begin{pmatrix} 0 \\ 1 \\ 0 \end{pmatrix} \to (-1/2, \sqrt{3}/6) \\ \begin{pmatrix} 0 \\ 0 \\ 1 \end{pmatrix} \to (0, -\sqrt{3}/3) \end{gathered} \tag{7.10}$$

These vectors, plotted in a plane, form the vertices of an equilateral triangle

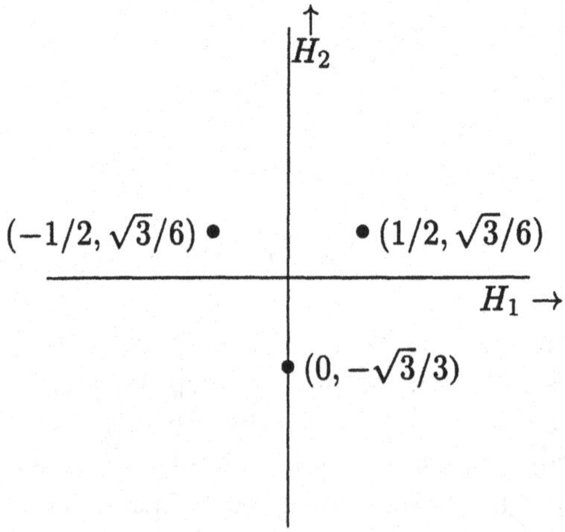

H_2 ↑

$(-1/2, \sqrt{3}/6)$ • • $(1/2, \sqrt{3}/6)$

H_1 →

• $(0, -\sqrt{3}/3)$

(7.11)

The roots are going to be differences of weights, because the corresponding generators must take us from one weight to another. It is not hard to see that the corresponding generators are those that have only one off-diagonal entry:

$$\begin{aligned}
\frac{1}{\sqrt{2}}(T_1 \pm iT_2) &= E_{\pm 1,0} \\
\frac{1}{\sqrt{2}}(T_4 \pm iT_5) &= E_{\pm 1/2,\pm\sqrt{3}/2} \\
\frac{1}{\sqrt{2}}(T_6 \pm iT_7) &= E_{\mp 1/2,\pm\sqrt{3}/2}
\end{aligned} \tag{7.12}$$

where the $\pm$ signs are correlated. The roots form a regular hexagon, plotted here along with the two elements of the Cartan subalgebra in the center:

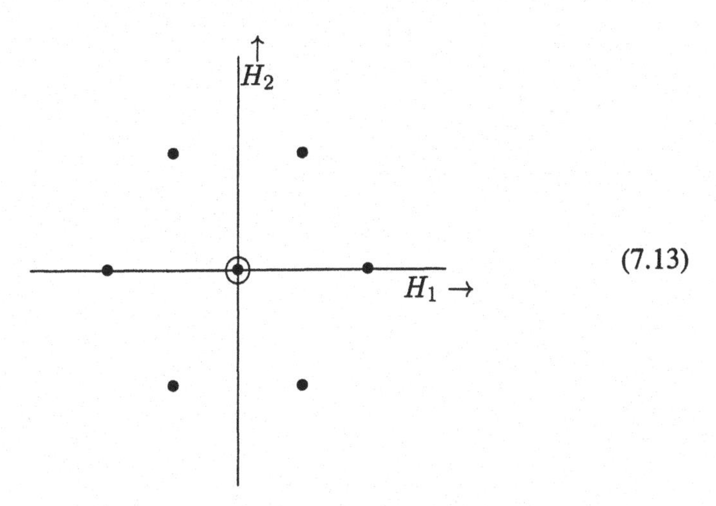

Problems

7.A. Calculate f_{147} and f_{458} in $SU(3)$.

7.B. Show that T_1, T_2 and T_3 generate an $SU(2)$ subalgebra of $SU(3)$. Every representation of $SU(3)$ must also be a representation of the subalgebra. However, the irreducible representations of $SU(3)$ are not necessarily irreducible under the subalgebra. How does the the representation generated by the Gell-Mann matrices transform under this subalgebra. That is, reduce, if necessary, the three dimensional representation into representations which are irreducible under the subalgebra and state which irreducible representations appear in the reduction. Then answer the same question for the adjoint representation of $SU(3)$.

7.C. Show that λ_2, λ_5 and λ_7 generate an $SU(2)$ subalgebra of $SU(3)$. Every representation of $SU(3)$ must also be a representation of the subalgebra. However, the irreducible representations of $SU(3)$ are not necessarily irreducible under the subalgebra. How does the representation generated by the Gell-Mann matrices transform under this subalgebra. That is, reduce, if necessary, the three dimensional representation into representations which are irreducible under the subalgebra and state which irreducible representations appear in the reduction. Then answer the same question for the adjoint representation of $SU(3)$.

Chapter 8

Simple Roots

What we need to complete the analogy between $SU(2)$ and an arbitrary simple Lie algebra is a notion of positivity for the weights. Then we can discuss things like raising and lowering operators, and the "highest weight" in a meaningful way. What we want is a definition that ensures that every non-zero weight is either positive or negative, and that if μ is positive, $-\mu$ is negative and vice versa.

8.1 Positive weights

It is easy to find such a scheme — indeed, in a multi-dimensional space, there are an infinite number. In some arbitrary basis for the Cartan subalgebra, the components, μ_1, μ_2, ..., of the weight are fixed. We will say that the weight is **positive** if its first non-zero component is positive and that the weight is **negative** if its first non-zero component is negative. While this depends on the arbitrary basis, it does have the properties we want. Eventually, we will see that the results will not depend on the basis, but for now, we will just fix it and forget it.

For example, in $SU(3)$, the 3 dimensional defining representation looks

 DOI: 10.1201/9780429499210-9

like this:

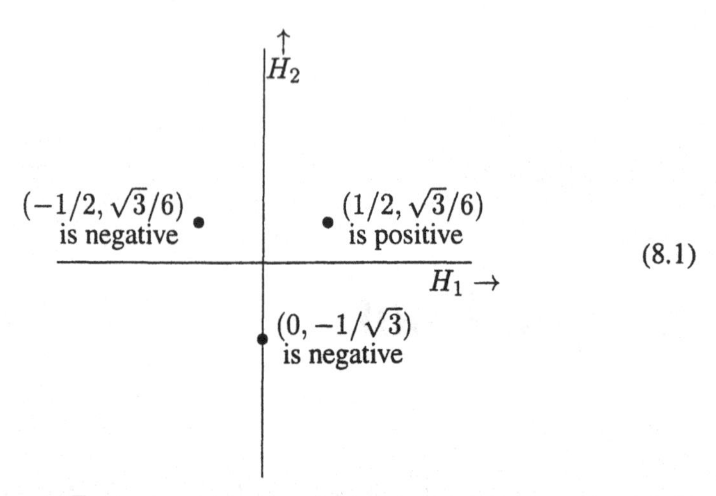

(8.1)

The weight $(0, -1/\sqrt{3})$ is negative because its first component is zero so the sign is determined by the sign of the second component.

With this definition, we can define an ordering in the obvious way:

$$\mu > \nu \text{ if } \mu - \nu \text{ is positive} \tag{8.2}$$

This allows us to talk about the highest weight in a representation.

In the adjoint representation, the positive roots correspond to raising operators and the negative roots to lowering operators. The highest weight of any representation has the property that we cannot raise it, so that all generators corresponding to positive roots must annihilate the corresponding state.

In the $SU(3)$ adjoint representation, in our usual basis, the positive roots are on the right and the negative on the left, as shown below:

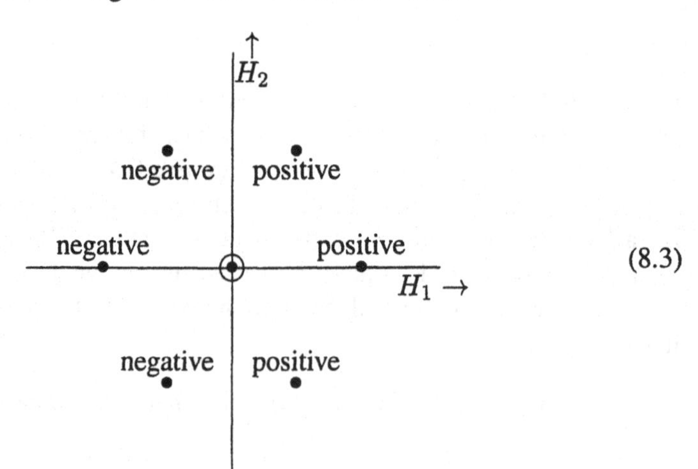

(8.3)

8.2 Simple roots

But we don't want to check all the roots if we don't have to. Clearly, some of the positive roots can be built out of others. So it makes sense to define **simple roots** as positive roots that cannot be written as a sum of other positive roots. We will then see that if a weight is annihilated by the generators of all the simple roots, it is the highest weight of an irreducible representation. Furthermore, from the geometry of the simple roots, it is possible to reconstruct the whole algebra. The logic of this is fun and worth understanding in detail.
1 — If α and β are different simple roots, then $\alpha - \beta$ is not a root. This is true because one of them, say β, is larger, so that $\beta - \alpha$ is positive. But then β is the sum of two positive roots, α and $\beta - \alpha$.
2 — Because $\alpha - \beta$ is not a root

$$E_{-\alpha}|E_\beta\rangle = E_{-\beta}|E_\alpha\rangle = 0 \tag{8.4}$$

Then in the master formula (6.36)

$$\frac{\alpha \cdot \beta}{\alpha^2} = -\frac{1}{2}(p - q)$$

the integer q is zero. Similarly in

$$\frac{\beta \cdot \alpha}{\beta^2} = -\frac{1}{2}(p' - q')$$

the integer q' is zero. Thus

$$\frac{\alpha \cdot \beta}{\alpha^2} = -\frac{p}{2} \qquad \frac{\beta \cdot \alpha}{\beta^2} = -\frac{p'}{2} \tag{8.5}$$

Knowing the integers p and p' for each simple root is equivalent to knowing the angles between the simple roots, and their relative lengths because

$$\cos\theta_{\alpha\beta} = -\frac{\sqrt{pp'}}{2} \qquad \frac{\beta^2}{\alpha^2} = \frac{p}{p'} \tag{8.6}$$

3 — The angle between any pair of simple roots satisfies

$$\frac{\pi}{2} \leq \theta < \pi \tag{8.7}$$

The first inequality follows from (8.6) because the cosine is less than or equal to zero. The second inequality follows because all the roots are positive. Simple multidimensional geometry then implies that the simple roots are linearly independent. Here's a proof — consider a linear combination of the simple roots,

$$\gamma = \sum_\alpha x_\alpha \alpha \tag{8.8}$$

If all the coefficients have the same sign, then clearly, γ cannot vanish unless all the coefficients vanish, because the α are all positive vectors. But if there are some coefficients of each sign, we can write

$$\gamma = \mu - \nu \tag{8.9}$$

where μ and ν are strictly positive vectors,

$$\mu = \sum_{x_\alpha > 0} x_\alpha \alpha \qquad \nu = -\sum_{x_\alpha < 0} x_\alpha \alpha \tag{8.10}$$

But the norm of γ cannot vanish because

$$(\mu - \nu)^2 = \mu^2 + \nu^2 - 2(\mu \cdot \nu) \geq \mu^2 + \nu^2 > 0 \tag{8.11}$$

where the last inequality follows from the fact $(\alpha \cdot \beta) \leq 0$ for any pair of simple roots, (8.6).

Thus no linear combination of the simple roots can vanish and they are linearly independent.

4 — Any positive root ϕ can be written as a linear combination of simple roots with non-negative integer coefficients, k_α

$$\phi = \sum_\alpha k_\alpha \alpha \tag{8.12}$$

This is just logic. If ϕ is simple, this is true. If not, we can split it into two positive roots and try again.

5 — The simple roots are not only linearly independent, they are complete, so the number of simple roots is equal to m, the rank of the algebra, the number of Cartan generators. If this were not true, then there would be some vector ξ orthogonal to all the simple roots, and therefore orthogonal to all the roots. But then

$$[\xi \cdot H, E_\phi] = 0 \quad \text{for all roots } \phi \tag{8.13}$$

Since $\xi \cdot H$ also commutes with the other Cartan generators, it commutes with all the generators and the algebra is not simple, contrary to assumption.

6 — Finally, we are in a position to construct the whole algebra from the simple roots. For now, we will simply show how to determine all the roots. We will find easier ways of doing this later, and also discuss how to construct the actual algebra.

We know that all the positive roots have the form

$$\phi_k = \sum_\alpha k_\alpha \alpha \tag{8.14}$$

for non-negative integers, k_α, where the integer k is

$$k = \sum_\alpha k_\alpha \tag{8.15}$$

If we can determine which ϕ_k are roots, we will have determined the roots in terms of the simple roots. It is straightforward to do this inductively, using the master formula.

All the ϕ_1's are roots because these are just the simple roots themselves. Suppose now we have determined the roots for $k \leq \ell$. Then we look at

$$E_\alpha |\phi_\ell\rangle \tag{8.16}$$

for all α, which gives roots of the form $\phi_{\ell+1}$. We can compute

$$\frac{2\alpha \cdot \phi_\ell}{\alpha^2} = -(p-q) \tag{8.17}$$

But we will always know q, because we will know the history of how ϕ_ℓ got built up by the action of the raising operators from smaller k. Thus we can determine p. If $p > 0$, then $\phi_\ell + \alpha$ is a root.

Let's illustrate this inductive procedure for $\ell = 1$. In this case, we always start with a simple root, $\phi_1 = \beta$ where β is a simple root. All the qs are zero so

$$\frac{2\alpha \cdot \phi_1}{\alpha^2} = \frac{2\alpha \cdot \beta}{\alpha^2} = -p \tag{8.18}$$

Thus if $\alpha \cdot \beta = 0$, then $p = 0$ and $\alpha + \beta$ is not a root. Otherwise $p > 0$ and $\alpha + \beta$ is a root.

The only way this procedure could fail to find a root is if there exists some positive root $\phi_{\ell+1}$ which is not the sum of a root ϕ_ℓ and some simple root. This is impossible, because if there were such a $\phi_{\ell+1}$, it would be annihilated by all the $E_{-\alpha}$ (because $E_{-\alpha}|\phi_{\ell+1}\rangle$ if non-zero would be a ϕ_ℓ state and we could apply E_α to it and get $|\phi_{\ell+1}\rangle$ back). Thus $|\phi_{\ell+1}\rangle$ would have to transform like the lowest weight state of all the $SU(2)$ subalgebras associated with the simple roots, which requires that the E_3 values $\alpha \cdot \phi_{\ell+1}/\alpha^2 \leq 0$ for all α. But then

$$\phi_{\ell+1}^2 = \sum_\alpha k_\alpha \alpha \cdot \phi_{\ell+1} \leq 0 \tag{8.19}$$

which is a contradiction. Thus we always find all the roots $\phi_{\ell+1}$ by acting on all the ϕ_ℓ with all the simple roots. For $SU(3)$, for example, the positive root $(1,0)$ is the sum of the other two, which are the simple roots

$$\alpha^1 = (1/2, \sqrt{3}/2) \qquad \alpha^2 = (1/2, -\sqrt{3}/2) \tag{8.20}$$

as shown

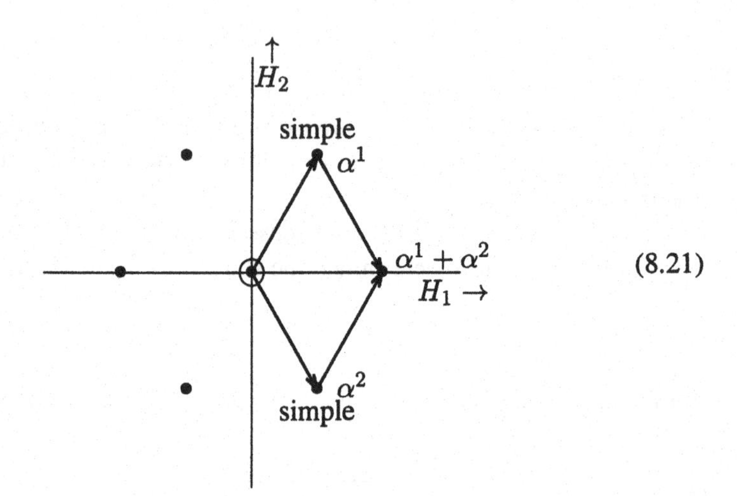

We have

$$\alpha^{1\,2} = \alpha^{2\,2} = 1 \qquad \alpha^1 \cdot \alpha^2 = -1/2 \tag{8.22}$$

thus

$$\frac{2\alpha^1 \cdot \alpha^2}{\alpha^{1\,2}} = \frac{2\alpha^1 \cdot \alpha^2}{\alpha^{2\,2}} = -1 \tag{8.23}$$

and thus $p = 1$ for both α^1 acting on $|\alpha^2\rangle$, and vice versa. Thus $\alpha^1 + \alpha^2$ is a root, but $2\alpha^1 + \alpha^2$ and $\alpha^1 + 2\alpha^2$ are not.

8.3 Constructing the algebra

The procedure outlined above can give us more than a listing of the roots. We can actually construct the entire algebra from the simple roots. Let us go back to the derivation of the master formula, where we found in (6.31) and (6.34)

$$\frac{\alpha \cdot \mu}{\alpha^2} + p = j \qquad \frac{\alpha \cdot \mu}{\alpha^2} - q = -j \tag{8.24}$$

This follows because a state $|\mu\rangle$ with weight μ in any irreducible representation must have some component that transforms under the largest spin representation of the $SU(2)$ algebra associated with α, generated by (from (6.20))

$$\begin{aligned} E^{\pm} &\equiv |\alpha|^{-1} E_{\pm\alpha} \\ E_3 &\equiv |\alpha|^{-2} \alpha \cdot H \end{aligned} \tag{8.25}$$

This largest spin is the j in (8.24). In general, $|\mu\rangle$ may also have lower spin components, but j is the highest one. It must be there so that $(E^+)^p |\mu\rangle$ transforms like a $|j, j\rangle$ state, which is non-zero, but annihilated by another E^+, and $(E^-)^m |\mu\rangle$ transforms like a $|j, -j\rangle$ state, annihilated by another E^-. Adding these two relations gives the master equation. But subtracting them gives

$$p + q = 2j \tag{8.26}$$

Thus if we know p and q, we know the highest spin representation that overlaps with the weight state.

If μ is a root in the adjoint representation, the situation is even simpler. Because we have already shown that each root appears only once in the adjoint representation, if we know p and q, and therefore j for a root β under the action of the $SU(2)$ associated with a simple root α, we can conclude that $|\beta\rangle$ is the state with $E_3 = \alpha \cdot \beta / \alpha^2$ in the spin j representation,

$$|\beta\rangle = |j, \alpha \cdot \beta / \alpha^2\rangle \tag{8.27}$$

It is completely determined up to a phase. Thus we also know exactly how $E^{\pm}$ acts, up to phases. Let's see how this enables us to construct the algebra in the example of $SU(3)$. The root diagram looks like this:

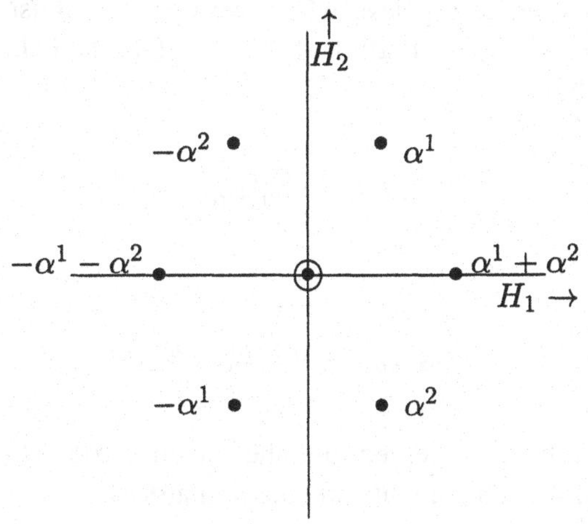

(8.28)

where

$$\alpha^{1^2} = \alpha^{2^2} = 1 \qquad \alpha^1 \cdot \alpha^2 = -1/2$$

$$\frac{\alpha^1 \cdot \alpha^2}{\alpha^{1^2}} = \frac{\alpha^1 \cdot \alpha^2}{\alpha^{2^2}} = -\frac{1}{2}$$

We already know that $p = 1$ and $q = 0$ for both α^1 acting on $|\alpha^2\rangle$, and vice versa. We already know how the Cartan generators commute with everything. So we just need the commutation relations of the raising and lowering operators. Let's begin by explicitly constructing the raising operator, $E_{\alpha^1+\alpha^2}$. Since $p + q = 1$, we have $j = 1/2$ and therefore, if we look at the α^1 raising operator

$$\begin{aligned} E^+|E_{\alpha^2}\rangle &= |\alpha^1|^{-1} E_{\alpha^1}|E_{\alpha^2}\rangle \\ &= E_{\alpha^1}|E_{\alpha^2}\rangle = |[E_{\alpha^1}, E_{\alpha^2}]\rangle \\ &= E_{\alpha^1}|1/2, -1/2\rangle = \frac{1}{\sqrt{2}}|1/2, 1/2\rangle \\ &= \frac{1}{\sqrt{2}}\eta|E_{\alpha^1+\alpha^2}\rangle \end{aligned} \tag{8.29}$$

where η is a phase. This may need some explanation. The $|E_{\alpha^2}\rangle$ state is properly normalized, by assumption, and it corresponds to the $|1/2, -1/2\rangle$ state under the α^1 $SU(2)$ (the E_3 value is $\alpha^1 \cdot \alpha^2/\alpha^{1^2} = -1/2$). Acting on it with the raising operator E^+ tells us what the correctly normalized $|1/2, 1/2\rangle$ state is. But this, up to a phase, which we called η, must be the properly normalized state $|E_{\alpha^1+\alpha^2}\rangle$. Putting all this together, and choosing $\eta = 1$ by convention, we have

$$|E_{\alpha^1+\alpha^2}\rangle = \sqrt{2}|[E_{\alpha^1}, E_{\alpha^2}]\rangle \tag{8.30}$$

and thus

$$E_{\alpha^1+\alpha^2} = \sqrt{2}\,[E_{\alpha^1}, E_{\alpha^2}] \tag{8.31}$$

Now that we have expressed the other positive root as a commutator of the simple roots, we can compute any commutator just using the Jacobi identity. For example

$$
\begin{aligned}
& [E_{-\alpha^1}, E_{\alpha^1+\alpha^2}] \\
&= \sqrt{2}\,[E_{-\alpha^1}, [E_{\alpha^1}, E_{\alpha^2}]] \\
&= \sqrt{2}\,[[E_{-\alpha^1}, E_{\alpha^1}], E_{\alpha^2}] \\
&= \sqrt{2}\left[-\alpha^1 \cdot H, E_{\alpha^2}\right] \\
&= -\sqrt{2}\alpha^1 \cdot \alpha^2 E_{\alpha^2} = \frac{1}{\sqrt{2}} E_{\alpha^2}
\end{aligned}
\tag{8.32}
$$

We already knew this, because we are just moving back down the $SU(2)$ representation. Here's another, slightly more interesting:

$$
\begin{aligned}
& [E_{-\alpha^2}, E_{\alpha^1+\alpha^2}] \\
&= \sqrt{2}\,[E_{-\alpha^2}, [E_{\alpha^1}, E_{\alpha^2}]] \\
&= \sqrt{2}\,[E_{\alpha^1}, [E_{-\alpha^2}, E_{\alpha^2}]] \\
&= \sqrt{2}\left[E_{\alpha^1}, -\alpha^2 \cdot H\right] \\
&= \sqrt{2}\left[\alpha^2 \cdot H, E_{\alpha^1}\right] \\
&= \sqrt{2}\alpha^1 \cdot \alpha^2 E_{\alpha^1} = -\frac{1}{\sqrt{2}} E_{\alpha^1}
\end{aligned}
\tag{8.33}
$$

The interesting thing here is the phase — which is determined to be a – sign.

8.4 Dynkin diagrams

A Dynkin diagram is a short-hand notation for writing down the simple roots. Each simple root is indicated by an open circle. Pairs of circles are connected by lines, depending on the angle between the pair of roots to which the circles correspond, as follows:

O≡O if the angle is 150° (8.34)

O=O if the angle is 135°

O–O if the angle is 120°

O O if the angle is 90°

The Dynkin diagram determines all the angles between pairs of simple roots. This doesn't quite fix the roots, because there may be more than one choice for the relative lengths. We will come back to that later.

$$\bigcirc \quad \text{is the diagram for } SU(2)$$

$$\bigcirc\!\!-\!\!\bigcirc \quad \text{is the diagram for } SU(3) \tag{8.35}$$

8.5 Example: G_2

Suppose that an algebra has simple roots

$$\alpha^1 = (0,1) \qquad \alpha^2 = (\sqrt{3}/2, -3/2) \tag{8.36}$$

This is an allowed pairing, because

$$\begin{gathered} {\alpha^1}^2 = 1 \qquad {\alpha^2}^2 = 3 \\ \alpha^1 \cdot \alpha^2 = -3/2 \\ \frac{2\alpha^1 \cdot \alpha^2}{{\alpha^1}^2} = -3 \qquad \frac{2\alpha^1 \cdot \alpha^2}{{\alpha^2}^2} = -1 \end{gathered} \tag{8.37}$$

The angle between the two roots is determined by

$$\cos\theta_{\alpha^1\alpha^2} = -\frac{\sqrt{3}}{2} \qquad \theta_{\alpha^1\alpha^2} = 150^\circ \tag{8.38}$$

Thus this corresponds to the Dynkin diagram

$$\bigcirc\!\!\equiv\!\!\bigcirc$$

This algebra is called G_2.

8.6 The roots of G_2

For E_{α^1} acting on $|\alpha^2\rangle$ we have $p = 3$. For E_{α^2} acting on $|\alpha^1\rangle$ we have $p = 1$. Thus

$$\alpha^1 + \alpha^2 \quad 2\alpha^1 + \alpha^2 \quad \text{and} \quad 3\alpha^1 + \alpha^2 \tag{8.39}$$

are all roots but

$$\alpha^1 + 2\alpha^2 \quad \text{and} \quad 4\alpha^1 + \alpha^2 \tag{8.40}$$

are not. In terms of ϕ_k in (8.14) we have

$$\phi_2 = \alpha^1 + \alpha^2 \quad \phi_3 = 2\alpha^1 + \alpha^2 \quad \phi_4 = 3\alpha^1 + \alpha^2 \tag{8.41}$$

We know that the ϕ_2 state is unique from the general properties of simple roots. The ϕ_3 state is unique because $\alpha^1 + 2\alpha^2$, the only other state that could be obtained by acting on ϕ_2, is not a root. To see whether there is another ϕ_4 state, we must check whether $2\alpha^1 + 2\alpha^2$ is a root, that is, whether it can be obtained by acting on ϕ_3 with a simple root, which must be α^2.

$$\frac{2\alpha^2 \cdot (2\alpha^1 + \alpha^2)}{\alpha^{2\,2}} = \frac{-6+6}{3} = 0 = -(p-q) \tag{8.42}$$

But $q = 0$ because $2\alpha^1$ is not a root, and thus $p = 0$, so $2\alpha^1 + 2\alpha^2$ is not a root. Actually, we could have come to this conclusion more simply by noting that $2\alpha^1 + 2\alpha^2 = 2(\alpha^1 + \alpha^2)$ which is twice the root $\alpha^1 + \alpha^2$, but we proved in the discussion after (6.26) that no multiple of a root can be a root.

Now to get the ϕ_5 states, note that we already know that $4\alpha^1 + \alpha^2$ is not a root, so we need only check $3\alpha^1 + 2\alpha^2$.

$$\frac{2\alpha^2 \cdot (3\alpha^1 + \alpha^2)}{\alpha^{2\,2}} = \frac{-9+6}{3} = -1 = -(p-q) \tag{8.43}$$

Again, $q = 0$, thus $p = 1$ and $\phi_5 = 3\alpha^1 + 2\alpha^2$ is a root. Because $p = 1$, we also know that $3\alpha^1 + 3\alpha^2$ is not a root, so to check for ϕ_6, we need only look at $4\alpha^1 + 2\alpha^2$.

$$\frac{2\alpha^1 \cdot (3\alpha^1 + 2\alpha^2)}{\alpha^{1\,2}} = \frac{6-6}{1} = 0 = -(p-q) \tag{8.44}$$

$q = 0$ because $2\alpha^1 + 2\alpha^2$ is not a root, so we are finished (again, we could has just used the fact that $2\alpha^1 + \alpha^2$ is a root to see that $4\alpha^1 + 2\alpha^2$ is not), and the roots look like this:

$-\alpha^2$ • H_2 ↑ • $3\alpha^1 + \alpha^2$

• α^1

$-\alpha^1 - \alpha^2$ • • $2\alpha^1 + \alpha^2$

$-3\alpha^1 - 2\alpha^2$ • ⊙ • $3\alpha^1 + 2\alpha^2$ H_1 → (8.45)

$-2\alpha^1 - \alpha^2$ • • $\alpha^1 + \alpha^2$

$-\alpha^1$ •

$-3\alpha^1 - \alpha^2$ • • α^2

What we did can be summarized in the following diagram:

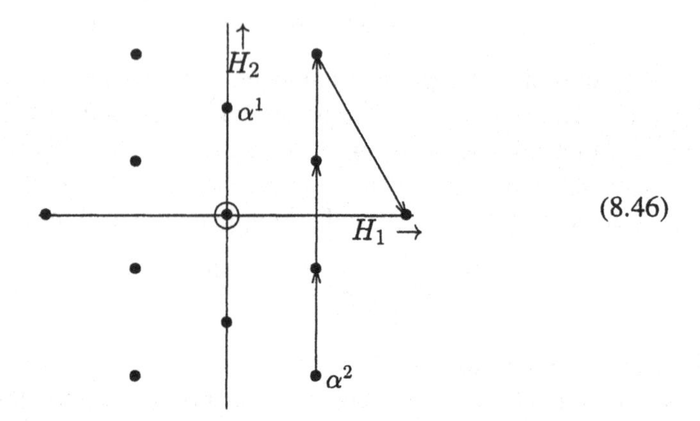

(8.46)

8.7 The Cartan matrix

There is a useful way of keeping track of the integers p^i and q^i associated with the action of a simple root α^i on a state $|\phi\rangle$ for a positive root ϕ that eliminates the need for tedious geometrical calculations. The idea is to label the roots directly by their $q^i - p^i$ values. The $q^i - p^i$ of any weight, μ, is simply twice its E_3 value, where E_3 is the Cartan generator of the $SU(2)$ associated with the simple root α^i, because

$$2E_3|\mu\rangle = \frac{2H \cdot \alpha^i}{\alpha^{i2}}|\mu\rangle = \frac{2\mu \cdot \alpha^i}{\alpha^{i2}}|\mu\rangle = \left(q^i - p^i\right)|\mu\rangle \,. \tag{8.47}$$

Because the α^i are complete and linearly independent, the $q^i - p^i$ values for the weights contain the same information as the values of the components of the weight vector, so we can use them to label the weights. The advantage of doing so is that it will make very transparent the structure of the representations under the $SU(2)$s generated by the simple roots.

Since a positive root, ϕ, can be written as $\phi = \sum_j k_j \alpha^j$, the master formula can be written as

$$\begin{aligned} q^i - p^i &= \frac{2\phi \cdot \alpha^i}{\alpha^{i2}} \\ &= \sum_j k_j \frac{2\alpha^j \cdot \alpha^i}{\alpha^{i2}} \\ &= \sum_j k_j A_{ji} \end{aligned} \tag{8.48}$$

where A is the **Cartan matrix**

$$A_{ji} \equiv \frac{2\alpha^j \cdot \alpha^i}{\alpha^{i2}} \tag{8.49}$$

The matrix element A_{ji} is the $q - p$ value for the simple root α^i acting on the state $|\alpha^j\rangle$, twice the E_3 value, thus all the entries of A are integers. The diagonal entries are all equal to 2. This is obvious from the definition, and also from the fact that the simple roots have $E_3 = 1$ because the $SU(2)$ generators themselves transform like the adjoint, spin 1 representation. The off-diagonal entries, all 0, -1, -2 or -3, record the angles between simple roots and their relative lengths — the same information as the Dynkin diagram, **and they tell us how the simple roots fit into representations of the $SU(2)$s associated with the other simple roots.** It is easy to see that the Cartan matrix is invertible because the α^j are complete and linearly independent. **Note that the jth row of the Cartan matrix consists of the $q_i - p_i$ values of the simple root α^j.**

For $SU(3)$, the Cartan matrix looks like

$$\begin{pmatrix} 2 & -1 \\ -1 & 2 \end{pmatrix} \tag{8.50}$$

For the G_2 algebra we have just analyzed, it looks like

$$\begin{pmatrix} 2 & -1 \\ -3 & 2 \end{pmatrix} \tag{8.51}$$

8.8 Finding all the roots

We now show how to use the Cartan matrix to simplify the procedure of building up all the roots from the simple roots. When we go from ϕ to $\phi + \alpha^j$ by the action of the raising operator E_{α^j}, this just changes k_j to $k_j + 1$, and thus $q^i - p^i$ to $q^i - p^i + A_{ji}$.

$$\begin{gathered} k_j \to k_j + 1 \\ q^i - p^i \to q^i - p^i + A_{ji} \end{gathered} \tag{8.52}$$

If we think of the $q^i - p^i$ as the elements of a row vector, this is equivalent to simply adding the jth row of the Cartan matrix, which is just the vector $q - p$ associated with the simple root α^j. This allows us to streamline the process of computing the roots. We will describe the procedure and illustrate it first for $SU(3)$. Start with the simple roots in the $q-p$ notation. We will put each in a rectangular box, and arrange them on a horizontal line, which represents the $k=1$ layer of positive roots — that is the simple roots.

$$k=1 \qquad \boxed{2\ -1} \quad \boxed{-1\ 2} \qquad \alpha^1, \alpha^2 \tag{8.53}$$

It is convenient to put a box with m zeros, representing the Cartan generators, on a line below, representing the $k=0$ layer.

$$\begin{array}{llll} k=1 & \boxed{2\ -1} \quad \boxed{-1\ 2} & & \alpha^1, \alpha^2 \\ \\ k=0 & \quad\quad\boxed{0\ \ 0} & & H_i \end{array} \tag{8.54}$$

Now for each element of each box we know the q^i value. For the ith element of α^i, $q^i = 2$, because the root is part of the $SU(2)$ spin 1 representation consisting of $E_{\pm\alpha^i}$ and $\alpha^i \cdot H$. For all the other elements, $q^j = 0$, because $\alpha^i - \alpha^j$ is not a root.

$$\begin{array}{llll} & q = 2\ \ 0 \quad\ \ 0\ \ 2 & & \\ k=1 & \boxed{2\ -1} \quad \boxed{-1\ 2} & & \alpha^1, \alpha^2 \\ \\ k=0 & \quad\quad\boxed{0\ \ 0} & & H_i \end{array} \tag{8.55}$$

Thus we can compute the corresponding p^i.

$$\begin{array}{llll} & p = 0\ \ 1 \quad\ \ 1\ \ 0 & & \\ k=1 & \boxed{2\ -1} \quad \boxed{-1\ 2} & & \alpha^1, \alpha^2 \\ \\ k=0 & \quad\quad\boxed{0\ \ 0} & & H_i \end{array} \tag{8.56}$$

Since the ith element of α^i is 2 (because it is a diagonal element of A), the corresponding p^i is zero (of course, since $2\alpha^i$ in not a root). For all the others, p is just minus the entry. For each non-zero p, we draw a line from the simple root to a new root with $k=2$, on a horizontal line above the $k=1$ line, obtained by adding the appropriate simple root. The line starts above the appropriate entry, so you can keep track of which root got added automatically. You can also draw such lines from the $k=0$ layer to the $k=1$ layer, and the lines for each root will have a different angle. You then try to put the boxes on the $k=2$ layer so that the lines associated with each root have the same angle they did between the 0 and 1 layer. These lines represent the action of the $SU(2)$ raising and lowering operators.

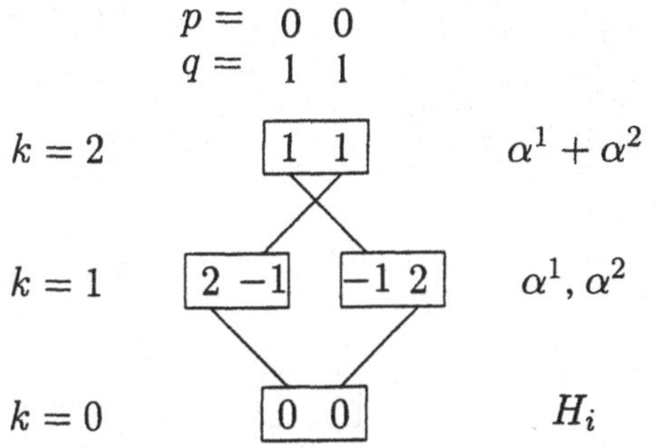

$$\begin{array}{lcc} & p = & 0\ 0 \\ & q = & 1\ 1 \\ k=2 & \boxed{1\ 1} & \alpha^1+\alpha^2 \\ k=1 & \boxed{2\ {-1}}\quad\boxed{-1\ 2} & \alpha^1,\alpha^2 \\ k=0 & \boxed{0\ 0} & H_i \end{array} \tag{8.57}$$

The procedure is now easy to iterate, because everything you need to know to go from $k = \ell$ to $k = \ell + 1$ is there in your diagram. At each stage, you compute p by subtracting the element of the vector from the corresponding q. For $SU(3)$, the procedure terminates at $k = 2$, because all the ps are zero.

Clearly, we could have continued this diagram farther down and shown the negative roots in the same way.

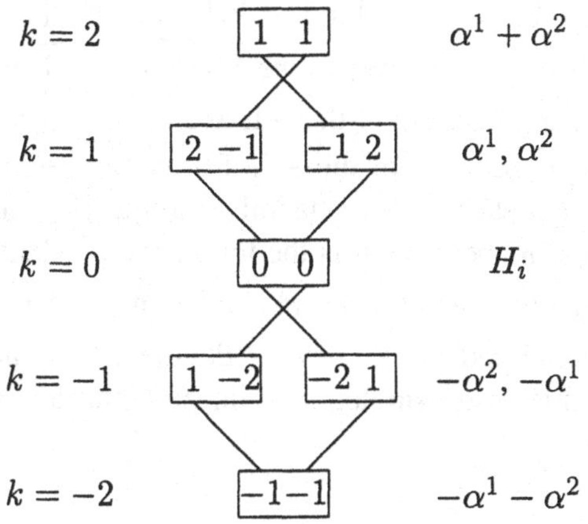

$$\begin{array}{lcc} k=2 & \boxed{1\ 1} & \alpha^1+\alpha^2 \\ k=1 & \boxed{2\ {-1}}\quad\boxed{-1\ 2} & \alpha^1,\alpha^2 \\ k=0 & \boxed{0\ 0} & H_i \\ k=-1 & \boxed{1\ {-2}}\quad\boxed{-2\ 1} & -\alpha^2,-\alpha^1 \\ k=-2 & \boxed{-1\ {-1}} & -\alpha^1-\alpha^2 \end{array} \tag{8.58}$$

8.9 The $SU(2)$s

The transformation properties of the roots under the two $SU(2)$s should be obvious from (8.58). In fact, instead of thinking about p and q, we can just see how each E_3 value fits into an $SU(2)$ representation. Then the process terminates as soon as all the $SU(2)$ representations are completed. This is equivalent to actually computing the ps and qs, because we got the master formula by thinking about this $SU(2)$ structure anyway, but it is much faster and easier.

Let us illustrate this with the diagram for G_2:

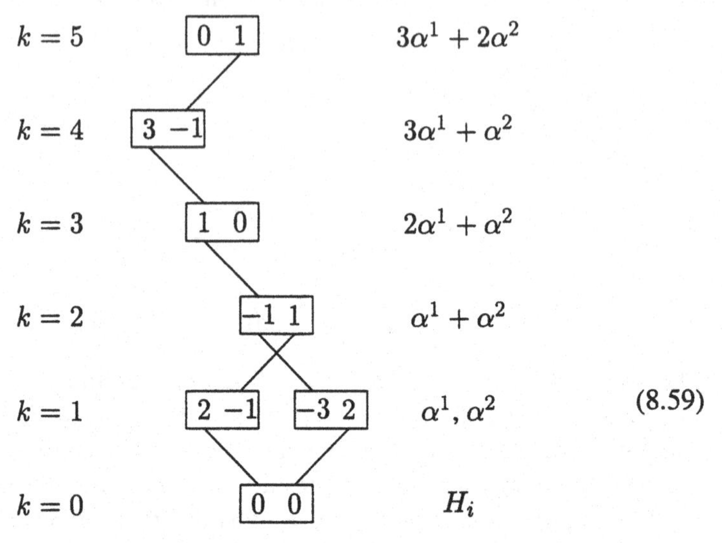

The argument in terms of the $SU(2)$ structure goes as follows. We know that the simple roots $\alpha^1 = \boxed{2\ -1}$ and $\alpha^2 = \boxed{-3\ 2}$ are the highest weights of spin 1 representations for their respective $SU(2)$s. $\boxed{2\ -1}$ must be the bottom of a doublet under α^2, because of the -1, and $\boxed{-3\ 2}$ is the bottom of a spin 3/2 quartet under α^1 because of the -3. So we just follow these up to the end, making sure that each root fits happily into representations under all the $SU(2)$s. $\boxed{-1\ 1}$ is fine because it is the top of the α^2 doublet and in the α^1 quartet. $\boxed{1\ 0}$ is fine because it is an α^2 singlet and in the α^1 quartet. $\boxed{3\ -1}$ finishes the α^1 quartet and starts a new α^2 doublet. And finally, $\boxed{0\ 1}$ finishes the α^2 doublet and is an α^1 singlet, so it doesn't start anything new, and the process terminates.

8.10 Constructing the G_2 algebra

We will do another example of this procedure later. Now let's stay with G_2 and construct the algebra. The two relevant raising operators are

$$E_1^+ = E_{\alpha^1} \qquad E_2^+ = \frac{1}{\sqrt{3}} E_{\alpha^2} \tag{8.60}$$

Start with the $|E_{\alpha^2}\rangle$ state. We know, because $p = 3$ and $q = 0$ or by looking at the roots in (8.59) that it is the lowest weight state in a spin 3/2 representation of the α^1 $SU(2)$. Call it

$$|E_{\alpha^2}\rangle = |3/2, -3/2, 1\rangle \tag{8.61}$$

Then applying the α^1 raising operator

$$\begin{aligned}
&|[E_{\alpha^1}, E_{\alpha^2}]\rangle \\
&= \sqrt{\frac{3}{2}}\,|3/2, -1/2, 1\rangle \\
&= \sqrt{\frac{3}{2}}\,|E_{\alpha^1+\alpha^2}\rangle
\end{aligned} \tag{8.62}$$

The last line is a definition, in that it involves a phase choice, but the construction using the raising operator guarantees that we have the normalization right. Applying the raising operator again gives

$$\begin{aligned}
&|[E_{\alpha^1}, [E_{\alpha^1}, E_{\alpha^2}]]\rangle \\
&= 2\sqrt{\frac{3}{2}}\,|3/2, 1/2, 1\rangle \\
&= \sqrt{6}\,|E_{2\alpha^1+\alpha^2}\rangle
\end{aligned} \tag{8.63}$$

and a third time gives

$$\begin{aligned}
&|[E_{\alpha^1}, [E_{\alpha^1}, [E_{\alpha^1}, E_{\alpha^2}]]]\rangle \\
&= \sqrt{\frac{3}{2}}\sqrt{6}\,|3/2, 3/2, 1\rangle \\
&= 3\,|E_{3\alpha^1+\alpha^2}\rangle \\
&= 3\,|1/2, -1/2, 2\rangle
\end{aligned} \tag{8.64}$$

where we have written the last line because this is also the lowest weight state of spin 1/2 representation of the α^2 $SU(2)$. Then applying the α^2 raising operator gives

$$\begin{aligned}
&|[E_{\alpha^2}, [E_{\alpha^1}, [E_{\alpha^1}, [E_{\alpha^1}, E_{\alpha^2}]]]]\rangle \\
&= 3\sqrt{3}\frac{1}{\sqrt{2}}\,|1/2, 1/2, 2\rangle \\
&= \frac{9}{\sqrt{6}}\,|E_{3\alpha^1+2\alpha^2}\rangle
\end{aligned} \tag{8.65}$$

Putting this all together, we have expressions for all the positive roots in terms of the generators associated with the simple roots:

$$
\begin{aligned}
E_{\alpha^1+\alpha^2} &= \sqrt{\frac{2}{3}}\,[E_{\alpha^1}, E_{\alpha^2}] \\
E_{2\alpha^1+\alpha^2} &= \sqrt{\frac{1}{6}}\,[E_{\alpha^1}, [E_{\alpha^1}, E_{\alpha^2}]] \\
E_{3\alpha^1+\alpha^2} &= \frac{1}{3}\,[E_{\alpha^1}, [E_{\alpha^1}, [E_{\alpha^1}, E_{\alpha^2}]]] \\
E_{3\alpha^1+2\alpha^2} &= \frac{\sqrt{6}}{9}\,[E_{\alpha^2}, [E_{\alpha^1}, [E_{\alpha^1}, [E_{\alpha^1}, E_{\alpha^2}]]]]
\end{aligned}
\tag{8.66}
$$

This is enough to determine all the commutation relations by just repeatedly using the Jacobi identity. For example, let's check that the α^1 lowering operator acts as we expect. For example

$$
\begin{aligned}
&[E_{-\alpha^1}, E_{\alpha^1+\alpha^2}] \\
&= \sqrt{\frac{2}{3}}\,[E_{-\alpha^1}, [E_{\alpha^1}, E_{\alpha^2}]] \\
&= \sqrt{\frac{2}{3}}\,[[E_{-\alpha^1}, E_{\alpha^1}], E_{\alpha^2}] \\
&= -\sqrt{\frac{2}{3}}\,[\alpha \cdot H, E_{\alpha^2}] \\
&= -\sqrt{\frac{2}{3}}\,\alpha^1 \cdot \alpha^2\, E_{\alpha^2} = \sqrt{\frac{3}{2}}\,E_{\alpha^2}
\end{aligned}
\tag{8.67}
$$

This is what we expect for a lowering operator acting on $|3/2, -1/2\rangle$ state. The other relations can be found similarly. The general form involves some multiple commutator of negative simple root generators with a multiple commutator of positive simple root generators. When these are rearranged using the Jacobi identity, the positive and negative generators always eat one another in pairs, so that in the end you get one of the positive root states, or one of the negative root states or a Cartan generator, so the algebra closes.

8.11 Another example: the algebra C_3

Let's look at the algebra corresponding to the following Dynkin diagram

$$\alpha^1 \quad \alpha^2 \quad \alpha^3 \qquad \bigcirc\!\!-\!\!\bigcirc\!\!=\!\!\bigcirc \tag{8.68}$$

where

$$\alpha^{1^2} = \alpha^{2^2} = 1 \qquad \alpha^{3^2} = 2 \tag{8.69}$$

The Cartan matrix is

$$\begin{pmatrix} 2 & -1 & 0 \\ -1 & 2 & -1 \\ 0 & -2 & 2 \end{pmatrix} \tag{8.70}$$

Then the construction of the positive roots goes as follows:[1]

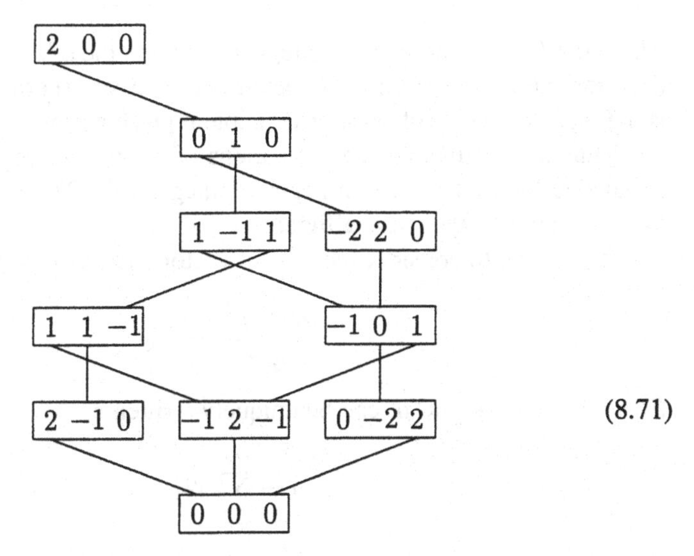

(8.71)

This algebra is called C_3. It is part of four infinite families of simple compact Lie algebras, called the classical groups. We will discuss them all later.

8.12 Fundamental weights

Suppose that the simple roots of some simple Lie algebra are α^j for $j =$ 1 to m. The highest weight, μ, of an arbitrary irreducible representation, D has the property that $\mu + \phi$ is not a weight for any positive root, ϕ. From the preceding discussion, it is clearly sufficient that $\mu + \alpha^j$ not be a weight in the representation for any j, because then

$$E_{\alpha^j}|\mu\rangle = 0 \; \forall j \tag{8.72}$$

which implies that all positive roots annihilate the state, because any positive root can be expressed as a multiple commutator of simple roots. We will see soon that this is an if and only if statement, because an entire irreducible

[1] Here, because the group is rank 3, we are projecting a three dimensional root space onto two dimensions — for groups of higher rank, these diagrams can get quite busy.

representation can be constructed by applying lowering operators to any state with this property. So if (8.72) is true, then μ is the highest weight of an irreducible representation. This means that for every E_{α^j} acting on $|\mu\rangle$, $p = 0$, and thus

$$\frac{2\alpha^j \cdot \mu}{\alpha^{j2}} = \ell^j \tag{8.73}$$

where the ℓ^j are non-negative integers. Because the α^js are linearly independent, the integers ℓ^j completely determine μ. Every set of ℓ^j gives a μ which is the highest weight of some irreducible representation.

Thus the irreducible representations of a rank m simple Lie algebra can be labeled by a set of m non-negative integers, ℓ^j. These integers are sometimes called the **Dynkin coefficients.**

It is useful to consider the weight vectors, μ^j, satisfying

$$\frac{2\alpha^j \cdot \mu^k}{\alpha^{j2}} = \delta_{jk} \tag{8.74}$$

Every highest weight, μ, can be uniquely written as

$$\mu = \sum_{j=1}^{m} \ell^j \mu^j \tag{8.75}$$

We can therefore build the representation with highest weight μ by constructing a tensor product of ℓ^1 representation of highest weight μ^1, ℓ^2 with highest weight μ^2, and so on. This representation will generally be reducible, but we can always then pick out the representation μ by applying lowering operators to $|\mu\rangle$.

The vectors μ^j are called the **fundamental weights** and the m irreducible representations that have these as highest weights are called the **fundamental representations**. We will sometimes denote them by D^j. Remember that the superscripts are just labels. The vectors also have vector indices. It's potentially confusing because both run from 1 to m.

There is more to say about the Dynkin coefficients. Since the fundamental weights form a complete set, we can expand any weight of any representation in terms of them, as in (8.75). Then we can run the argument backwards and get (8.73) which implies that for a general μ,

$$\ell^j = q^j - p^j \tag{8.76}$$

that is ℓ^j is the $q^j - p^j$ value for the simple root α^j. Thus the matrix elements of the vectors we were manipulating in constructing the positive roots of various algebras were just the Dynkin coefficients of the roots (though of course, for a general weight or root, the Dynkin coefficients will not necessarily be positive). In particular, the highest box in the construction is just the highest weight in the adjoint representation. The rows of the Cartan matrix are the Dynkin coefficients of the simple roots. Later, we will use a similar analysis to discuss arbitrary representations.

8.13 The trace of a generator

This is a good place to prove a theorem that we will see often in examples, and that will play a crucial role in the discussion of unification of forces:

Theorem 8.9 *The trace of any generator of any representation of a compact simple Lie group is zero.*

Proof: It suffices to prove this in the standard basis that we have developed in chapter 6 and this chapter, because the trace is invariant under similarity transformations. In the weight basis, every generator is a linear combination of Cartan generators and raising and lowering operators. The trace of raising or lowering operators is zero because they have no diagonal elements. The Cartan generators can be written as linear combinations of $\vec{\alpha}^i \cdot \vec{H}$, because the simple roots, $\vec{\alpha}^i$, are complete. But each $\vec{\alpha}^i \cdot \vec{H}$ is proportional to the generator of an $SU(2)$ subalgebra and its trace is zero because every $SU(2)$ representation is symmetrical about 0 — the spin runs from $-j$ to j. Thus the Cartan generators have zero trace and the theorem is proved.

Problems

8.A. Find the simple roots and fundamental weights and the Dynkin diagram for the algebra discussed in problem (6.C).

8.B. Consider the algebra generated by σ_a and $\sigma_a \eta_1$ where σ_a and η_a are independent Pauli matrices. Show that this algebra generates a group which is semisimple but not simple. Nevertheless, you can define simple roots. What does the Dynkin diagram look like?

8.C. Consider the algebra corresponding to the following Dynkin diagram

$\alpha^1 \quad \alpha^2 \quad \alpha^3$

○—○═○

where

$$\alpha^{1^2} = \alpha^{2^2} = 2 \qquad \alpha^{3^2} = 1$$

Note that this is similar to C_3 in (8.68), but the lengths (and relative lengths) are different. Find the Cartan matrix and find the Dynkin coefficients of all of the positive roots, using the diagrammatic construction described in this chapter. Don't forget to put the lines in the right place — this will make it harder to get confused.

Chapter 9

More $SU(3)$

In this chapter, we study other irreducible representations of $SU(3)$ and in the process, learn some useful general things about the irreducible representations of Lie algebras.

9.1 Fundamental representations of $SU(3)$

Label the $SU(3)$ simple roots as

$$\begin{aligned} \alpha^1 &= (1/2, \sqrt{3}/2) \\ \alpha^2 &= (1/2, -\sqrt{3}/2) \end{aligned} \tag{9.1}$$

Then we find the μ^i, the highest weights of the fundamental representations, by requiring that $\mu^i \cdot \alpha^j = 0$ if $i \neq j$

$$\begin{aligned} &\mu^i = (a^i, b^i) \\ &\mu^i \cdot \alpha^j = (a^i \mp \sqrt{3}\, b^i)/2 \\ &\Rightarrow a^i = \pm\sqrt{3}\, b^i \end{aligned} \tag{9.2}$$

and then normalizing to satisfy (8.74) which gives

$$\begin{aligned} \mu^1 &= (1/2, \sqrt{3}/6) \\ \mu^2 &= (1/2, -\sqrt{3}/6) \end{aligned} \tag{9.3}$$

 DOI: 10.1201/9780429499210-10

μ^1 is the highest weight of the defining representation generated by the Gell-Mann matrices —

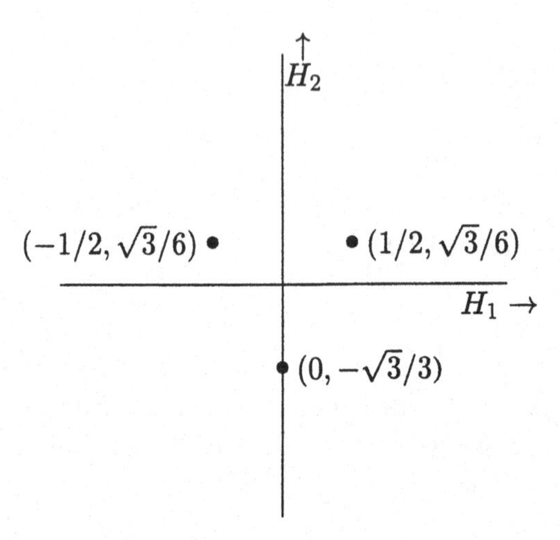

We can also write it in the $q-p$ notation. In fact, we can use the properties of the lowering operators in the $q-p$ notation to construct all weights by working down from the highest weight. μ^1 is associated with the vector (1,0), and in the graphical notation in which the simple roots look like

$$\begin{array}{ccc} \alpha^1 & & \alpha^2 \\ \boxed{2\ -1} & & \boxed{-1\ 2} \\ & \searrow\ \swarrow & \\ & \boxed{0\ \ 0} & \end{array} \tag{9.4}$$

it looks like this

$$\begin{array}{ccc} \boxed{1\ \ 0} & & \\ & \searrow & \\ & \boxed{-1\ 1} & \\ & \swarrow & \\ \boxed{0\ -1} & & \end{array} \tag{9.5}$$

The highest weight is the top of an α^1 doublet. Subtracting α^1 gives $\boxed{1\ 0} - \boxed{2\ -1} = \boxed{-1\ 1}$, thus because the second entry is a 1, and there is no state above it in the α^2 direction, it is the top of an α^2 doublet. Subtracting α^2 gives $\boxed{-1\ 1} - \boxed{-1\ 2} = \boxed{0\ -1}$, so the series terminates and the weights are

$$\mu^1 \qquad \mu^1 - \alpha^1 \qquad \mu^1 - \alpha^1 - \alpha^2 \tag{9.6}$$

There are many ways of obtaining the representation μ^2, or (0,1). The most straightforward way to construct it from scratch is to use the $q-p$ notation, starting with the highest weight, and working downwards in the $SU(2)$ representations:

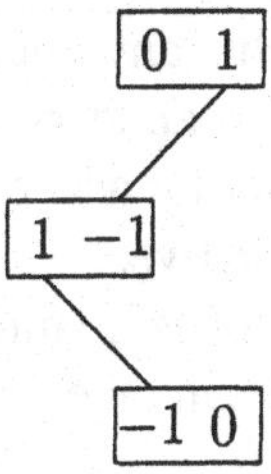

$$\boxed{0\ 1} \to \boxed{1\ -1} \to \boxed{-1\ 0} \tag{9.7}$$

so the weights are

$$\mu^2 \qquad \mu^2-\alpha^2 \qquad \mu^2-\alpha^2-\alpha^1 \tag{9.8}$$

or

$$(1/2,-\sqrt{3}/6) \quad (0,\sqrt{3}/3) \quad (-1/2,-\sqrt{3}/6) \tag{9.9}$$

so in the H_i plane, these look like

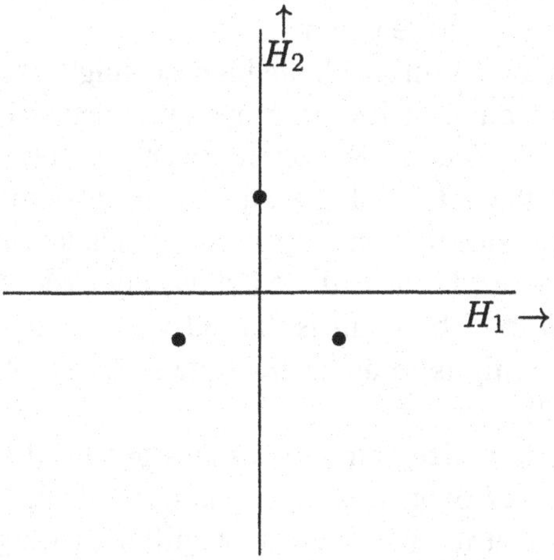

$\uparrow H_2$

$H_1 \rightarrow$

(9.10)

9.2 Constructing the states

Do all of these weights correspond to unique states so that the representation is 3 dimensional? We strongly suspect so, since that was the case for the (1,0) representation, and these two are clearly closely related. But we would like to be able to answer the question in general, so let's see how we might answer it for a general representation. Suppose we have an irreducible representation with highest weight μ. Then there will be one or more states, $|\mu\rangle$, and all the states in the representation can clearly be written in the form

$$E_{\phi_1} E_{\phi_2} \cdots E_{\phi_n} |\mu\rangle \tag{9.11}$$

where the ϕ_i are any roots. But any such state with a positive ϕ can be dropped, because if a positive ϕ appears, we can move it to the right using the commutation relations until it acts on $|\mu\rangle$, which give zero. So we can take all the ϕ_i in (9.11) to be negative.

But since every negative root is a sum over the simple roots with non-positive integer coefficients, we can take only states of the form

$$E_{-\alpha^{\beta_1}} E_{-\alpha^{\beta_2}} \cdots E_{-\alpha^{\beta_n}} |\mu\rangle \tag{9.12}$$

where α^{β_i} are simple roots without losing any of the states.

Now it is clear if it wasn't before that the highest weight state is unique. If there were two states with the highest weight, the representation would break up into two independent sets of the form (9.12) above, and it would not be irreducible.

In addition, any state that can be obtained in only one way by the action of simple lowering operators on $|\mu\rangle$ is unique. This shows that all the states in the (0,1) representation are unique.

This technique of starting with the highest weight state and builiding the other states by the action of lowering operators actually allows us to build the representation explicitly. We can compute the norms of the states (see problem 9.A) and the action of the raising and lowering operators on the states is built in by construction. We will not actually use this to build the representation matrices, because there are usually easier ways, but we could do it if we wanted to. The point is that all we need to do to construct the representation explicitly is to understand the structure of the space spanned by the kets in (9.12).

The first thing to notice is that two states produced by a different set of $E_{-\alpha^j}$s in (9.12) are orthogonal. This is clear because the linear independence of the α^js implies that the two states have different weights. But then they are orthogonal because they are eigenstates of the hermitian Cartan generators with different eigenvalues.

The norm of any state can be easily computed using the commutation relations of the simple root generators. The norm has the form

$$\langle\mu| E_{\alpha^{\beta_n}} \cdots E_{\alpha^{\beta_2}} E_{\alpha^{\beta_1}} E_{-\alpha^{\beta_1}} E_{-\alpha^{\beta_2}} \cdots E_{-\alpha^{\beta_n}} |\mu\rangle \tag{9.13}$$

The raising operators, starting with $E_{\alpha^{\beta_1}}$ can be successively moved to the right until they annihilate $|\mu\rangle$. On the way, they commute with the lowering operators, unless the roots are a positive and negative pair, in which case the commutator is a linear combination of Cartan generators, which can be evaluated in terms of μ and the α^{β_j}s, leaving a matrix element with one less raising and lowering operator. This is a tedious, but completely systematic procedure — you can easily write a recursive program to do it explicitly.

If two states in (9.12) have the same set of β_js but in a different order, then they have the same weight. In this case, we must compute their norms and matrix elements to understand the structure of the space. The matrix elements can be calculated in the same way as the norm in (9.13) because they have a similar structure:

$$\langle\mu|E_{\alpha^{\gamma_n}}\cdots E_{\alpha^{\gamma_2}}E_{\alpha^{\gamma_1}}E_{-\alpha^{\beta_1}}E_{-\alpha^{\beta_2}}\cdots E_{-\alpha^{\beta_n}}|\mu\rangle \tag{9.14}$$

where γ_j and β_j are two lists of the same set of integers but in a different order. Again, simply moving the raising operators to the right gives a completely systematic way of computing the matrix element.

Once you have the norms and the matrix elements, you can construct an orthonormal basis in the Hilbert space using the Gram-Schmidt procedure. For example, suppose that one weight space involves two states of the form (9.12), $|A\rangle$ and $|B\rangle$. Then if

$$\langle A \mid B\rangle \cdot \langle B \mid A\rangle = \langle A \mid A\rangle \cdot \langle B \mid B\rangle \tag{9.15}$$

the two states are linearly dependent, and there is a single basis state that we can take to be

$$\frac{|A\rangle}{\sqrt{\langle A \mid A\rangle}}. \tag{9.16}$$

While if

$$\langle A \mid B\rangle \cdot \langle B \mid A\rangle \neq \langle A \mid A\rangle \cdot \langle B \mid B\rangle \tag{9.17}$$

then the two states are linearly independent and we can choose an orthnormal basis of the form

$$\begin{gathered}\frac{|A\rangle}{\sqrt{\langle A \mid A\rangle}} \text{ and} \\ \frac{|B\rangle\langle A \mid A\rangle - |A\rangle\langle A \mid B\rangle}{\sqrt{(\langle A \mid A\rangle)\cdot(\langle A \mid B\rangle\cdot\langle B \mid A\rangle - \langle A \mid A\rangle\cdot\langle B \mid B\rangle)}}.\end{gathered} \tag{9.18}$$

An orthonormal basis is all you need to complete the construction of the space. Again, I want to emphasize that while this procedure is tedious and boring, it is completely systematic. You could program your computer to do it, and thus construct any representation completely from the highest weight and the simple roots.

9.3 The Weyl group

We could also have shown that the states in the μ^2 representation are not degenerate by understanding the obvious symmetry of the representations. The symmetry arises because there is an $SU(2)$ associated with each root direction, and all $SU(2)$ representations are symmetrical under the reflection $E_3 \to -E_3$. In particular, if μ is a weight, and $E_3 = \alpha \cdot H/\alpha^2$ is the E_3 associated with the root α, then

$$E_3|\mu\rangle = \frac{\alpha \cdot \mu}{\alpha^2}|\mu\rangle \tag{9.19}$$

and the reflection symmetry implies that $\mu - (q-p)\alpha$ (where $(q-p) = 2(\alpha \cdot \mu/\alpha^2)$) is a weight with the opposite E_3 value

$$E_3|\mu - (q-p)\alpha\rangle = -\frac{\alpha \cdot \mu}{\alpha^2}|\mu - (q-p)\alpha\rangle \tag{9.20}$$

Note that the vector average of μ and its reflection, $\mu - (q-p)\alpha$, is $\mu - (q-p)\alpha/2$ which is in the hyperplane perpendicular to α, because

$$\alpha \cdot (\mu - (q-p)\alpha/2) = \alpha \cdot \mu - (q-p)\alpha^2/2 = 0 \tag{9.21}$$

as illustrated below:

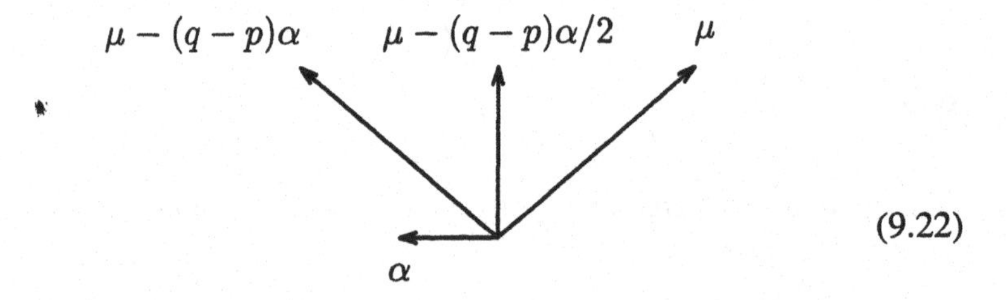

(9.22)

All such reflections for all roots are transformations on the weight space that leave the roots unchanged. We can obtain additional transformations that leave the roots unchanged by combining two or more such reflections.[1] The set of all transformations obtained in this way forms a transformation group called the **Weyl group** of the algebra. The individual reflections are called Weyl reflections. The Weyl group is a simple way of understanding the hexagonal and triangular structures that appear in $SU(3)$ representations.

[1]For example, in $SU(3)$, if you first reflect in the plane perpendicular to α^1 and then in the plane perpendicular to α^2, the result in a rotation by 120°.

9.4 Complex conjugation

Notice that the weights of the second fundamental representation are just the negatives of the weights of the first.

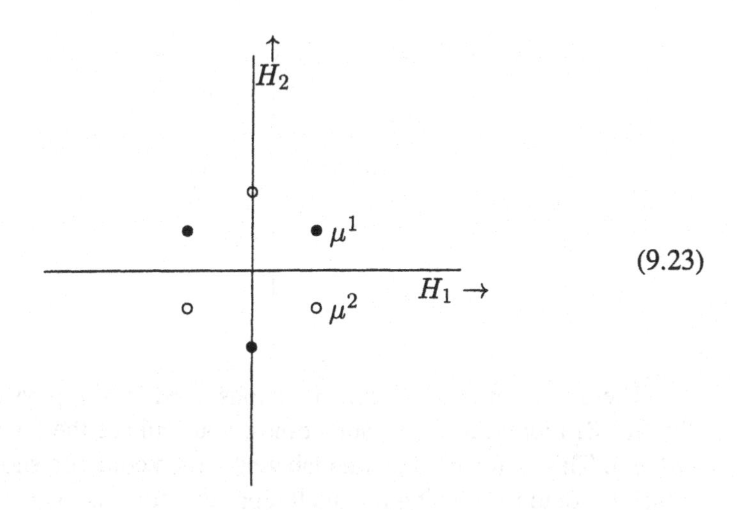

(9.23)

This means that the two representations are related by complex conjugation. If T_a are the generators of some representation, D, of some Lie Algebra, the objects $-T_a^*$ have the same commutation relations (because the structure constants are real) —

$$\begin{aligned} ([T_a, T_b])^* &= (i\, f_{abc} T_c)^* \\ \left[T_a^*, T_b^*\right] &= -i\, f_{abc} T_c^* \\ \left[-T_a^*, -T_b^*\right] &= i\, f_{abc}(-T_c^*) \end{aligned} \tag{9.24}$$

It is called the **complex conjugate of the representation** D, and is sometimes denoted by $\overline{D}$. D is said to be a **real representation** if it is equivalent to its complex conjugate. If not, it is a **complex representation**. The Cartan generators of the complex conjugate representation are $-H_i^*$. Because H_i is hermitian, H_i^* has the same eigenvalues as H_i. Thus if μ is a weight in D, $-\mu$ is a weight in $\overline{D}$. In particular, the highest weight of $\overline{D}$ is minus the lowest weight of D.

Obviously, in $SU(3)$, the representation (0,1) is the complex conjugate of (1,0). The lowest weight of (1,0) is $-\mu^2$, the negative of the highest weight of (0,1) and vice versa.

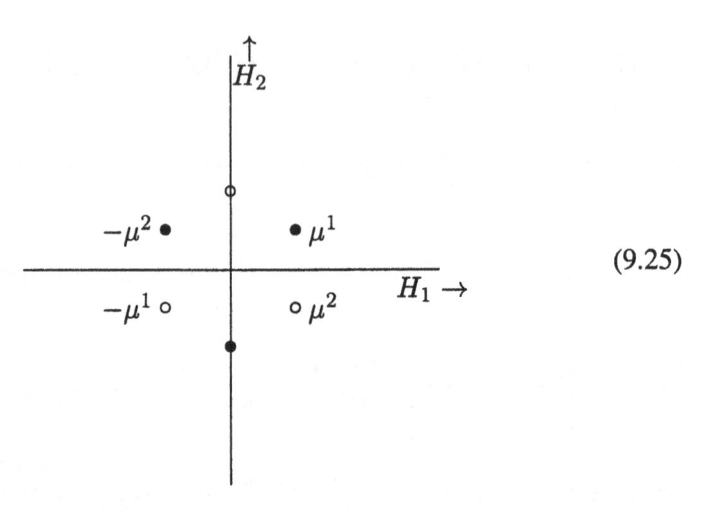

(9.25)

There are various different notations used in the physics literature for the $SU(3)$ representations. Sometimes, you will see the Dynkin coefficients (q_1, q_2). This is always a unique labeling. But a common shorthand notation, which is not too confusing for small representations is to give their dimension and to distinguish between a representation and its complex conjugate with a bar, so that (for example),

$$(1,0) \equiv 3 \qquad (0,1) \equiv \overline{3} \tag{9.26}$$

In general, the complex conjugate of (n,m) is (m,n). This is clear if you think about the highest and lowest weights. Because the lowest weight of $(1,0)$ is minus the highest weight of $(0,1)$, and vice versa, we have

$$\begin{array}{lll} (n,m) & \text{has highest weight} & n\mu^1 + m\mu^2 \\ (n,m) & \text{has lowest weight} & -n\mu^2 - m\mu^1 \\ & \Rightarrow & \\ (m,n) & \text{has highest weight} & n\mu^2 + m\mu^1 \end{array} \tag{9.27}$$

Representations of the form (n,n) are real.

9.5 Examples of other representations

Consider the representation (2,0). It has highest weight

$$2\mu^1 = (1, 1/\sqrt{3}) \tag{9.28}$$

In the $q-p$ notation, of (9.4) and (9.5) it looks like this

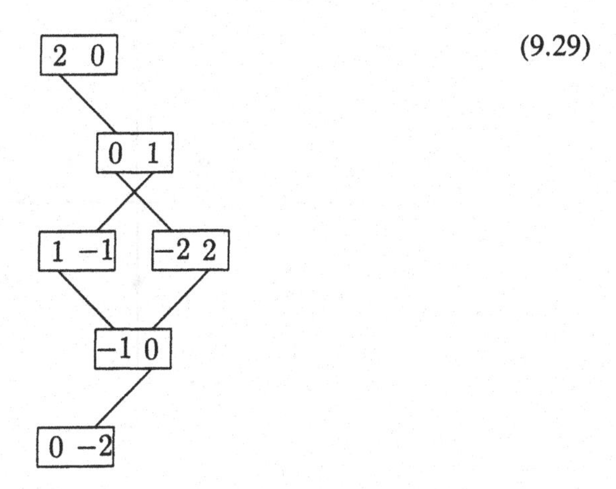

(9.29)

or in the Cartan space

H_2

$2\mu^1 - \alpha^1 - \alpha^2$

$2\mu^1$

$H_1 \rightarrow$

$2\mu^1 - \alpha^1$

$2\mu^1 - 2\alpha^1$

(9.30)

All the weights correspond to unique states. $2\mu^1 - \alpha^1$, $2\mu^1 - 2\alpha^1$ and $2\mu^1 - \alpha^1 - \alpha^2$ are reached by unique paths from the highest weight, and the others are related to these by Weyl reflections. Thus this is a 6-dimension representation — $(2,0) \equiv 6$. It is evidently complex. Its complex conjugate, $(0,2) \equiv \overline{6}$ is inverted in the origin in the Cartan plane, and reflected about a vertical line in the $q-p$ picture:

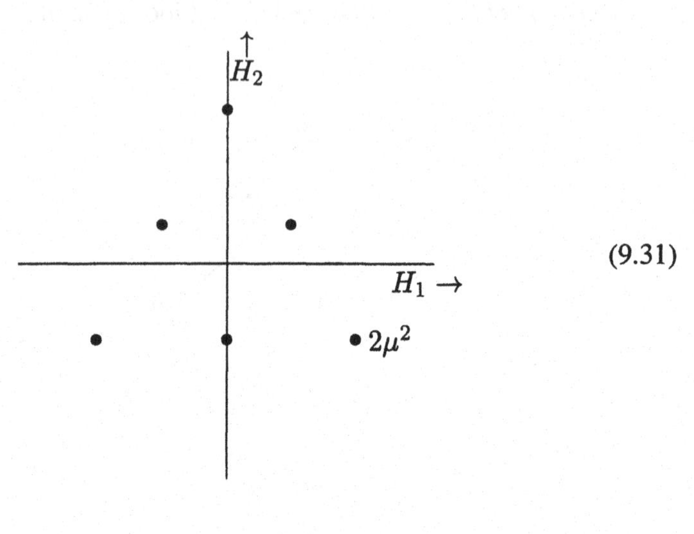

(9.31)

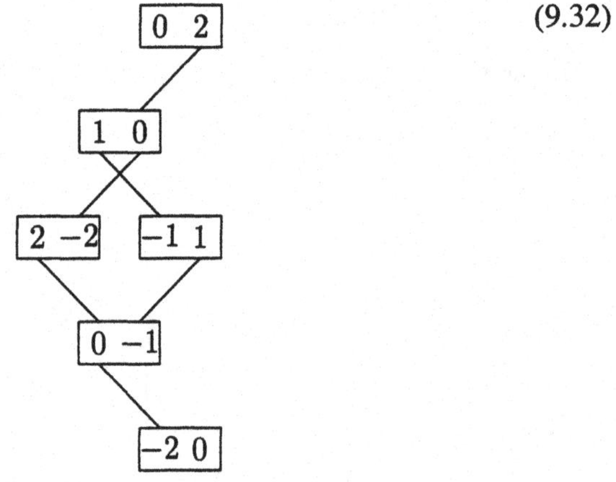

(9.32)

Now consider the representation (1,1). Note that

$$\mu^1 + \mu^2 = \alpha^1 + \alpha^2 \tag{9.33}$$

but $\alpha^1 + \alpha^2$ is the highest weight of the adjoint representation (8.58) We know that the zero weight is doubly degenerate — corresponding to the two states of the Cartan subalgebra. The two ways of getting to zero from the highest weights give different states. You can show that

$$E_{-\alpha^1} E_{-\alpha^2} |\mu^1 + \mu^2\rangle \not\propto E_{-\alpha^2} E_{-\alpha^1} |\mu^1 + \mu^2\rangle \tag{9.34}$$

You will do this in problem (9.A).

Now look at (3,0), with highest weight $3\mu^1$

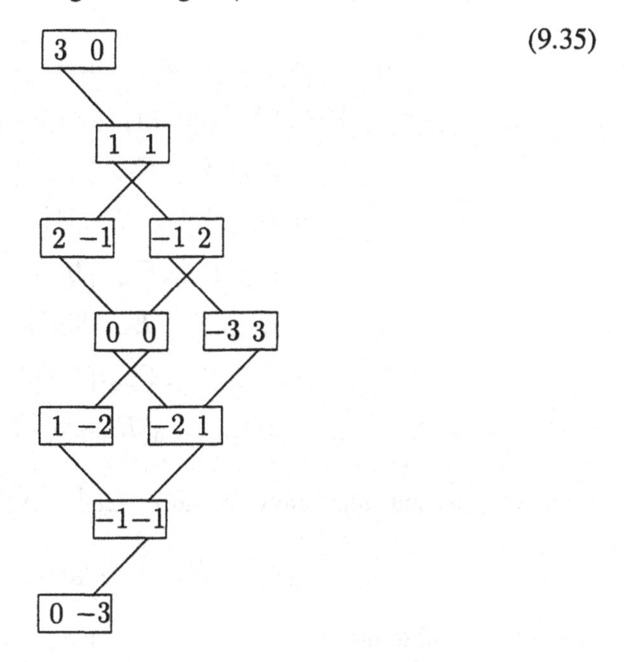

(9.35)

The weight diagram is

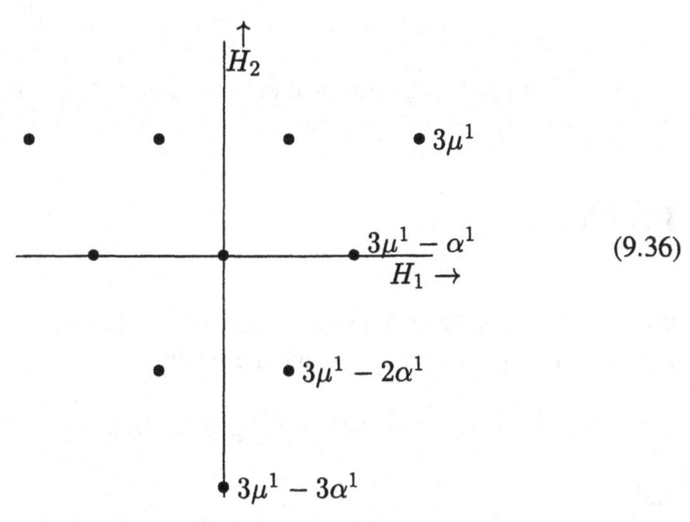

(9.36)

All the states are obviously unique except the state in the center. But it turns out to be unique also. The two vectors

$$
\begin{aligned}
&E_{-\alpha^1}E_{-\alpha^2}E_{-\alpha^1}|3\mu^1\rangle \\
&E_{-\alpha^2}E_{-\alpha^1}E_{-\alpha^1}|3\mu^1\rangle
\end{aligned}
\tag{9.37}
$$

are linearly dependent, because we can write

$$\begin{aligned}
&E_{-\alpha^2}E_{-\alpha^1}E_{-\alpha^1}|3\mu^1\rangle \\
&= [E_{-\alpha^2}, E_{-\alpha^1}]\, E_{-\alpha^1}|3\mu^1\rangle \\
&+E_{-\alpha^1}E_{-\alpha^2}E_{-\alpha^1}|3\mu^1\rangle \\
&= E_{-\alpha^1}\,[E_{-\alpha^2}, E_{-\alpha^1}]\,|3\mu^1\rangle \\
&+E_{-\alpha^1}E_{-\alpha^2}E_{-\alpha^1}|3\mu^1\rangle \\
&= E_{-\alpha^1}E_{-\alpha^2}E_{-\alpha^1}|3\mu^1\rangle \\
&+E_{-\alpha^1}E_{-\alpha^2}E_{-\alpha^1}|3\mu^1\rangle \\
&= 2E_{-\alpha^1}E_{-\alpha^2}E_{-\alpha^1}|3\mu^1\rangle
\end{aligned} \tag{9.38}$$

where the second step follows because $-2\alpha^1 - \alpha^2$ is not a root, and thus

$$[[E_{-\alpha^2}, E_{-\alpha^1}]\,, E_{-\alpha^1}] = 0 \tag{9.39}$$

and the second to last because $3\mu^1 - \alpha^2$ is not a root, and thus

$$[E_{-\alpha^2}, E_{-\alpha^1}]\,|3\mu^1\rangle = E_{-\alpha^2}E_{-\alpha^1}|3\mu^1\rangle \tag{9.40}$$

Thus this is a ten dimensional representation. It is sometimes called just the 10. Its complex conjugate, the $(0,3)$ representation, is the $\overline{10}$.

Problems

9.A. If $|\mu\rangle$ is the state of the highest weight ($\mu = \mu^1 + \mu^2$) of the adjoint representation of $SU(3)$, show that the states

$$|A\rangle = E_{-\alpha^1}\, E_{-\alpha^2}\,|\mu\rangle$$

and

$$|B\rangle = E_{-\alpha^2}\, E_{-\alpha^1}\,|\mu\rangle$$

are linearly independent. Hint: Calculate the matrix elements $\langle A \mid A\rangle$, $\langle A \mid B\rangle$, and $\langle B \mid B\rangle$. Show that $|A\rangle$ and $|B\rangle$ are linearly dependent if and only if

$$\langle A \mid A\rangle\langle B \mid B\rangle = \langle A \mid B\rangle\langle B \mid A\rangle$$

9.B. Consider the following matrices defined in the six dimensional tensor product space of the $SU(3)$ λ_a matrices and the Pauli matrices σ_a:

$$\frac{1}{2}\lambda_a\sigma_2 \text{ for } a = 1, 3, 4, 6 \text{ and } 8$$
$$\frac{1}{2}\lambda_a \text{ for } a = 2, 5 \text{ and } 7$$

Show that these generate a reducible representation and reduce it.

9.C. Decompose the tensor product of 3×3 using highest weight techniques.

Chapter 10

Tensor Methods

Tensors are great tools for doing practical calculations in $SU(3)$ and in many other groups as well. As you will see, the idea of tensors is closely related to the idea of a wave function in quantum mechanics.

10.1 lower and upper indices

The idea starts by labeling the states of the 3 as

$$\begin{aligned} |1/2, \sqrt{3}/6\rangle &\equiv |_1\rangle \\ |-1/2, \sqrt{3}/6\rangle &\equiv |_2\rangle \\ |0, -1/\sqrt{3}\rangle &\equiv |_3\rangle \end{aligned} \tag{10.1}$$

The 1, 2 and 3 are to remind you of the fact that the eigenvectors of the H_1 and H_2 matrices corresponding to these weights are vectors with a single non-zero entry in the first, second and third position. We have also written the indices below the line for a reason that will become clear shortly.

If we define a set of matrices with one upper and one lower index, as follows

$$[T_a]^i_j = \frac{1}{2}[\lambda_a]_{ij} \tag{10.2}$$

then the triplet of states, $|_i\rangle$, transforms under the algebra as

$$T_a|_i\rangle = |_j\rangle[T_a]^j_i \tag{10.3}$$

The important thing to notice is that the sum over j involves one upper and one lower index.

DOI: 10.1201/9780429499210-11

Label the states of the $\overline{3}$ as

$$\begin{aligned} &|-1/2, -\sqrt{3}/6\rangle \equiv |^1\rangle \\ &|1/2, -\sqrt{3}/6\rangle \equiv |^2\rangle \\ &|0, 1/\sqrt{3}\rangle \equiv |^3\rangle \end{aligned} \tag{10.4}$$

Then

$$T_a|^i\rangle = -|^j\rangle [T_a]^i_j \tag{10.5}$$

This is true because the $\overline{3}$ is the complex conjugate representation, generated by

$$-[T_a^*]^j_i = -[T_a^T]^j_i = -[T_a]^i_j \tag{10.6}$$

Now I can define, as usual, a state in the tensor product of n 3s and m $\overline{3}$s.

$$|^{i_1\cdots i_m}_{j_1\cdots j_n}\rangle = |^{i_1}\rangle \cdots |^{i_m}\rangle |_{j_1}\rangle \cdots |_{j_n}\rangle \tag{10.7}$$

It transforms as

$$\begin{aligned} &T_a|^{i_1\cdots i_m}_{j_1\cdots j_n}\rangle \\ &= \sum_{\ell=1}^{n} |^{i_1\cdots i_m}_{j_1\cdots j_{\ell-1}kj_{\ell+1}\cdots j_n}\rangle \, [T_a]^k_{j_\ell} \\ &- \sum_{\ell=1}^{m} |^{i_1\cdots i_{\ell-1}ki_{\ell+1}\cdots i_m}_{j_1\cdots j_n}\rangle \, [T_a]^{i_\ell}_k \end{aligned} \tag{10.8}$$

This distinction between upper and lower indices is useful because $SU(3)$ has two different kinds of 3-dimensional representations — the 3 and $\overline{3}$. We need some way to distinguish them. Raising and lowering the indices is just a handy notational device.

10.2 Tensor components and wave functions

Now consider an arbitrary state in this tensor product space

$$|v\rangle = |^{i_1\cdots i_m}_{j_1\cdots j_n}\rangle \, v^{j_1\cdots j_n}_{i_1\cdots i_m} \tag{10.9}$$

v is called a **tensor**. A tensor is just a "wave-function", because we can find v by taking the matrix element of $|v\rangle$ with the tensor product state.

$$v^{j_1\cdots j_n}_{i_1\cdots i_m} = \langle^{i_1\cdots i_m}_{j_1\cdots j_n}|v\rangle \tag{10.10}$$

The correspondence here is exactly like the relation between the space wave function of a particle in quantum mechanics and the state describing the particle in the Hilbert space.

$$\psi(x) = \langle x|\psi\rangle \tag{10.11}$$

The tensor v is characterized by its **tensor components,** $v^{j_1\cdots j_n}_{i_1\cdots i_m}$. Now we can think of the action of the generators on $|v\rangle$ as an action on the tensor components, as follows:

$$T_a|v\rangle = |T_a v\rangle \tag{10.12}$$

where

$$\begin{aligned} &T_a v^{j_1\cdots j_n}_{i_1\cdots i_m} \\ &= \sum_{\ell=1}^{n} [T_a]^{j_\ell}_k \, v^{j_1\cdots k\cdots j_n}_{i_1\cdots i_m} \\ &- \sum_{\ell=1}^{m} [T_a]^k_{i_\ell} \, v^{j_1\cdots j_n}_{i_1\cdots k\cdots i_m} \end{aligned} \tag{10.13}$$

This defines the action of the generators on an arbitary tensor!

10.3 Irreducible representations and symmetry

We can now use the highest weight procedure to pick out the states in the tensor product corresponding to the irreducible representation (n,m). Because $|_1\rangle$ is the highest weight of the (1,0) representation, and $|^2\rangle$ is the highest weight of the (0,1) representation, the state with highest weight in (n,m) is

$$|^{222\cdots}_{111\cdots}\rangle \tag{10.14}$$

It corresponds to the tensor v_H with components

$$v_H{}^{j_1\cdots j_n}_{i_1\cdots i_m} = N\,\delta_{j_1 1}\cdots\delta_{j_n 1}\,\delta_{i_1 2}\cdots\delta_{i_m 2} \tag{10.15}$$

Now we can construct all the states in (n,m) by acting on the tensor v_H with lowering operators. The important point is that v_H has two properties that are preserved by the transformation $v_H \to T_a v_H$.

1. v_H is symmetric in the upper indices, and symmetric in the lower indices.

2. v_H satisfies

$$\delta^{i_1}_{j_1}\, v_H{}^{j_1 \cdots j_n}_{i_1 \cdots i_m} = 0 \tag{10.16}$$

The first is preserved because the generators act the same way on all the upper indices and the same way on all the lower indices. The second is preserved because of the minus sign in (10.13) and the tracelessness of the T_as. All the states in (n, m) therefore correspond to tensors of this form, symmetric in both upper and lower indices, and traceless (which just means that it satisfies (10.16) which is a generalization of the condition of tracelessness for a matrix). It turns out that the correspondence also goes the other way. Every such tensor gives a state in (n, m).

10.4 Invariant tensors

The $\delta^{i_1}_{j_1}$ is called an **invariant tensor**. An invariant tensor is one that does not change under an $SU(3)$ transformation. The change in a tensor is proportional to the action of some linear combination of generators on it, but

$$[T_a \delta]^i_j = [T_a]^i_k\, \delta^k_j - [T_a]^k_j\, \delta^i_k = [T_a]^i_j - [T_a]^i_j = 0 \tag{10.17}$$

thus $\delta^{i_1}_{j_1}$ doesn't change under any $SU(3)$ transformation. There are two other invariant tensors in $SU(3)$ — the completely antisymmetric tensors, ϵ^{ijk} and ϵ_{ijk}. These are invariant because of the tracelessness of T_a. Consider

$$[T_a\, \epsilon]^{ijk} = [T_a]^i_\ell\, \epsilon^{\ell jk} + [T_a]^j_\ell\, \epsilon^{i\ell k} + [T_a]^k_\ell\, \epsilon^{ij\ell} \tag{10.18}$$

This is completely antisymmetric, so we can look at the 123 component —

$$\begin{aligned} [T_a\ \epsilon]^{123} &= [T_a]^1_\ell\, \epsilon^{\ell 23} + [T_a]^2_\ell\, \epsilon^{1\ell 3} + [T_a]^3_\ell\, \epsilon^{12\ell} \\ &= [T_a]^1_1\, \epsilon^{123} + [T_a]^2_2\, \epsilon^{123} + [T_a]^3_3\, \epsilon^{123} = 0 \end{aligned} \tag{10.19}$$

Thus ϵ^{ijk} is invariant.

10.5 Clebsch-Gordan decomposition

We can use tensors to decompose tensor products explicitly. Suppose that u is a tensor with n upper indices and m lower indices, and v is a tensor with p upper indices and q lower indices. Then it follows from the definition of a tensor that the product, $u \otimes v$ defined by the product of the tensor components

$$[u \otimes v]^{i_1 \cdots i_n i'_1 \cdots i'_p}_{j_1 \cdots j_m j'_1 \cdots j'_q} = u^{i_1 \cdots i_n}_{j_1 \cdots j_m}\ v^{i'_1 \cdots i'_p}_{j'_1 \cdots j'_q} \tag{10.20}$$

is the tensor that describes the tensor product state $|u\rangle \otimes |v\rangle$. Thus we can analyze tensor products by manipulating tensors — the wave-functions of the corresponding states. The general strategy for doing these decompositions is to make irreducible representations out of the product of tensors, and then express the original product as a sum of terms proportional to various irreducible combinations. The advantage of this procedure is that we are directly manipulating wave functions, which is often what we want to know about.

Consider, for example, $3 \otimes 3$. If we have two 3s, u^i and v^j, we can write the product as

$$u^i\, v^j = \frac{1}{2}\left(u^i\, v^j + u^j\, v^i\right) + \frac{1}{2}\epsilon^{ijk}\, \epsilon_{k\ell m}\, u^\ell\, v^m \tag{10.21}$$

The first term,

$$\frac{1}{2}\left(u^i\, v^j + u^j\, v^i\right) \tag{10.22}$$

transforms like a 6. This contains the highest weight state $u^1\, v^1$. We have added the $u^j\, v^i$ term to make it completely symmetric in the two upper indices, and thus irreducible. The lower index object

$$\epsilon_{k\ell m} u^\ell\, v^m \tag{10.23}$$

having only one lower index, transforms like a $\overline{3}$. Thus we have explicitly decomposed the tensor product into a sum of 6 and a $\overline{3}$. Not only does this show that

$$3 \otimes 3 = 6 \oplus \overline{3} \tag{10.24}$$

or

$$(1,0) \otimes (1,0) = (2,0) \oplus (0,1) \tag{10.25}$$

it shows us how to actually build wave functions with the required symmetry properties. Later, we will see how this makes some kinds of calculations easy.

Note also how as in (10.23) whenever a tensor with more than one upper index is not completely symmetric, we can trade two upper indices for one lower index using the ϵ tensor.

Next, let's look at $3 \otimes \overline{3}$, a product of u^i (a 3) and v_j (a $\overline{3}$). We can write

$$u^i\, v_j = \left(u^i\, v_j - \frac{1}{3}\delta^i_j\, u^k\, v_k\right) + \frac{1}{3}\delta^i_j\, u^k\, v_k \tag{10.26}$$

The first term in parentheses is traceless, and transforms like the 8, while the tensor with no indices, $u^k v_k$, transforms like the trivial representation, (0,0), or 1. Thus

$$3 \otimes \overline{3} = 8 \oplus 1 \tag{10.27}$$

or

$$(1,0)\otimes(0,1)=(1,1)\oplus(0,0) \tag{10.28}$$

Notice the role of the invariant tensor, δ^i_j. One way of understanding why only traceless tensors are irreducible is that if a tensor is not traceless, the δ tensor can be used to construct a non-zero tensor with fewer indices, and thus explicitly reduce it.

One more example — u^i (a 3) times v^j_k (an 8).

$$\begin{aligned}
&u^i\,v^j_k = \\
&\frac{1}{2}\left(u^i\,v^j_k + u^j\,v^i_k - \frac{1}{4}\delta^i_k\,u^\ell\,v^j_\ell - \frac{1}{4}\delta^j_k\,u^\ell\,v^i_\ell\right) \\
&+\frac{1}{4}\epsilon^{ij\ell}\left(\epsilon_{\ell mn}\,u^m\,v^n_k + \epsilon_{kmn}\,u^m\,v^n_\ell\right) \\
&+\frac{1}{8}\left(3\delta^i_k\,u^\ell\,v^j_\ell - \delta^j_k\,u^\ell\,v^i_\ell\right)
\end{aligned} \tag{10.29}$$

where the first term on the right hand side is a (2,1) (or 15 — it's 15 dimensional), the second term is a $\overline{6}$, and the last is a 3.

$$3\otimes 8 = 15\oplus\overline{6}\oplus 3. \tag{10.30}$$

or

$$(1,0)\otimes(1,1)=(2,1)\oplus(0,2)\oplus(1,0). \tag{10.31}$$

10.6 Triality

Notice that $(n-m)\mod 3$ is conserved in these tensor product decompositions. This is true because the invariant tensors all have $(n-m)\mod 3=0$, so there is no way to change this quantity. It is called **triality**.

10.7 Matrix elements and operators

The bra state $\langle v|$ is

$$\langle v| = v^{j_1\cdots j_n\,*}_{i_1\cdots i_m}\,\langle^{i_1\cdots i_m}_{j_1\cdots j_n}| \tag{10.32}$$

The bra transforms under the algebra with an extra minus sign. For example, the triplet $\langle_i|$ transforms as

$$-\langle_i|T_a = -\langle_i|T_a|_j\rangle\langle_j| = -[T_a]^i_j\langle_j| \tag{10.33}$$

In words, this says that the bra with a lower index transforms as if it had an upper index. This is because of the complex conjugation involved in going from a bra to a ket. Similarly, the bra with an upper index transforms as if it has a lower index. This suggests that we define the tensor corresponding to a bra state with the upper and lower indices interchanged so that the contraction of indices works in (10.33). Thus we say that the tensor components of the bra tensor $\langle v|$ are

$$\overline{v}^{i_1\cdots i_m}_{j_1\cdots j_n} \equiv v^{j_1\cdots j_n}_{i_1\cdots i_m}{}^* \tag{10.34}$$

so that (10.32) becomes

$$\langle v| = \overline{v}^{i_1\cdots i_m}_{j_1\cdots j_n} \langle {}^{i_1\cdots i_m}_{j_1\cdots j_n}| \tag{10.35}$$

Then when the state $\langle v|$ is transformed by $-\langle v|T_a$, the tensor $\overline{v}$ is transformed by $T_a\overline{v}$.

For example, consider the matrix element

$$\begin{aligned}\langle u|v\rangle &= \overline{u}^{k_1\cdots k_m}_{\ell_1\cdots \ell_n} v^{j_1\cdots j_n}_{i_1\cdots i_m} \langle {}^{k_1\cdots k_m}_{\ell_1\cdots \ell_n}|{}^{i_1\cdots i_m}_{j_1\cdots j_n}\rangle \\ &= \overline{u}^{k_1\cdots k_m}_{\ell_1\cdots \ell_n} v^{j_1\cdots j_n}_{i_1\cdots i_m} \delta^{i_1}_{k_1}\cdots\delta^{i_m}_{k_m}\delta^{\ell_1}_{j_1}\cdots\delta^{\ell_n}_{j_n} \\ &= \overline{u}^{i_1\cdots i_m}_{j_1\cdots j_n} v^{j_1\cdots j_n}_{i_1\cdots i_m}\end{aligned} \tag{10.36}$$

The indices are all repeated and summed over (**contracted** for short), which they must be because the matrix element is invariant.

10.8 Normalization

A corollary to (10.36) is that if the state $|v\rangle$ is normalized, satisfying $\langle v \mid v\rangle = 1$, then the tensor components must satisfy a normalization condition

$$\sum_{\substack{i_1\cdots i_m \\ j_1\cdots j_n}} | v^{j_1\cdots j_n}_{i_1\cdots i_m} |^2 = 1 \tag{10.37}$$

For example, the tensor v_H in (10.15) satisfies (10.37), because only a single term contributes in the sum. But a tensor of the form

$$N(2\,\delta^j_1\delta^1_k - \delta^j_2\delta^2_k - \delta^j_3\delta^3_k) \tag{10.38}$$

describes a state with norm

$$|N|^2\sum_{j,k}(2\,\delta^j_1\delta^1_k - \delta^j_2\delta^2_k - \delta^j_3\delta^3_k)^2 = |N|^2\Big(\overbrace{4}^{j=k=1} + \overbrace{1}^{j=k=2} + \overbrace{1}^{j=k=3}\Big) = 6\,|N|^2 \tag{10.39}$$

10.9 Tensor operators

We can extend the concept of tensors to include the coefficients of tensor operators in an obvious way. For example, if a set of operators, O, transform according to the (n,m) representation of $SU(3)$, we can form the general linear combination

$$W = w^{j_1\cdots j_n}_{i_1\cdots i_m}\, O^{i_1\cdots i_m}_{j_1\cdots j_n} \tag{10.40}$$

then if W transforms by commutation with the generators, the coefficients, $w^{j_1\cdots j_n}_{i_1\cdots i_m}$, transform like (n,m), (10.13).

10.10 The dimension of (n,m)

A very simple application of tensor methods is the calculation of the dimension of the irreducible representation (n,m). The dimension of the representation is the number of independent components in the tensor. We know that the tensor has n upper and m lower indices, and is separately symmetric in each. The number of independent components of a object symmetric in n indices each of which run from 1 to 3 is equal to the number of way of separating n identical objects with two identical partitions — which is

$$\binom{n+2}{2} = \frac{(n+2)!}{n!\,2!} = \frac{(n+2)(n+1)}{2} \tag{10.41}$$

Thus if there were no other constraint, the number of independent coefficients would be

$$B(n,m) = \frac{(n+2)(n+1)}{2}\frac{(m+2)(m+1)}{2} \tag{10.42}$$

However, the tensor is also required to be traceless. This says that the object we get by taking the trace vanishes, and it is symmetric in $n-1$ upper indices and $m-1$ lower indices. Thus this imposes $B(n-1,m-1)$ constraints, so the total is

$$\begin{aligned} D(n,m) &= B(n,m) - B(n-1,m-1) \\ &= \frac{(n+2)(n+1)}{2}\frac{(m+2)(m+1)}{2} \\ &\quad - \frac{(n+1)n}{2}\frac{(m+1)m}{2} \\ &= \frac{(n+1)(m+1)[(n+2)(m+2)-nm]}{4} \\ &= \frac{(n+1)(m+1)(n+m+2)}{2} \end{aligned} \tag{10.43}$$

You can check that this formula works for the small irreducible representations that we have discussed.

10.11 * The weights of (n, m)

We can now put together various pieces of what we know about the irreducible representation (n, m) of $SU(3)$. A general irreducible representation of $SU(3)$, (n, m), has the form of a hexagon, illustrated below for $n = 8$ and $m = 4$.

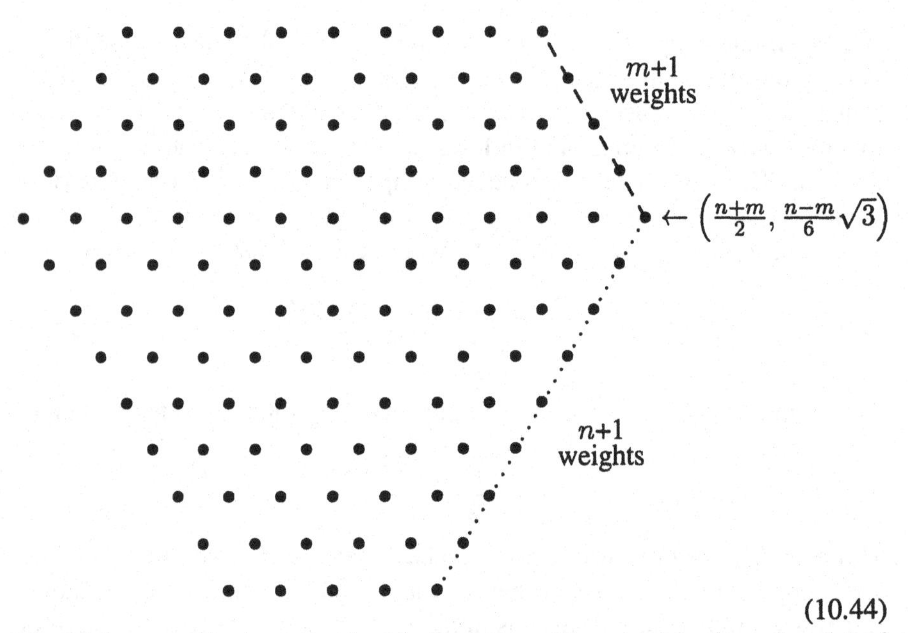

(10.44)

The highest weight is marked with the arrow. The weights along the dotted line can be reached from the highest weight by repeated application of the lowering operator $E_{-\alpha^1}$. Thus counting the highest weight, there are $n + 1$ weights along the dotted side of the hexagon. The weights along the dashed line can be reached from the highest weight by repeated application of the lowering operator $E_{-\alpha^2}$. Thus counting the highest weight, there are $m + 1$ weights along the dashed side of the hexagon. If either n or m is 0, the hexagon degenerates in to a triangle.

The Weyl reflection symmetries guarantee that the hexagon is symmetrical. Thus the three dashed sides are equivalent in the diagram below, as are the three dotted sides.

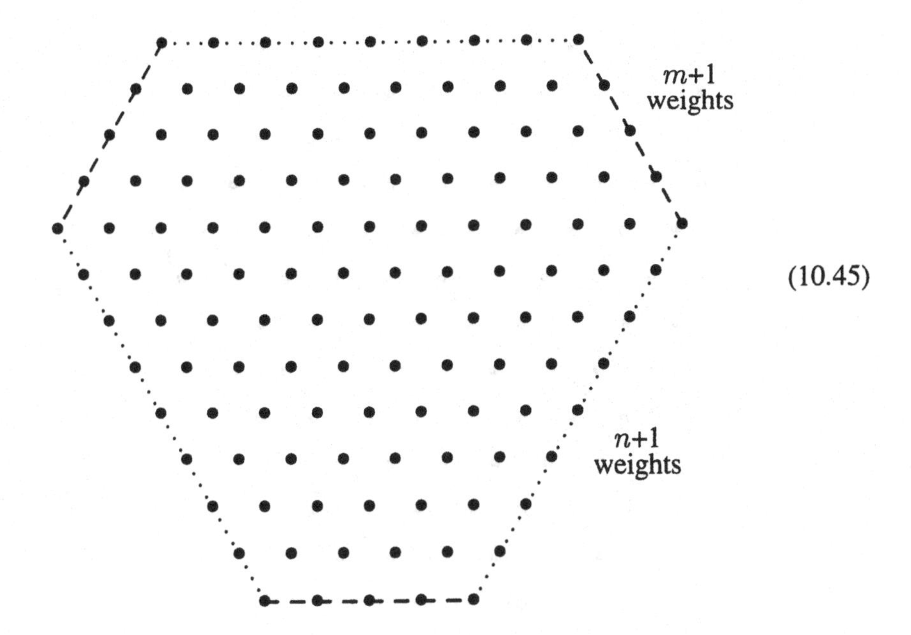

(10.45)

The dashed and dotted sides together make up an outer layer of weights, all of which correspond to unique states in the irreducible representation because they are equivalent by Weyl reflection to states that can be reached uniquely by simple lowering operators from the highest weight. Obviously, we can define a next layer of weights by considering the hexagon immediately inside this one, and we can continue this process until each weight is assigned to a "layer".

We will now show how to prove the following result:

Theorem 10.10 *As you go in from one layer to the next, the degeneracy of the weight space for states in the layer increases by one each time, until you get to a triangular layer. From then on, the degeneracy remains constant.*

Theorem 10.10 implies that the degeneracies of the layers in our example are as shown in the following figure.

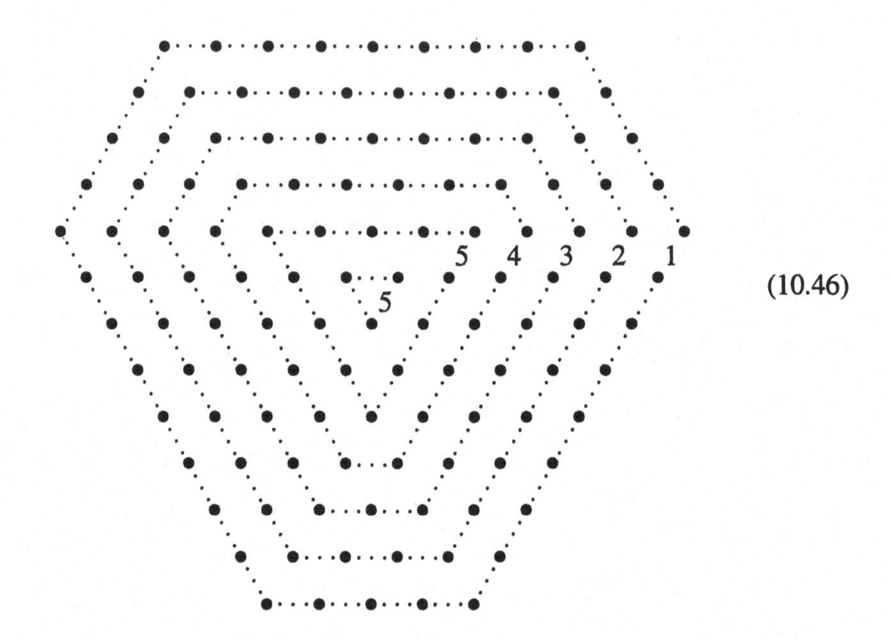

(10.46)

We will prove theorem 10.10 by counting the independent components of the tensors which represent these states. So we must first understand in detail how to go from weights to tensor components. We begin by considering tensors without the constraint of tracelessness. Such a tensor is simply the most general object completely symmetric in its upper and lower indices. We will then use the strategy of the previous section (on the dimension of (n, m)) to deal with the trace constraint. The advantage of this, as we saw above, is that it is easy to count the independent states because the tensor components are completely determined by the number of upper and lower components of each value. Now consider a tensor symmetric in n upper indices and in m lower indices. This corresponds to a reducible $SU(3)$ representation, but it contains the general irreducible representation (n, m) that we wish to study. The weight diagram for this reducible representation will be the same as (10.44), but the degeneracies of the states corresponding to each weight may be different (and larger). The outer layer, however, is nondegenerate even for this more general tensor, because it contains no tensor components which have the same value of any upper and lower index, so there is no trace constraint. As we go around this outer layer, along each side, one index changes to another as illustrated in the following figure (where $\bar{j}$ indicates a lower j index).

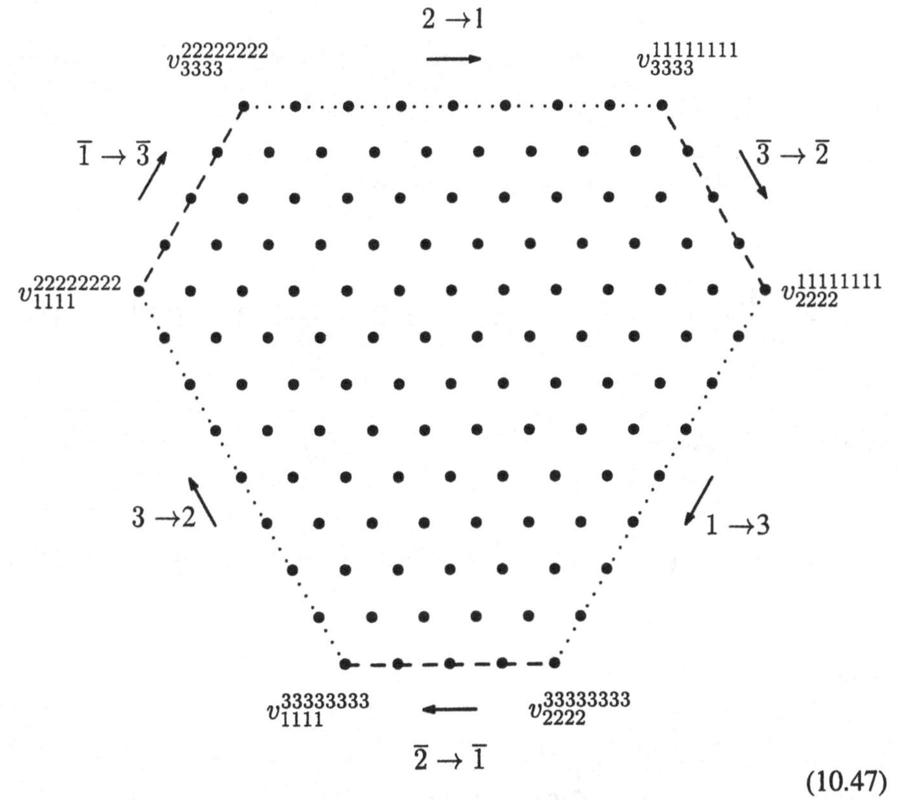

(10.47)

To deal with the inner layers, we first need to understand the connection between the tensor components and the weights. Consider a particular component of the general tensor with n upper and m lower indices. We will assume, for convenience that $n \geq m$ (as is the case for the example in the figures). This is not a serious restriction, because if it is not true for a representation, then it is true for the complex conjugate representation obtained by changing the signs of all the weights (with all degeneracies remaining unchanged). Thus if we understand the representations for $n \geq m$, we actually understand them all. Define $z(j)$ to be the number of upper components with the value j, and $\bar{z}(j)$ to be the number of lower components with the value j. Thus for example, for the highest weight of our tensor, $z(1) = n$, $\bar{z}(2) = m$, and the others are all zero. Note that the three quantities

$$z(1) - \bar{z}(1)\,, \qquad z(2) - \bar{z}(2)\,, \qquad z(3) - \bar{z}(3)\,, \tag{10.48}$$

are determined by the weight associated with the component and the value of $n - m$. By straightforward calculation, the 1-component of the weight vector satisfies

$$\mu_1 = \frac{1}{2}[z(1) - \overline{z}(1)] - \frac{1}{2}[z(2) - \overline{z}(2)]\,. \tag{10.49}$$

The 2-component of the weight vector satisfies

$$\mu_2 = \frac{\sqrt{3}}{6}[z(1) - \overline{z}(1)] + \frac{\sqrt{3}}{6}[z(2) - \overline{z}(2)] - \frac{\sqrt{3}}{3}[z(3) - \overline{z}(3)]\,. \tag{10.50}$$

And $n - m$ satisfies

$$n - m = [z(1) - \overline{z}(1)] + [z(2) - \overline{z}(2)] + [z(3) - \overline{z}(3)]\,. \tag{10.51}$$

Thus

$$\begin{aligned} z(1) - \overline{z}(1) &= \frac{1}{6}[6\mu_1 + 2\sqrt{3}\,\mu_2 + 2(n-m)] \\ z(2) - \overline{z}(2) &= \frac{1}{6}[-6\mu_1 + 2\sqrt{3}\,\mu_2 + 2(n-m)] \\ z(3) - \overline{z}(3) &= \frac{1}{3}[(n-m) - 2\sqrt{3}\,\mu_2] \end{aligned} \tag{10.52}$$

Actually, rather than using (10.52), it will be easier just to follow what is happening to the $z(j) - \overline{z}(j)$ by thinking about what happens to the indices as you move around the weight diagram, but the important point is that these differences of upper and lower indices are all fixed.

Now suppose that we move into the k-th inner layer by taking k steps to the left. Assume that $k < m$ so we are still in a hexagonal layer (the number of weights along the upper-right-hand edge of the layer decreases by one for each increase in k, thus $k = m$ is the corner of the first triangular layer — we will come back to this later). Then take j steps down to a weight on the k-th inner layer, as shown below (for $k = 3$ and $j = 2$):

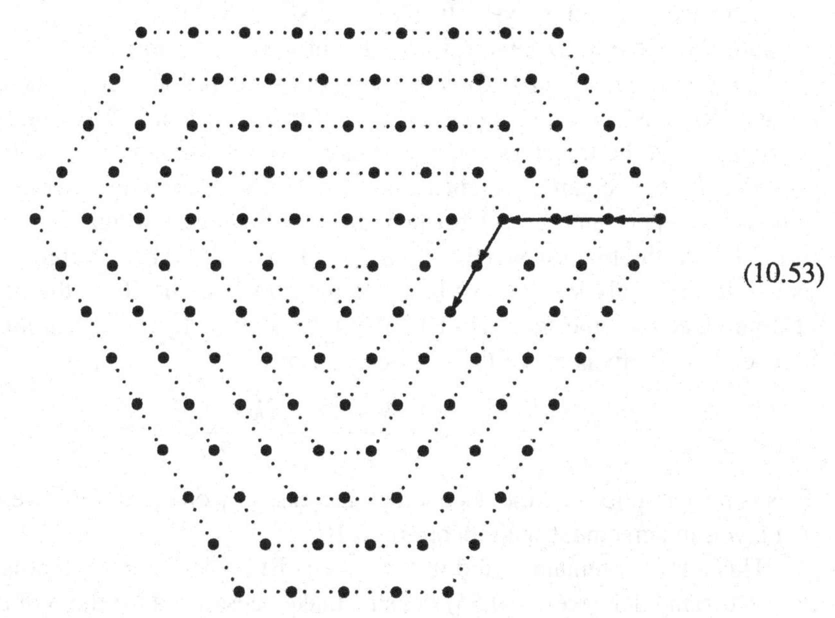

(10.53)

As we move in horizontally, in each step either $1 \to 2$, or $\overline{2} \to \overline{1}$, or $1 \to 3$ and $\overline{2} \to \overline{3}$ (see (10.47), and thus $z(1) - \overline{z}(1)$ is decreased by one and $z(2) - \overline{z}(2)$ is increased by one. Similarly, as we go down the kth layer, at each step $z(1) - \overline{z}(1)$ is decreased by one and $z(3) - \overline{z}(3)$ is increased by one. And thus

$$
\begin{aligned}
z(1) - \overline{z}(1) &= n - k - j \\
z(2) - \overline{z}(2) &= -(m - k) \\
z(3) - \overline{z}(3) &= j
\end{aligned}
\tag{10.54}
$$

Now we can count the number of components. (10.54) implies that the tensor component can be written with

$n-k-j$ 1s, $m-k$ $\overline{2}$s, j 3s, and k j-$\overline{j}$ pairs, where j is 1, 2 or 3. (10.55)

The j-$\overline{j}$ pairs do not contribute to the $z(j) - \overline{z}(j)$ values —they do not affect the weight, but they give different independent tensor components. Thus the number of ways we can get this weight is the number of different ways we can choose these k pairs, which is

$$
\binom{k+2}{2} = \frac{(k+2)(k+1)}{2} \tag{10.56}
$$

Note that it only depends on k — that is it only depends on the layer, not on where we are in the layer. It is easy to see that this is also true for the weights above the horizontal. The argument is similar and we leave it as an exercise for the reader. (10.56) is the number of independent components for the general tensor with n upper and m lower components. The number of components in the traceless tensor is this minus the number of independent components of the same weight in the trace. The trace is a general tensor with $n-1$ upper and $m-1$ lower indices. It's highest weight is one step to the left of the highest weight of (n,m). From the highest weight of this tensor, it takes only $k-1$ steps to get to the same weight. Thus the number of components of the trace is (10.56) for $k \to k-1$, and the number of independent components of the traceless tensor is

$$\binom{k+2}{2} - \binom{k+1}{2} = \frac{(k+2)(k+1)}{2} - \frac{(k+1)k}{2} = k+1 \,. \tag{10.57}$$

This simply implies that the degeneracy increases by one each time we move in a layer, in agreement with theorem 10.10.

This analysis remains valid in to $k = m$. But when $k > m$, that is after the first triangular layer, (10.55) doesn't make sense because the number of $\overline{2}$s cannot be negative. Instead, the component must look like

$$n-k-j \text{ 1s, } k-m \text{ 2s, } j \text{ 3s, and } m \text{ } j\text{-}\overline{j} \text{ pairs,} \tag{10.58}$$
where j is $1, 2$ or 3.

Thus from this layer on, m plays the role of k, and the degeneracy remains at $m+1$ the rest of the way in. This completes the proof of theorem 10.10.

10.12 Generalization of Wigner-Eckart

Consider a matrix element of a tensor operator between tensor states

$$\langle u|W|v\rangle \tag{10.59}$$

This is really a shorthand for a very large number of possible matrix elements, because each representation may have many independent components (the number being just the dimension of the representation). Suppose that the representations of v, u and W are irreducible. Then we expect that the symmetry will ensure that there are relations among the various possible matrix elements. For example, if these were $SU(2)$ tensor operators, we know that all the matrix elements would be determined in terms of a single number. To determine the consequences of the symmetry in this case, we imagine decomposing the state $W|v\rangle$ into irreducible representations. Then we know there can be no contribution to the matrix element except from representations that transform under the same irreducible representation as u. This follows from Shur's lemma. Also because of Shur's lemma, the contribution from each appearance of the representation of u is determined by a single unknown number, because the matrix element is proportional to the identity in the space of the u representation. Thus there will be one (and only one) unknown constant in the matrix element for each time the representation of u appears in the tensor product $W \otimes v$. However, there is an important difference between this and the case of $SU(2)$. Here u may appear more than once in the decomposition of $W|v\rangle$.

Tensor analysis will make it relatively easy to write an expression for the matrix element, (10.59), that automatically incorporates all of the constraints that follow from the symmetry. Because of (10.9), (10.35), and (10.40), the matrix element will be proportional to the tensor components of $|v\rangle$

$$v^{i_1\cdots i_n}_{j_1\cdots j_m} \tag{10.60}$$

and of W

$$W^{k_1\cdots k_a}_{\ell_1\cdots \ell_b} \tag{10.61}$$

and proportional to the conjugate components of $\langle u|$

$$\overline{u}^{i'_1\cdots i'_{n'}}_{j'_1\cdots j'_{m'}} \tag{10.62}$$

The entire matrix element must be invariant. Which means that the answer for the matrix element must have all the indices either contracted with each other or with invariant tensors (that is δs and ϵs). Another way to say the same thing is that the matrix element

$$\begin{aligned} &\langle {}^{i'_1\cdots i'_{n'}}_{j'_1\cdots j'_{m'}} | O^{\ell_1\cdots \ell_b}_{k_1\cdots k_a} | {}^{j_1\cdots j_m}_{i_1\cdots i_n} \rangle \\ &= \Gamma^{\ell_1\cdots \ell_b j_1\cdots j_m j'_1\cdots j'_{m'}}_{i_1\cdots i_n k_1\cdots k_a i'_1\cdots i'_{n'}} \end{aligned} \tag{10.63}$$

must itself be an invariant tensor — a sum of terms made of δs and ϵs.

Here is an example that will be of some interest later. Suppose that u, W and v are all 8s. This is particularly easy to analyze, because $\overline{u}$, W and v can all be thought of as matrices, with the upper index referring to the row, and the lower index to the column. Then, for example,

$$W^i_j \, v^j_k = [W\, v]^i_k \tag{10.64}$$

Now a term with all indices contracted is one in which the matrices are combined into a number by matrix multiplication, and taking traces. There are precisely two such terms:

$$\text{Tr}(\overline{u}\,W\,v) \quad \text{and} \quad \text{Tr}(\overline{u}\,v\,W) \tag{10.65}$$

There are two terms because the tensor product $8 \otimes 8$ contains 8 twice. Thus in this case the matrix element is

$$\lambda_1\,\text{Tr}(\overline{u}\,W\,v) + \lambda_2\,\text{Tr}(\overline{u}\,v\,W) \tag{10.66}$$

where λ_1 and λ_2 are constants that must be fixed by the physics, rather than the group theory (like the reduced matrix elements in the Wigner-Eckart theorem for $SU(2)$). This means that all $8^3 = 512$ matrix elements are determined in terms of these two parameters! Furthermore, we know how all the matrix elements are related simply by matrix multiplication, without any fancy tables or whatnot.

Here's another example — suppose u and v are 10s, and W is an 8. The matrix element will be proportional to the tensor components of $|v\rangle$

$$v^{i_1 i_2 i_3} \tag{10.67}$$

and of W

$$W^k_\ell \tag{10.68}$$

and proportional to the conjugate components of $\langle u|$

$$\overline{u}_{j_1 j_2 j_3} \tag{10.69}$$

In this case, it is easy to see that there is only one way to put the indices together (because of the total symmetry of $\overline{u}$ and v)

$$\overline{u}_{i_1 i_2 k} W^k_\ell v^{i_1 i_2 \ell} \tag{10.70}$$

Thus the matrix element is determined by a single reduced matrix element

$$\langle u|W|v\rangle = \lambda_1\,\overline{u}_{i_1 i_2 k} W^k_\ell v^{i_1 i_2 \ell} \tag{10.71}$$

10.13 * Tensors for $SU(2)$

You may be wondering why we have spent so much time on the description of tensors for $SU(3)$, while we barely mentioned them for $SU(2)$. The answer to this question is really more history and sociology than it is math or physics. The description of irreducible $SU(2)$ representations in term of tensors is simply not used very much, perhaps because it is not needed. Everyone knows how to deal with the $SU(2)$ irreducible representations in other ways. Nevertheless, tensor analysis is useful for $SU(2)$, and it may be instructive to see how it works.

Tensor analysis for $SU(2)$ is even simpler than for $SU(3)$ because there is only one fundamental representation for $SU(2)$ — the doublet. Thus we can look only at tensors with upper indices, and there is no trace constraint. The ϵ symbol now has two indices, rather than three in $SU(3)$. So while we could use the invariant ϵ tensor to lower two upper indices in $SU(3)$, in $SU(2)$, the ϵ just eats two upper indices completely. This implies that the irreducible representations correspond to completely symmetric tensors, because if a tensor with n indices is not completely symmetric, then we can get a non-zero tensor with $n-2$ indices by contracting two indices with ϵ, and that would reduce the representation. Obviously, the completely symmetric tensor with n indices corresponds to the irreducible spin-$n/2$ representation, because the highest weight state, in which all the indices are 1, has $J_3 = n/2$.

Because the $SU(2)$ irreducible representations correspond to such simple tensors, we can write them down explicitly. Consider the tensor with $2s$ indices, corresponding to the spin-s representation, and let us pick out the component with $J_3 = m$. Because each 1 index carries $J_3 = 1/2$ and each 2 index carries $J_3 = -1/2$, the $J_3 = m$ state must correspond to a tensor with $s+m$ 1s and $s-m$ 2s. Let us define the tensor

$$v_{s,m}^{j_1 \cdots j_{2s}} \tag{10.72}$$

to be a completely symmetric tensor which is equal to 1 if $s+m$ indices are 1 and zero otherwise. There are

$$\binom{2s}{s+m} = \frac{(2s)!}{(s+m)!\,(s-m)!} \tag{10.73}$$

ways to pick the $s+m$ 1s out of the $2s$ indices. Thus $v_{s,m}$ is a sum of $\binom{2s}{s+m}$ terms. For example,

$$v_{3/2,1/2}^{j_1\,j_2\,j_3} = \delta_1^{j_1}\,\delta_1^{j_2}\,\delta_2^{j_3} + \delta_1^{j_1}\,\delta_2^{j_2}\,\delta_1^{j_3} + \delta_2^{j_1}\,\delta_1^{j_2}\,\delta_1^{j_3} \tag{10.74}$$

The state $|v_{s,m}\rangle$ is not properly normalized. The squared norm gets a contribution 1 from each of the $\binom{2s}{s+m}$ terms in (10.72). Thus the properly normalized state is

$$|s,m\rangle = \binom{2s}{s+m}^{-1/2} |v_{s,m}\rangle \tag{10.75}$$

It is easy to check that this definition behaves properly under the $SU(2)$ raising and lowering operators. We will next illustrate its use.

10.14 * Clebsch-Gordan coefficients from tensors

Consider the product

$$v_{s_1,m_1}^{i_1\cdots i_{2s_1}} \cdot v_{s_2,m_2}^{j_1\cdots j_{2s_2}} \tag{10.76}$$

From the general properties of tensors, we know that it transforms like a component in the tensor product of s_1 and s_2. The corresponding normalized state is easy to write down, because (10.76) is simply a product:

$$|s_1, s_2, m_1, m_2\rangle = \binom{2s_1}{s_1+m_1}^{-1/2} \binom{2s_2}{s_2+m_2}^{-1/2} |v_{s_1,m_1} \cdot v_{s_2,m_2}\rangle \tag{10.77}$$

The tensor product (10.76) lives in the space of tensors which are symmetric in the $2s_1$ i indices and separately symmetric in the $2s_2$ j indices. What we would like to do is to decompose this tensor product space into subspaces that transform irreducibly under $SU(2)$. We can do this by combining tensors corresponding to the various spin states in $s_1 \otimes s_2$ with invariant tensors to produce tensors with the appropriate symmetry properties.

The simplest example of this is spin $s_1 + s_2$. We know that we can write the states of spin $s_1 + s_2$ in terms of a completely symmetric tensor with $2s_1 + 2s_2$ indices,

$$|s_1+s_2, m_1+m_2\rangle = \binom{2s_1+2s_2}{s_1+s_2+m_1+m_2}^{-1/2} |v_{s_1+s_2,m_1+m_2}\rangle \tag{10.78}$$

Since $v_{s_1+s_2,m_1+m_2}$ is completely symmetric, it is symmetric in its first $2s_1$ and its last $2s_2$ indices, and is thus in the same space as the tensor product, (10.76). The scalar product of the two states, (10.77) and (10.78) is the answer to the question "What is the amplitude to find the state of spin $s_1 + s_2$ in the tensor product state?" In other words, it is the Clebsch-Gordan coefficient,

$$\langle s_1 + s_2, m_1 + m_2 \mid s_1, s_2, m_1, m_2\rangle \tag{10.79}$$

But we can also easily calculate the matrix element directly from the definitions because

$$\begin{aligned} &\langle v_{s_1+s_2,m_1+m_2} \mid v_{s_1,m_1} \cdot v_{s_2,m_2}\rangle \\ &= \sum_{\substack{i_1\cdots i_{2s_1} \\ j_1\cdots j_{2s_2}}} \left(v_{s_1+s_2,m_1+m_2}^{i_1\cdots i_{2s_1} j_1\cdots j_{2s_2}}\right)^* v_{s_1,m_1}^{i_1\cdots i_{2s_1}} \cdot v_{s_2,m_2}^{j_1\cdots j_{2s_2}} \end{aligned} \tag{10.80}$$

The complex conjugation is unnecessary, because the tensors are real. The sum looks formidable, but actually it is easy to do. Each configuration of 1s and 2s for the tensors on the right-hand-side will correspond to some configuration for $v_{2s_1+2s_2}$, which contains all possible configurations with the right total number of 1s and 2s. Thus the sum in (10.80) gives

$$\binom{2s_1}{s_1+m_1}\binom{2s_2}{s_2+m_2} \tag{10.81}$$

and the result for the Clebsch-Gordan coefficient is

$$\begin{aligned} &\langle s_1+s_2, m_1+m_2 \mid s_1, s_2, m_1, m_2 \rangle \\ &= \binom{2s_1}{s_1+m_1}^{1/2}\binom{2s_2}{s_2+m_2}^{1/2}\binom{2s_1+2s_2}{s_1+s_2+m_1+m_2}^{-1/2} \end{aligned} \tag{10.82}$$

You can check in special cases that this is right using raising and lowering operators, but tensors gave us the general result rather easily.

10.15 * Spin $s_1 + s_2 - 1$

The component of the tensor product, (10.77), in spin $s_1 + s_2 - 1$ will be proportional to the tensor $v_{s_1+s_2-1,m_1+m_2}$. To compute the Clebsch-Gordan coefficient, we must find the normalized state proportional to this tensor in the space of the tensor product tensor, (10.76). That is, we must construct a tensor proportional to $v_{s_1+s_2-1,m_1+m_2}$, but with $2s_1 + 2s_2$ indices and symmetric in the first $2s_1$ indices, and the last $2s_2$ indices. To add two indices, we need the invariant tensor ϵ^{ij}. Thus we can form the tensor

$$\epsilon^{i_1 j_1} v_{s_1+s_2-1,m_1+m_2}^{i_2\cdots i_{2s_1} j_2 \cdots j_{2s_2}} \tag{10.83}$$

Now we have to symmetrize in the i indices and symmetrize in the j indices. Let us call the symmetrized tensor

$$\mathcal{V}_{s_1+s_2-1,m_1+m_2,2s_1,2s_2}^{i_1\cdots i_{2s_1} j_1 \cdots j_{2s_2}} \tag{10.84}$$

Each term in $\mathcal{V}$ has a single $\epsilon^{i_r j_s}$ multiplied by $v_{s_1+s_2-1,m_1+m_2}$ with all the other indices. There are $4s_1 s_2$ terms in $\mathcal{V}$, corresponding to the $2s_1$ possible values of r and the $2s_2$ values of s.

The tensor $\mathcal{V}$ is not properly normalized. Call the normalization factor N. Then the Clebsch-Gordan coefficient is

$$\begin{aligned}
&\langle s_1+s_2-1, m_1+m_2 \mid s_1, s_2, m_1, m_2\rangle \\
&= N \begin{pmatrix} 2s_1 \\ s_1+m_1 \end{pmatrix}^{-1/2} \begin{pmatrix} 2s_2 \\ s_2+m_2 \end{pmatrix}^{-1/2} \\
&\sum_{\substack{i_1\cdots i_{2s_1} \\ j_1\cdots j_{2s_2}}} \left(\mathcal{V}^{i_1\cdots i_{2s_1} j_1\cdots j_{2s_2}}_{s_1+s_2-1, m_1+m_2, 2s_1, 2s_2}\right)^* v_{s_1,m_1}^{i_1\cdots i_{2s_1}} \cdot v_{s_2,m_2}^{j_1\cdots j_{2s_2}}
\end{aligned} \tag{10.85}$$

The sum in (10.85) is easy to simplify because each of the $4s_1s_2$ terms in $\mathcal{V}$ gives the same result. This is true because these terms were constructed by symmetrizing in the i and j indices, but the tensor product is already symmetric in the i and j indices. And as before, the complex conjugation is irrelevant. Thus we can rewrite (10.85) as

$$\begin{aligned}
&\langle s_1+s_2-1, m_1+m_2 \mid s_1, s_2, m_1, m_2\rangle \\
&= N \begin{pmatrix} 2s_1 \\ s_1+m_1 \end{pmatrix}^{-1/2} \begin{pmatrix} 2s_2 \\ s_2+m_2 \end{pmatrix}^{-1/2} 4s_1s_2 \\
&\sum_{\substack{i_1\cdots i_{2s_1} \\ j_1\cdots j_{2s_2}}} \epsilon^{i_1j_1} \cdot v_{s_1+s_2-1, m_1+m_2}^{i_2\cdots i_{2s_1} j_2\cdots j_{2s_2}} \cdot v_{s_1,m_1}^{i_1\cdots i_{2s_1}} \cdot v_{s_2,m_2}^{j_1\cdots j_{2s_2}} \\
&= N \begin{pmatrix} 2s_1 \\ s_1+m_1 \end{pmatrix}^{-1/2} \begin{pmatrix} 2s_2 \\ s_2+m_2 \end{pmatrix}^{-1/2} 4s_1s_2 \\
&\cdot\left[\begin{pmatrix} 2s_1-1 \\ s_1+m_1-1 \end{pmatrix}\begin{pmatrix} 2s_2-1 \\ s_2+m_2 \end{pmatrix} - \begin{pmatrix} 2s_1-1 \\ s_1+m_1 \end{pmatrix}\begin{pmatrix} 2s_2-1 \\ s_2+m_2-1 \end{pmatrix}\right]
\end{aligned} \tag{10.86}$$

where the last line follows by simply expanding the two non-zero values of $\epsilon^{i_1j_1}$ in the sum to get

$$\sum_{\substack{i_2\cdots i_{2s_1} \\ j_2\cdots j_{2s_2}}} v_{s_1+s_2-1, m_1+m_2}^{i_2\cdots i_{2s_1} j_2\cdots j_{2s_2}} \cdot \left(v_{s_1,m_1}^{1\,i_2\cdots i_{2s_1}} \cdot v_{s_2,m_2}^{2\,j_2\cdots j_{2s_2}} - v_{s_1,m_1}^{2\,i_2\cdots i_{2s_1}} \cdot v_{s_2,m_2}^{1\,j_2\cdots j_{2s_2}}\right) \tag{10.87}$$

To calculate N in (10.85), note that we can expand one of the $\mathcal{V}$s, and each of the terms will give the same result because of the symmetry of the remaining $\mathcal{V}$, so

$$\begin{aligned}
1/N^2 &= \sum_{\substack{i_1\cdots i_{2s_1} \\ j_1\cdots j_{2s_2}}} \left|\mathcal{V}^{i_1\cdots i_{2s_1} j_1\cdots j_{2s_2}}_{s_1+s_2-1, m_1+m_2, 2s_1, 2s_2}\right|^2 \\
&= 4s_1s_2 \sum_{\substack{i_1\cdots i_{2s_1} \\ j_1\cdots j_{2s_2}}} \epsilon^{i_1j_1}\, v_{s_1+s_2-1, m_1+m_2}^{i_2\cdots i_{2s_1} j_2\cdots j_{2s_2}}\, \mathcal{V}^{i_1\cdots i_{2s_1} j_1\cdots j_{2s_2}}_{s_1+s_2-1, m_1+m_2, 2s_1, 2s_2}
\end{aligned} \tag{10.88}$$

Now we can do the sum over i_1 and j_1 in (10.88) by expanding $\mathcal{V}$. What makes this doable is the fact that

$$\sum_{i_1,j_1} \epsilon^{i_1 j_1} \mathcal{V}^{i_1\cdots i_{2s_1}\, j_1\cdots j_{2s_2}}_{s_1+s_2-1,m_1+m_2,2s_1,2s_2} = C\, v^{i_2\cdots i_{2s_1}\, j_2\cdots j_{2s_2}}_{s_1+s_2-1,m_1+m_2} \tag{10.89}$$

for some constant C, so that

$$\sum_{\substack{i_1\cdots i_{2s_1}\\ j_1\cdots j_{2s_2}}} \epsilon^{i_1 j_1}\, v^{i_2\cdots i_{2s_1}\, j_2\cdots j_{2s_2}}_{s_1+s_2-1,m_1+m_2}\, \mathcal{V}^{i_1\cdots i_{2s_1}\, j_1\cdots j_{2s_2}}_{s_1+s_2-1,m_1+m_2,2s_1,2s_2}$$
$$= C \sum_{\substack{i_2\cdots i_{2s_1}\\ j_2\cdots j_{2s_2}}} \left| v^{i_2\cdots i_{2s_1}\, j_2\cdots j_{2s_2}}_{s_1+s_2-1,m_1+m_2} \right|^2 = C \begin{pmatrix} 2s_1+2s_2-2 \\ s_1+s_2-1+m_1+m_2 \end{pmatrix} \tag{10.90}$$

To compute C, note that the term in $\mathcal{V}$ proportional to $\epsilon^{i_1 j_1}$ gives a factor of 2 because

$$\sum_{i_1\, j_1} \left|\epsilon^{i_1 j_1}\right|^2 = 2 \tag{10.91}$$

The terms proportional to $\epsilon^{i_1 j_m}$ for $m \neq 1$ give 1 each because

$$\sum_{i_1} \epsilon^{i_1 j_1}\, \epsilon^{i_1 j_m} = \delta^{j_1 j_m} \tag{10.92}$$

The sum over j_1 then simply reproduces $v^{i_2\cdots i_{2s_1}\, j_2\cdots j_{2s_2}}_{s_1+s_2-1,m_1+m_2}$. Since there are $2s_2 - 1$ possible values for j_m, such terms give $2s_2 - 1$ total contribution to C. The terms proportional to $\epsilon^{i_m j_1}$ for $m \neq 1$ contribute another $2s_1 - 1$. The other terms contribute nothing. Thus $C = (2+2s_1-1+2s_2-1) = 2s_1+2s_2$, and

$$1/N^2 = 4\, s_1 s_2 (2s_1 + 2s_2) \begin{pmatrix} 2s_1+2s_2-2 \\ s_1+s_2-1+m_1+m_2 \end{pmatrix} \tag{10.93}$$

and putting all this together gives

$$\langle s_1+s_2-1, m_1+m_2 \mid s_1, s_2, m_1, m_2 \rangle$$
$$= \begin{pmatrix} 2s_1 \\ s_1+m_1 \end{pmatrix}^{-1/2} \begin{pmatrix} 2s_2 \\ s_2+m_2 \end{pmatrix}^{-1/2}$$
$$\cdot \begin{pmatrix} 2s_1+2s_2-2 \\ s_1+s_2-1+m_1+m_2 \end{pmatrix}^{-1/2} \left(\frac{2\, s_1 s_2}{s_1+s_2} \right)^{1/2}$$
$$\cdot \left[\begin{pmatrix} 2s_1-1 \\ s_1+m_1-1 \end{pmatrix} \begin{pmatrix} 2s_2-1 \\ s_2+m_2 \end{pmatrix} - \begin{pmatrix} 2s_1-1 \\ s_1+m_1 \end{pmatrix} \begin{pmatrix} 2s_2-1 \\ s_2+m_2-1 \end{pmatrix} \right] \tag{10.94}$$

10.16 * Spin $s_1 + s_2 - k$

The analysis of the general case is similar to that in the previous section. The total spin can be written as $s_1 + s_2 - k$ for k between 0 and $|s_1 - s_2|$. The component of the tensor product, (10.77), in spin $s_1 + s_2 - k$ is proportional to the tensor $v_{s_1+s_2-k,m_1+m_2}$. To compute the Clebsch-Gordan coefficient, we must find the normalized state proportional to this tensor in the space of the tensor product tensor, (10.76), so we must construct a tensor proportional to $v_{s_1+s_2-k,m_1+m_2}$, but with $2s_1 + 2s_2$ indices and symmetric in the first $2s_1$ indices, and the last $2s_2$ indices. For this, we need k invariant tensors ϵ^{ij} to form the tensor

$$\epsilon^{i_1 j_1} \cdots \epsilon^{i_k j_k} v_{s_1+s_2-k,m_1+m_2}^{i_{k+1}\cdots i_{2s_1} j_{k+1}\cdots j_{2s_2}} \tag{10.95}$$

Now we symmetrize in the i indices and symmetrize in the j indices. The symmetrized tensor is

$$\mathcal{V}_{s_1+s_2-k,m_1+m_2,2s_1,2s_2}^{i_1\cdots i_{2s_1} j_1\cdots j_{2s_2}} \tag{10.96}$$

Each term in $\mathcal{V}$ has k $\epsilon^{i_r j_s}$ multiplied by $v_{s_1+s_2-k,m_1+m_2}$ with the other indices. We can count the number of terms in $\mathcal{V}$ as follows. The i index in the first ϵ can have any of $2s_1$ values, and the k index any of $2s_2$ values. Then the i index in the second ϵ can have any of $2s_1 - 1$ values, and the k index any of $2s_2 - 1$ values. And so on for k terms. However, this overcounts by a factor of $k!$, because the order of the ϵs doesn't matter. Thus the number of terms is

$$\mu \equiv \frac{(2s_1)!}{(2s_1-k)!} \frac{(2s_2)!}{(2s_2-k)!} \frac{1}{k!} \tag{10.97}$$

Again, let N be the normalization factor required to properly normalize the tensor $\mathcal{V}$. Then the Clebsch-Gordan coefficient is

$$\begin{aligned} &\langle s_1+s_2-k, m_1+m_2 \mid s_1, s_2, m_1, m_2 \rangle \\ &= N \begin{pmatrix} 2s_1 \\ s_1+m_1 \end{pmatrix}^{-1/2} \begin{pmatrix} 2s_2 \\ s_2+m_2 \end{pmatrix}^{-1/2} \\ &\sum_{\substack{i_1\cdots i_{2s_1} \\ j_1\cdots j_{2s_2}}} \left(\mathcal{V}_{s_1+s_2-k,m_1+m_2,2s_1,2s_2}^{i_1\cdots i_{2s_1} j_1\cdots j_{2s_2}} \right)^* v_{s_1,m_1}^{i_1\cdots i_{2s_1}} \cdot v_{s_2,m_2}^{j_1\cdots j_{2s_2}} \end{aligned} \tag{10.98}$$

As above, the sum in (10.98) can be simplified because each of the μ terms in $\mathcal{V}$ gives the same result. The complex conjugation is irrelevant as usual .

Thus we can rewrite (10.98) as

$$
\begin{aligned}
&\langle s_1 + s_2 - k, m_1 + m_2 \mid s_1, s_2, m_1, m_2 \rangle \\
&= N \binom{2s_1}{s_1+m_1}^{-1/2} \binom{2s_2}{s_2+m_2}^{-1/2} \mu \\
&\sum_{\substack{i_1\cdots i_{2s_1}\\ j_1\cdots j_{2s_2}}} \epsilon^{i_1 j_1} \ldots \epsilon^{i_k j_k} \cdot v^{i_{k+1}\cdots i_{2s_1} j_{k+1}\cdots j_{2s_2}}_{s_1+s_2-k, m_1+m_2} \cdot v^{i_1\cdots i_{2s_1}}_{s_1,m_1} \cdot v^{j_1\cdots j_{2s_2}}_{s_2,m_2} \\
&= N \binom{2s_1}{s_1+m_1}^{-1/2} \binom{2s_2}{s_2+m_2}^{-1/2} \mu \\
&\cdot \sum_{\ell=0}^{k} (-1)^\ell \binom{k}{\ell} \binom{2s_1-k}{s_1+m_1-k+\ell} \binom{2s_2-k}{s_2+m_2-\ell}
\end{aligned}
\tag{10.99}
$$

where in the last line we have expanded all the ϵs.

To calculate N in (10.98), note that

$$
\begin{aligned}
1/N^2 &= \sum_{\substack{i_1\cdots i_{2s_1}\\ j_1\cdots j_{2s_2}}} \left| \mathcal{V}^{i_1\cdots i_{2s_1} j_1\cdots j_{2s_2}}_{s_1+s_2-k, m_1+m_2, 2s_1, 2s_2} \right|^2 \\
&= \mu\, \epsilon^{i_1 j_1} \ldots \epsilon^{i_k j_k}\, v^{i_{k+1}\cdots i_{2s_1} j_{k+1}\cdots j_{2s_2}}_{s_1+s_2-k, m_1+m_2}\, \mathcal{V}^{i_1\cdots i_{2s_1} j_1\cdots j_{2s_2}}_{s_1+s_2-k, m_1+m_2, 2s_1, 2s_2}
\end{aligned}
\tag{10.100}
$$

Now, again, we can do the sum over i_1 and j_1 by expanding $\mathcal{V}$. The result will be proportional to $\mathcal{V}$ with two fewer indices

$$
\sum_{i_1, j_1} \epsilon^{i_1 j_1}\, \mathcal{V}^{i_1\cdots i_{2s_1}\, j_1\cdots j_{2s_2}}_{s_1+s_2-k, m_1+m_2, 2s_1, 2s_2} = C\, \mathcal{V}^{i_2\cdots i_{2s_1}\, j_2\cdots j_{2s_2}}_{s_1+s_2-k, m_1+m_2, 2s_1-1, 2s_2-1}
\tag{10.101}
$$

because the sum over i_1 and j_1 doesn't affect the symmetry in the other indices. We can find the coefficient C in (10.101) by picking out all contributions to any single term in the expansion of the right-hand-side, for example,

$$
\epsilon^{i_2 j_2} \ldots \epsilon^{i_k j_k}\, v^{i_{k+1}\cdots i_{2s_1} j_{k+1}\cdots j_{2s_2}}_{s_1+s_2-k, m_1+m_2}
\tag{10.102}
$$

We will now look at all terms in $\mathcal{V}_{s_1+s_2-k, m_1+m_2, 2s_1, 2s_2}$ that contribute to the particular term (10.102) in $\mathcal{V}_{s_1+s_2-k, m_1+m_2, 2s_1-1, 2s_2-1}$ and find their contributions to the coefficient C.

The term

$$
\epsilon^{i_1 j_1} \epsilon^{i_2 j_2} \ldots \epsilon^{i_k j_k}\, v^{i_{k+1}\cdots i_{2s_1} j_{k+1}\cdots j_{2s_2}}_{s_1+s_2-k, m_1+m_2}
\tag{10.103}
$$

contributes 2.

Each of the $2s_2 - k$ terms of the form

$$\epsilon^{i_1 j_m} \epsilon^{i_2 j_2} \dots \epsilon^{i_k j_k} v_{s_1+s_2-k,m_1+m_2}^{i_{k+1}\cdots i_{2s_1} j_{k+1}\cdots j_{m-1}\, j_1\, j_{m+1}\cdots j_{2s_2}} \tag{10.104}$$

for $m > k$ contributes 1.

Each of the $2s_1 - k$ terms of the form

$$\epsilon^{i_m j_1} \epsilon^{i_2 j_2} \dots \epsilon^{i_k j_k} v_{s_1+s_2-k,m_1+m_2}^{i_{k+1}\cdots i_{m-1}\, i_1\, i_{m+1}\cdots i_{2s_1} j_{k+1}\cdots j_{2s_2}} \tag{10.105}$$

for $m > k$ contributes 1.

Each of the $k - 1$ terms of the form

$$\epsilon^{i_1 j_m} \epsilon^{i_m j_1} \epsilon^{i_2 j_2} \dots \epsilon^{i_{m-1} j_{m-1}} \epsilon^{i_{m+1} j_{m+1}} \dots \epsilon^{i_k j_k} v_{s_1+s_2-k,m_1+m_2}^{i_{k+1}\cdots i_{2s_1} j_{k+1}\cdots j_{2s_2}} \tag{10.106}$$

for $2 \leq m \leq k$ contributes 1.

Putting all this together gives $C = (2s_1 + 2s_2 - k + 1)$ and thus

$$\begin{aligned} &\sum_{i_1,j_1} \epsilon^{i_1 j_1} \mathcal{V}_{s_1+s_2-k,m_1+m_2,2s_1,2s_2}^{i_1\cdots i_{2s_1}\, j_1\cdots j_{2s_2}} \\ &= (2s_1 + 2s_2 - k + 1)\, \mathcal{V}_{s_1+s_2-k,m_1+m_2,2s_1-1,2s_2-1}^{i_2\cdots i_{2s_1}\, j_2\cdots j_{2s_2}} \end{aligned} \tag{10.107}$$

Now we iterate the procedure, and do the sum over i_2 and j_2 in the same way, then i_3 and j_3 and so on. In general

$$\begin{aligned} &\sum_{i_{p+1},j_{p+1}} \epsilon^{i_{p+1} j_{p+1}} \mathcal{V}_{s_1+s_2-k,m_1+m_2,2s_1-p,2s_2-p}^{i_{p+1}\cdots i_{2s_1}\, j_{p+1}\cdots j_{2s_2}} \\ &\quad = C_p\, \mathcal{V}_{s_1+s_2-k,m_1+m_2,2s_1-p-1,2s_2-p-1}^{i_{p+2}\cdots i_{2s_1}\, j_{p+2}\cdots j_{2s_2}} \end{aligned} \tag{10.108}$$

where $C_0 = C$ and

$$\mathcal{V}_{s_1+s_2-k,m_1+m_2,2s_1-k,2s_2-k}^{i_{k+1}\cdots i_{2s_1}\, j_{k+1}\cdots j_{2s_2}} = v_{s_1+s_2-k,m_1+m_2}^{i_{k+1}\cdots i_{2s_1}\, j_{k+1}\cdots j_{2s_2}} \tag{10.109}$$

As we did before for C_0, to calculate C_p, we can look at all contributions to a single term in the expansion of the right-hand-side, for example,

$$\epsilon^{i_{p+2} j_{p+2}} \dots \epsilon^{i_k j_k} v_{s_1+s_2-k,m_1+m_2}^{i_{k+1}\cdots i_{2s_1} j_{k+1}\cdots j_{2s_2}} \tag{10.110}$$

We will now look at all terms in $\mathcal{V}_{s_1+s_2-k,m_1+m_2,2s_1-p,2s_2-p}$ that contribute to the particular term (10.110) in $\mathcal{V}_{s_1+s_2-k,m_1+m_2,2s_1-p-1,2s_2-p-1}$ and find their contributions to the coefficient C_p.

The term

$$\epsilon^{i_{p+1} j_{p+1}} \epsilon^{i_{p+2} j_{p+2}} \dots \epsilon^{i_k j_k} v_{s_1+s_2-k,m_1+m_2}^{i_{k+1}\cdots i_{2s_1} j_{k+1}\cdots j_{2s_2}} \tag{10.111}$$

contributes 2.

Each of the $2s_2 - k$ terms of the form

$$\epsilon^{i_{p+1}j_m}\epsilon^{i_{p+2}j_{p+2}}\ldots\epsilon^{i_kj_k}v_{s_1+s_2-k,m_1+m_2}^{i_{k+1}\cdots i_{2s_1}j_{k+1}\cdots j_{m-1}\,j_{p+1}\,j_{m+1}\cdots j_{2s_2}} \tag{10.112}$$

for $m > k$ contributes 1.

Each of the $2s_1 - k$ terms of the form

$$\epsilon^{i_mj_{p+1}}\epsilon^{i_{p+2}j_{p+2}}\ldots\epsilon^{i_kj_k}v_{s_1+s_2-k,m_1+m_2}^{i_{k+1}\cdots i_{m-1}\,i_{p+1}\,i_{m+1}\cdots i_{2s_1}j_{k+1}\cdots j_{2s_2}} \tag{10.113}$$

for $m > k$ contributes 1.

Each of the $k - p - 1$ terms of the form

$$\epsilon^{i_{p+1}j_m}\epsilon^{i_mj_{p+1}}\epsilon^{i_{p+2}j_{p+2}}\ldots\epsilon^{i_{m-1}j_{m-1}}\epsilon^{i_{m+1}j_{m+1}}\ldots\epsilon^{i_kj_k}v_{s_1+s_2-k,m_1+m_2}^{i_{k+1}\cdots i_{2s_1}j_{k+1}\cdots j_{2s_2}} \tag{10.114}$$

for $2 \le m \le k$ contributes 1.

Thus

$$C_p = 2s_1 + 2s_2 - k - p + 1 \tag{10.115}$$

and

$$\begin{aligned}
&\sum_{\substack{i_1\cdots i_{2s_1}\\ j_1\cdots j_{2s_2}}} \epsilon^{i_1j_1}\ldots\epsilon^{i_kj_k}v_{s_1+s_2-k,m_1+m_2}^{i_{k+1}\cdots i_{2s_1}j_{k+1}\cdots j_{2s_2}}\,\mathcal{V}_{s_1+s_2-k,m_1+m_2,2s_1,2s_2}^{i_1\cdots i_{2s_1}j_1\cdots j_{2s_2}}\\
&= \left(\prod_{p=0}^{k-1} C_p\right) \sum_{\substack{i_{k+1}\cdots i_{2s_1}\\ j_{k+1}\cdots j_{2s_2}}} \left|v_{s_1+s_2-k,m_1+m_2}^{i_{k+1}\cdots i_{2s_1}j_{k+1}\cdots j_{2s_2}}\right|^2\\
&= \frac{(2s_1+2s_2-k+1)!}{(2s_1+2s_2-2k+1)!}\binom{2s_1+2s_2-2k}{s_1+s_2+k+m_1+m_2}
\end{aligned} \tag{10.116}$$

and

$$1/N^2 = \mu\,\frac{(2s_1+2s_2-k+1)!}{(2s_1+2s_2-2k+1)!}\binom{2s_1+2s_2-2}{s_1+s_2-1+m_1+m_2} \tag{10.117}$$

Now we have all the pieces, and can put them together into the Clebsch-

Gordan coefficient for $|s_1+s_2-k, m_1+m_2\rangle$ in $|s_1,m_1\rangle \times |s_2,m_2\rangle$,

$$
\begin{aligned}
&\langle s_1+s_2-k, m_1+m_2 \mid s_1, s_2, m_1, m_2\rangle \\
&= \left[\binom{2s_1+2s_2-2k}{s_1+s_2+m_1+m_2-k}\binom{2s_1}{s_1+m_1}\binom{2s_2}{s_2+m_2}\right]^{-1/2} \\
&\quad \cdot \left(\frac{(2s_1)!\,(2s_2)!}{k!}\right)^{1/2}\left(\frac{(2s_1+2s_2-2k+1)!}{(2s_1+2s_2-k+1)!}\right)^{1/2} \\
&\quad \cdot \sum_{\ell=0}^{k}(-1)^{\ell}\binom{k}{\ell}\binom{2s_1-k}{s_1+m_1-k+\ell}\binom{2s_2-k}{s_2+m_2-\ell} \\
&= \left[\binom{2s_1+2s_2-2k}{s_1+s_2+m_1+m_2-k}\binom{2s_1}{s_1+m_1}\binom{2s_2}{s_2+m_2}\right]^{-1/2} \\
&\quad \cdot \left[\binom{2s_1}{k}\binom{2s_2}{k}\right]^{1/2}\left[\binom{2s_1+2s_2-k+1}{k}\right]^{-1/2} \\
&\quad \cdot \sum_{\ell=0}^{k}(-1)^{\ell}\binom{k}{\ell}\binom{2s_1-k}{s_1+m_1-k+\ell}\binom{2s_2-k}{s_2+m_2-\ell}
\end{aligned}
\tag{10.118}
$$

This relation is a bit complicated for humans to use, but it is easy to code it into a computer program that computes arbitrary Clebsch-Gordan coefficients, and it is nice example of the power of tensor methods.

Problems

10.A. Decompose the product of tensor components $u^i\, v^{jk}$ where $v^{jk} = v^{kj}$ transforms like a 6 of $SU(3)$.

10.B. Find the matrix elements $\langle u|T_a|v\rangle$ where T_a are the $SU(3)$ generators and $|u\rangle$ and $|v\rangle$ are tensors in the adjoint representation of $SU(3)$ with components u^i_j and v^i_j. Write the result in terms of the tensor components and the λ_a matrices of (7.4).

10.C. In the 6 of $SU(3)$, for each weight find the corresponding tensor component v^{ij}.

***10.D.**

a. Use the $SU(2)$ tensor methods described in this chapter to redo problem V.C.

b. Check equation (10.86) by using the highest weight procedure to find the state $|3/2, 1/2\rangle$ in the tensor product of spin 3/2 states and spin 1 states and compare with the result from (10.86) for the Clebsch-Gordan coefficient

$$\langle 3/2, 1, 3/2, 1/2 \mid 1, 3/2, 0, 1/2\rangle$$

***10.E.** Consider πP scattering, ignoring isospin breaking. There are three easily accessible processes (because it is hard to make π^0 beams or neutron targets):

$$\begin{aligned} \pi^+ p \rightarrow \pi^+ p &\text{ with amplitude } A_{+p} = \langle \pi^+ p | H_I | \pi^+ p\rangle \\ \pi^- p \rightarrow \pi^- p &\text{ with amplitude } A_{-p} = \langle \pi^- p | H_I | \pi^- p\rangle \\ \pi^- p \rightarrow \pi^0 n &\text{ with amplitude } A_{0n} = \langle \pi^0 n | H_I | \pi^- p\rangle \end{aligned} \tag{1}$$

where H_I is the interaction Hamiltonian, which is approximately $SU(3)$ invariant. If we describe the pion and nucleon wave functions in terms of $SU(2)$ tensors, $\pi^{jk} = \pi^{kj}$ for the pions and N^j for the nucleons, we can write the most general amplitude for this process as follows:

$$\langle \pi N | H_I | \pi N\rangle = A_1 \, \overline{\pi}_{jk} \, \overline{N}_\ell \, \pi^{jk} \, N^\ell + A_2 \, \overline{\pi}_{jk} \, \overline{N}_\ell \, \pi^{j\ell} \, N^k \tag{2}$$

a. Use (2) and the techniques described in the notes on tensor analysis for $SU(2)$ to write the three scattering amplitudes in (1) in terms of A_1 and A_2 and find the relation among the three amplitudes in (1).

b. The scattering amplitudes in (1) can also be written in terms of $I = 3/2$ and $I = 1/2$ amplitudes by decomposing the πN states into irreducible $I = 3/2$ and $I = 1/2$ representations and using Schur's lemma. Find these amplitudes in terms of A_1 and A_2.

Chapter 11

Hypercharge and Strangeness

Let's now go back to the 1950s to the discovery of strange particles. By this time, isospin was well established as an approximate symmetry of the strong interactions, broken by the weak and electromagnetic interactions. Experimenters began to notice a strange new class of particles which were produced in pairs in scattering of strongly interacting particles, and decayed much more slowly back into ordinary particles. It was eventually realized that a lot of the physics could be understood if there were an additive quantum number, called **strangeness** that was conserved by the strong (and electromagnetic) interactions, but not by weak interactions. Call it S. Particles which carry a non-zero value of S are called **strange particles.**

11.1 The eight-fold way

The lightest strange particles are the K mesons, an isospin doublet with $S = 1$

$$\begin{array}{lll} K^+ & \text{with} & I_3 = 1/2 \\ K^0 & \text{with} & I_3 = -1/2 \end{array} \tag{11.1}$$

and their antiparticles with $S = -1$

$$\begin{array}{lll} \overline{K}^0 & \text{with} & I_3 = 1/2 \\ K^- & \text{with} & I_3 = -1/2 \end{array} \tag{11.2}$$

Like the πs, the Ks have spin zero and baryon number zero.

There are also strange **baryons**, particles with baryon number $+1$. With

DOI: 10.1201/9780429499210-12

$S = -1$ there is an **isotriplet** (that is an isospin 1 representation)

$$\begin{aligned} \Sigma^+ \quad &\text{with} \quad I_3 = 1 \\ \Sigma^0 \quad &\text{with} \quad I_3 = 0 \\ \Sigma^- \quad &\text{with} \quad I_3 = -1 \end{aligned} \tag{11.3}$$

and an **isosinglet** (isospin 0)

$$\Lambda \quad \text{with} \quad I_3 = 0 \tag{11.4}$$

With $S = -2$ there is an **isodoublet** (isospin 1/2)

$$\begin{aligned} \Xi^0 \quad &\text{with} \quad I_3 = 1/2 \\ \Xi^- \quad &\text{with} \quad I_3 = -1/2 \end{aligned} \tag{11.5}$$

All of these particles satisfy

$$Q = T_3 + Y/2 \tag{11.6}$$

where Q is the electric charge in units of the proton charge, T_3 is the third component of isospin, and Y is a quantum number called **hypercharge**, defined by

$$Y = B + S \tag{11.7}$$

If we plot T_3 versus Y for the baryons, we get a plot that suggests the 8 dimensional adjoint representation of $SU(3)$:

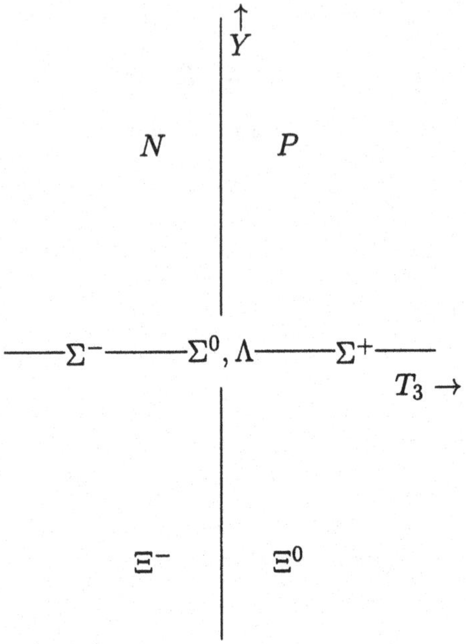

$$
\begin{array}{ccccc}
 & & \uparrow Y & & \\
 & N & & P & \\
\Sigma^- & & \Sigma^0, \Lambda & & \Sigma^+ \quad T_3 \rightarrow \\
 & \Xi^- & & \Xi^0 &
\end{array}
\tag{11.8}
$$

To make the hexagon regular, we can choose

$$
H_1 = T_3 \qquad H_2 = \sqrt{3}\,Y/2 \tag{11.9}
$$

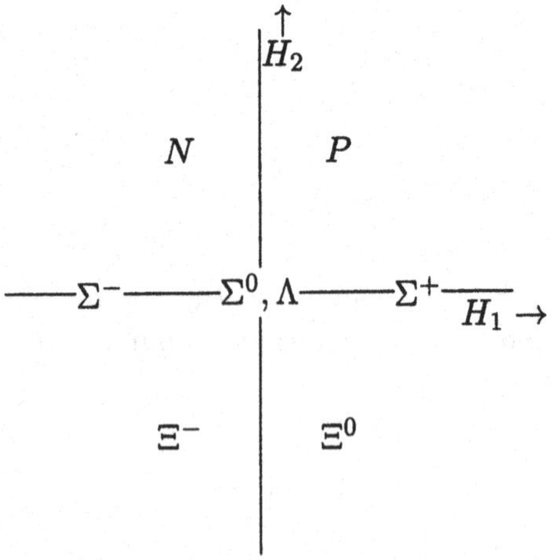

$$
\begin{array}{ccccc}
 & & \uparrow H_2 & & \\
 & N & & P & \\
\Sigma^- & & \Sigma^0, \Lambda & & \Sigma^+ \quad H_1 \rightarrow \\
 & \Xi^- & & \Xi^0 &
\end{array}
\tag{11.10}
$$

Something similar works for the light spin 0 mesons if we add the somewhat heavier η meson that was not discovered when $SU(3)$ was first discussed.

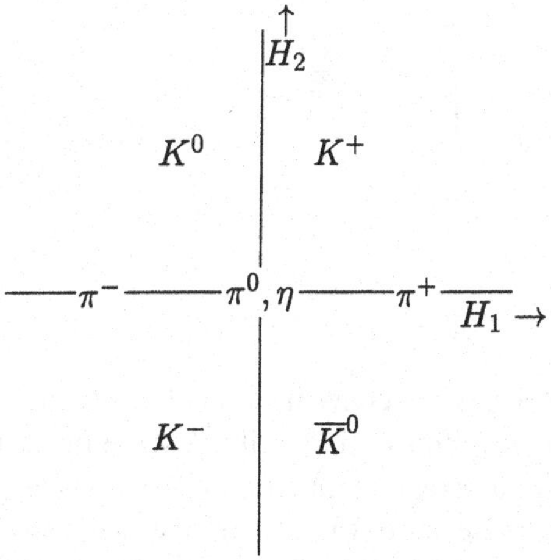

$$\begin{array}{ccccc} & & \uparrow H_2 & & \\ & K^0 & & K^+ & \\ \text{---}\pi^-\text{---} & & \pi^0,\eta & & \text{---}\pi^+\text{---}\; H_1 \rightarrow \\ & K^- & & \overline{K}^0 & \end{array} \tag{11.11}$$

It turns out that the $SU(3)$ symmetry suggested by these plots is not a symmetry of what were thought of as the strong interactions in the early 1960. But then **Gell-Mann** realized that you could imagine dividing the strong interactions into two parts:

very strong interactions, invariant under and $SU(3)$ symmetry under which the light baryons and mesons transform like 8s and;

medium strong interactions, which break $SU(3)$, but conserve isospin generated by T_1, T_2 and T_3, and hypercharge generated by $2T_8/\sqrt{3}$. (11.12)

Then he showed that $SU(3)$ symmetry could be used in perturbation theory, like isospin, to understand a lot about the strong interactions.

11.2 The Gell-Mann Okubo formula

$SU(3)$ is much more than just taxonomy. Consider the baryon masses. We can write the baryon states as

$$B^i_j \, |^j_i\rangle \tag{11.13}$$

where B^i_j is a tensor (wave function) that labels the particular baryon we are interested in. For example, the proton state corresponds to

$$B^i_j = \delta_{i1}\,\delta_{j3} \tag{11.14}$$

We can survey all the states at once by writing the following matrix:

$$B^i_j = \begin{pmatrix} \frac{\Sigma^0}{\sqrt{2}} + \frac{\Lambda}{\sqrt{6}} & \Sigma^+ & P \\ \Sigma^- & \frac{-\Sigma^0}{\sqrt{2}} + \frac{\Lambda}{\sqrt{6}} & N \\ \Xi^- & \Xi^0 & \frac{-2\Lambda}{\sqrt{6}} \end{pmatrix}_{ij} \tag{11.15}$$

Now

$$B^i_j \left|{}^j_i\right\rangle = P|P\rangle + N|N\rangle + \cdots \tag{11.16}$$

The entries of the matrix are convenient labels for keeping track of the states — P is the proton wave function, Λ is the Λ wave function, and so on.

If $SU(3)$ were an exact symmetry, all the particles of the **octet** would have the same mass. Because it is broken, the masses will be different. But if, as Gell-Mann assumed in (11.12), the $SU(3)$ conserving very strong interactions are much more important than the $SU(3)$ breaking medium strong interactions, the differences between different particle masses within the representations, the **mass splittings**, should be small compared to the average mass. Explicitly, the mass is the energy of a particle state at rest, so we want to calculate the matrix element of the Hamiltonian between the various single particle states. We will ignore the weak and electromagnetic interactions for now. The weak interactions give a negligible contribution to masses. The electromagnetic interactions are more interesting, because while their effect is small, it is significant because it violates isospin symmetry. We will come back to this later. Then the matrix element we want is

$$\langle B|H_S|B\rangle = \langle B|H_{VS}|B\rangle + \langle B|H_{MS}|B\rangle \tag{11.17}$$

where H_{VS} commutes with the $SU(3)$ generators and H_{MS} commutes with the isospin and hypercharge generators, but not with the others. The first term contributes a common mass to each particle in the $SU(3)$ representation. In general, we cannot say anything very interesting about the second term. However, there is reason to believe that H_{MS} transforms like the 8 component of an octet of $SU(3)$. Actually, when Gell-Mann first did the calculation, he didn't have any reason to believe this. But it is the simplest possibility consistent with the isospin and hypercharge conservation assumed in (11.12). The point is that we know the strong interaction Hamiltonian commutes with the isospin and hypercharge generators. Thus it must be made up of tensor operators that have components with $I = 0$ and $Y = 0$. The 8 is the smallest nontrivial representation that has a state with $I = Y = 0$. The $I = Y = 0$ state corresponds to the generator T_8. So Gell-Mann just assumed that H_{MS}

transformed like the T_8 component of an 8 in order to be able to use the Wigner-Eckart theorem to get a result. In particular, he assumed

$$H_{MS} = [T_8]^i_j \, O^j_i \tag{11.18}$$

where O is a tensor operator transforming like an 8 under $SU(3)$,

$$\left[T_a, O^j_i\right] = O^j_k \, [T_a]^k_i - O^k_i \, [T_a]^j_k \tag{11.19}$$

In (11.18) we are using the matrix element $[T_8]^i_j$ as a tensor component to pick out the component of the operator that commutes with $\vec{I}$ and Y.

Now we can use tensor methods to compute the matrix element

$$\langle B|H_{MS}|B\rangle \tag{11.20}$$

This is a calculation we have already done. The two independent $SU(3)$ invariant combinations of the tensor coefficients are

$$\begin{aligned} \overline{B}^i_j \, [T_8]^j_k \, B^k_i &= \text{Tr}\left(B^\dagger \, T_8 \, B\right) \\ \overline{B}^i_j \, B^j_k \, [T_8]^k_i &= \text{Tr}\left(B^\dagger \, B \, T_8\right) \end{aligned} \tag{11.21}$$

where we have used the fact that in matrix notation,

$$\overline{B} \to B^\dagger = B^{*T} \tag{11.22}$$

because $\overline{B}$ involves complex conjugation and interchanging of upper and lower indices, which is equivalent to transposition.

Thus purely from $SU(3)$ symmetry and the assumption that H_{MS} is a component of an octet we know that

$$\begin{aligned} &\langle B|H_{MS}|B\rangle \\ &= X \, \text{Tr}\left(B^\dagger \, B \, T_8\right) + Y \, \text{Tr}\left(B^\dagger \, T_8 \, B\right) \\ &= X \left([B^\dagger \, B]^1_1 + [B^\dagger \, B]^2_2 - 2[B^\dagger \, B]^3_3\right) / \sqrt{12} \\ &+ Y \left([B \, B^\dagger]^1_1 + [B \, B^\dagger]^2_2 - 2[B \, B^\dagger]^3_3\right) / \sqrt{12} \end{aligned} \tag{11.23}$$

So we need to compute the diagonal elements of $B^\dagger B$ and $B\,B^\dagger$

$$\begin{aligned}
[B^\dagger B]_{11} &= \left|\frac{\Sigma^0}{\sqrt{2}} + \frac{\Lambda}{\sqrt{6}}\right|^2 + |\Sigma^-|^2 + |\Xi^-|^2 \\
[B^\dagger B]_{22} &= |\Sigma^+|^2 + \left|\frac{-\Sigma^0}{\sqrt{2}} + \frac{\Lambda}{\sqrt{6}}\right|^2 + |\Xi^0|^2 \\
[B^\dagger B]_{33} &= |p|^2 + |n|^2 + \left|\frac{-2\Lambda}{\sqrt{6}}\right|^2 \\
[B\,B^\dagger]_{11} &= \left|\frac{\Sigma^0}{\sqrt{2}} + \frac{\Lambda}{\sqrt{6}}\right|^2 + |\Sigma^+|^2 + |p|^2 \\
[B\,B^\dagger]_{22} &= |\Sigma^-|^2 + \left|\frac{-\Sigma^0}{\sqrt{2}} + \frac{\Lambda}{\sqrt{6}}\right|^2 + |n|^2 \\
[B\,B^\dagger]_{33} &= |\Xi^-|^2 + |\Xi^0|^2 + \left|\frac{-2\Lambda}{\sqrt{6}}\right|^2
\end{aligned} \tag{11.24}$$

Thus

$$\begin{aligned}
&\langle B|H_{MS}|B\rangle \\
&= X\left(|\Sigma|^2 + |\Xi|^2 - |\Lambda|^2 - 2|N|^2\right)/\sqrt{12} \\
&+Y\left(|\Sigma|^2 + |N|^2 - |\Lambda|^2 - 2|\Xi|^2\right)/\sqrt{12}
\end{aligned} \tag{11.25}$$

where we sum over particle types in each isospin representation. Now it is easy to find the contribution to each particle mass by picking out the right tensor coefficient. Adding the common mass M_0 from the very strong interactions, we have

$$\begin{aligned}
M_N &= M_0 - 2X/\sqrt{12} + Y/\sqrt{12} \\
M_\Sigma &= M_0 + X/\sqrt{12} + Y/\sqrt{12} \\
M_\Lambda &= M_0 - X/\sqrt{12} - Y/\sqrt{12} \\
M_\Xi &= M_0 + X/\sqrt{12} - 2Y/\sqrt{12}
\end{aligned} \tag{11.26}$$

Thus we have expressed the four masses in terms of three parameters. We know nothing (from the symmetry) about the values of M_0, X, and Y, but we can eliminate these dynamical quantities and get one relation that follows purely from the symmetry —

$$2(M_N + M_\Xi) = 3M_\Lambda + M_\Sigma \tag{11.27}$$

This is the **Gell-Mann-Okubo formula**. And it works very well. For example, it implies

$$M_\Lambda = \frac{1}{3}(2(M_N + M_\Xi) - M_\Sigma) \tag{11.28}$$

Putting in (in MeV)

$$M_N = 940 \qquad M_\Sigma = 1190 \qquad M_\Xi = 1320 \tag{11.29}$$

gives

$$M_\Lambda = 1110 \qquad \text{(exp. 1115)} \tag{11.30}$$

compared to the experimental value of 1115, the difference is less than 1%. Considering that isospin breaking is bigger than this, it is much better than we have any right to expect.

11.3 Hadron resonances

Particles like the baryons and mesons that participate in the strong interactions are generically called **hadrons**. The baryons and mesons that we have discussed are the lightest hadrons. But there are also an enormous number of excited states of these light states that can be produced in particle collisions but decay back into the light states so quickly that they appear only as enhancements in the scattering cross-section. The first hadron resonance to be discovered was the Δ, which shows up as a very large enhancement in the πP scattering cross-section at about 1230 MeV for angular momentum 3/2. The resonance appears in all the charge states, from $+2$ to -1, so this is a spin 3/2, isospin 3/2 state. It is part of a 10 of $SU(3)$. All the other states in the 10 have now been observed.

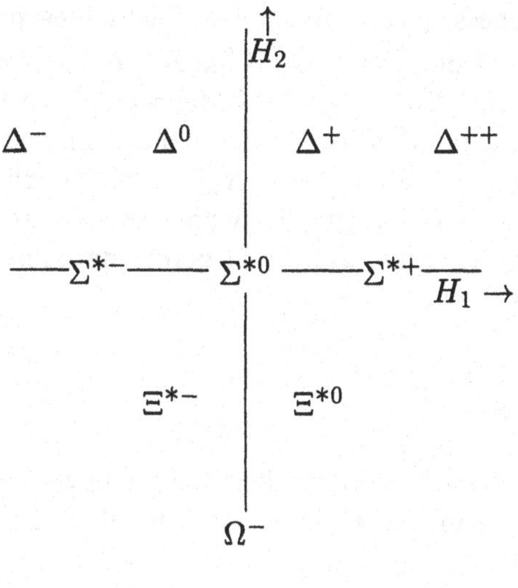

$$\begin{array}{ccccccc} & & & \uparrow H_2 & & & \\ \Delta^- & & \Delta^0 & & \Delta^+ & & \Delta^{++} \\ & \Sigma^{*-} & & \Sigma^{*0} & & \Sigma^{*+} & H_1 \rightarrow \\ & & \Xi^{*-} & & \Xi^{*0} & & \\ & & & \Omega^- & & & \end{array} \tag{11.31}$$

When Gell-Mann first discussed $SU(3)$, the Ω^- had not yet been observed. Gell-Mann was able to predict not only its existence, but also its mass. Let us repeat his calculation.

Again, we assume that H_{MS} transforms like the 8 components of an octet, and compute

$$\langle B^*|H_{MS}|B^*\rangle \tag{11.32}$$

where we have called the decuplet wavefunction B^* and

$$|B^*\rangle = B^{*jk\ell}|_{jk\ell}\rangle \tag{11.33}$$

is the general decuplet state. We have already done the tensor analysis, and in this case, we can immediately write down the result just by thinking. Since there is only one reduced matrix element, the matrix elements of all octet operators are proportional (component by component). Thus the matrix element we want is proportional to the matrix element of the generator, T_8, and thus to the hypercharge, Y. This means that we predict equal spacing for the isospin representations

$$M_{\Sigma^*} - M_\Delta = M_{\Xi^*} - M_{\Sigma^*} = M_{\Omega^-} - M_{\Xi^*} \tag{11.34}$$

Experimentally, in MeV,

$$M_\Delta = 1230 \qquad M_{\Sigma^*} = 1385 \qquad M_{\Xi^*} = 1530 \tag{11.35}$$

The spacings are nearly equal, and the average is about 150, thus we expect the Ω^- at about 1680. Gell-Mann was even able to predict the fate of the Ω^-. With the predicted mass, it could not decay into two lighter particles conserving strangeness and baryon number. The lightest pair of particles with baryon number 1 and strangeness -3 is the K^- and Ξ^0 (or $\overline{K}^0$ and Ξ^-) with a total mass of about 1815. Thus Gell-Mann predicted that the Ω^- would look not like a resonance, but like a weakly decaying particle, decaying into states with strangeness -2, $\Xi\pi$ and ΛK^-. Sure enough, the Ω^- was seen (first in bubble chamber photographs) with a mass of 1672 MeV. This was very convincing evidence that $SU(3)$ is a good approximate symmetry of the strong interactions.

11.4 Quarks

Today, this may seem rather trivial, because we now know that all of these strongly interacting particles are built out of the three light quarks, u, d and

s, transforming like the 3 of $SU(3)$.

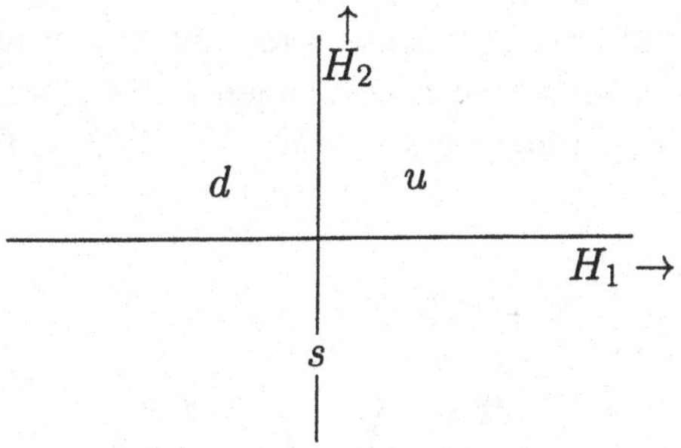

$$\begin{array}{c} \uparrow \\ H_2 \\ d \quad\quad u \\ \hline \quad\quad\quad H_1 \rightarrow \\ s \end{array} \tag{11.36}$$

The baryons can be built of three quarks because

$$3 \otimes 3 \otimes 3 = 10 \oplus 8 \oplus 8 \oplus 1 \tag{11.37}$$

To see this, note that $3 \otimes 3 = 6 \oplus \overline{3}$, so

$$3 \otimes 3 \otimes 3 = (6 \otimes 3) \oplus (\overline{3} \otimes 3) \tag{11.38}$$

and $6 \otimes 3$ looks like

$$\begin{aligned} u^{ij}\, v^k &= \frac{1}{3}\left(u^{ij}\, v^k + u^{ik}\, v^j + u^{kj}\, v^i\right) \\ &+ \frac{1}{3}\left(\epsilon^{ik\ell}\epsilon_{\ell mn}\, u^{mj}\, v^n + \epsilon^{jk\ell}\epsilon_{\ell mn}\, u^{im}\, v^n\right) \end{aligned} \tag{11.39}$$

which is $10 \oplus 8$. Thus both the 10 and the 8 of baryons that we have already seen can be built out of 3 quarks (the 1 also appears, in higher angular momentum states).

The corresponding antiquarks transform like a $\overline{3}$

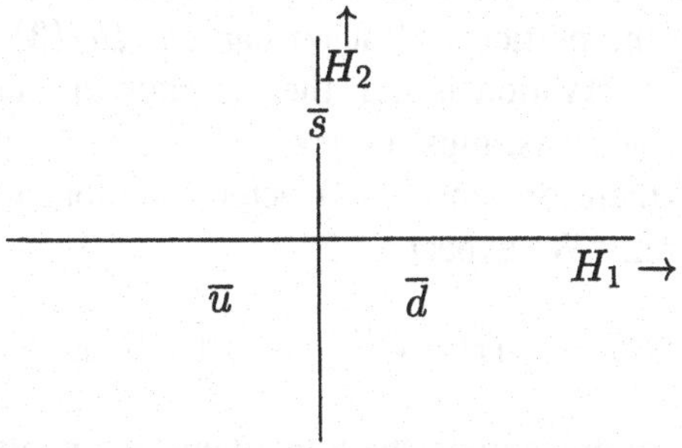

$$\begin{array}{c} \uparrow \\ H_2 \\ \overline{s} \\ \hline \quad\quad\quad H_1 \rightarrow \\ \overline{u} \quad\quad \overline{d} \end{array} \tag{11.40}$$

The mesons are built out of quark plus antiquark. Since $3 \otimes \overline{3} = 8 \oplus 1$, this is either an octet like the π, K, η states we have already seen, or a singlet, like the η' (actually, because of the medium strong interactions, the η and η' mix slightly).

The quarks have spin 1/2, and as we will see later, carry another property, color, that is essential for the understanding of the strong interactions. They also have baryon number 1/3 because three quarks are required to make a baryon. The u and d quarks have zero strangeness, and thus their hypercharge is 1/3. The s quark has strangeness -1, and $Y = -2/3$. For the quarks, the electric charge is

$$\begin{aligned} Q &= T_3 + Y/2 \\ &= \begin{pmatrix} \frac{1}{2} & 0 & 0 \\ 0 & -\frac{1}{2} & 0 \\ 0 & 0 & 0 \end{pmatrix} + \begin{pmatrix} \frac{1}{6} & 0 & 0 \\ 0 & \frac{1}{6} & 0 \\ 0 & 0 & -\frac{1}{3} \end{pmatrix} \\ &= \begin{pmatrix} \frac{2}{3} & 0 & 0 \\ 0 & -\frac{1}{3} & 0 \\ 0 & 0 & -\frac{1}{3} \end{pmatrix} \end{aligned} \tag{11.41}$$

Quarks were originally introduced as a mathematical device, a shorthand for doing $SU(3)$ calculations. Today, though we know that quarks are real and we have a detailed understanding of many aspects of the strong interactions. Still, however, there is much that we cannot calculate. We are often forced to fall back on symmetry arguments.

I can't resist doing one more example of an $SU(3)$ relation from the early days. The octet of spin 1/2 baryons have magnetic moments. Unlike the electron magnetic moment, however, the baryon moments cannot be calculated just from their masses and quantum electrodynamics. They depend on the internal structure of the particles. But we can use $SU(3)$ to say a lot about them. The crucial observation is that the operator that describes the magnetic moment, whatever it is, must be proportional to Q, the electric charge of the quarks. It is therefore an $SU(3)$ octet operator, and we can use the Wigner-Eckart theorem. We expect

$$\mu(B) = \alpha\,\mathrm{Tr}(B\,B^\dagger\,Q) + \beta\,\mathrm{Tr}(B^\dagger\,B\,Q) \tag{11.42}$$

Thus we expect 6 relations among the 8 magnetic moments (there is actually a 9th, because it is a **transition magnetic moment** that is responsible for the electromagnetic decay, $\Sigma^0 \to \Lambda\gamma$). In fact, all the magnetic moments can be calculated in terms of $\mu(P)$ and $\mu(N)$. These predictions were first worked out by Sidney Coleman and Shelly Glashow in 1961.

Problems

11.A. What would the Gell-Mann-Okubo argument tell you about the masses of particles transforming like a 6 of $SU(3)$?

11.B. Compare the probability for Δ^+ production in $\pi^0 P \to \Delta^+$ with the probability for Σ^{*0} production in $K^- P \to \Sigma^{*0}$, assuming $SU(3)$ symmetry of the S-matrix.

11.C. Use the $SU(3)$ argument discussed in the chapter to repeat the calculation of Coleman and Glashow, predicting all the spin 1/2 baryon magnetic moments in terms of $\mu(P)$ and $\mu(N)$.

Chapter 12

Young Tableaux

We discussed Young tableaux in connection with the irreducible representations of the symmetric groups. We will now see that they are useful for dealing with irreducible representations of Lie groups. We will begin by discussing this for $SU(3)$, but the real advantage is that it generalizes to $SU(n)$.

12.1 Raising the indices

The crucial observation is that the $\overline{3}$ representation is an antisymmetric combination of two 3s, so we really do not need the second fundamental representation to construct higher representations. We can write an arbitrary representation as a tensor product of 3s with appropriate symmetry. In fact, as we have seen, Young tableaux correspond to irreducible representations of the permutation group, and the connection with the irreducible representations of $SU(3)$ (and $SU(N)$, as we will see later) is that the irreducible representations of $SU(3)$ transform irreducibly under permutation of the labels of the indices.

Consider a general representation, (n,m). It is a tensor (in the old language) with components

$$A^{i_1\cdots i_n}_{j_1\cdots j_m} \tag{12.1}$$

separately symmetric in upper and lower indices, and traceless. We can raise all the lower indices with ϵ tensors to get

$$\begin{aligned} &a^{i_1\cdots i_n k_1\ell_1\cdots k_m\ell_m} \\ &= \epsilon^{j_1k_1\ell_1}\ldots\epsilon^{j_mk_m\ell_m}A^{i_1\cdots i_n}_{j_1\cdots j_m} \end{aligned} \tag{12.2}$$

Clearly, it is antisymmetric in each pair, $k_i \leftrightarrow \ell_i$, and symmetric in the exchange of pairs $k_i,\ell_i \leftrightarrow k_j\ell_j$.

DOI: 10.1201/9780429499210-13

Now for each such tensor, we can associate a Young tableau:

$$\begin{array}{|c|c|c|c|c|c|}\hline k_1 & \cdots & k_m & i_1 & \cdots & i_n \\ \hline \ell_1 & \cdots & \ell_m \\ \cline{1-3}\end{array} \tag{12.3}$$

What we would like to do is to find a rule that associates with the Young tableau the specific symmetry of the tensor (12.2). We can do that by thinking about the highest weight of the representation, (n, m). Because the lowering operators preserve the symmetry, if we find the symmetry of the tensor components describing the highest weight, all the states will have that symmetry. The highest weight is associated with the components in which all the is are 1, and all the k, ℓ pairs are 1,3. All of these can be obtained by antisymmetrizing the k, ℓ pairs from the component in which all the ks are 1, and all the ℓs are 3. But this one component is symmetric under arbitrary permutations of the is and ks, and separately symmetric under permutations of the ℓs. Thus we will obtain a tensor with the right symmetry if we start with an arbitrary tensor with $n+2m$ components, and first symmetrize all the is and ks, and separately the ℓs, and then antisymmetrize in every k, ℓ pair. In the Young tableau language, this is very easy to state. We first symmetrize in the components in the rows, then antisymmetrize in the components in the columns. The result is symmetric in the is and in the k, ℓ pairs as (12.2) must be. But it also has a property that is the analog of tracelessness. Because we have raised the indices with ϵs, the condition of tracelessness becomes

$$\epsilon_{i_1 k_1 \ell_1} \, a^{i_1 \cdots i_n k_1 \ell_1 \cdots k_m \ell_m} = 0 \tag{12.4}$$

This vanishes for a tensor with the symmetry properties just described because of the symmetrization of the components in the rows.

Thus a Young tableau like the one above is a rule for symmetrizing a tensor to project out a specific irreducible representation. For example, if $a^{j_1 j_2 k_1}$ is a general tensor with three upper indices, but no special symmetry property, the Young diagram

$$\begin{array}{|c|c|}\hline j_1 & j_2 \\ \hline k_1 \\ \cline{1-1}\end{array} \tag{12.5}$$

produces the tensor

$$\begin{aligned} &a^{j_1 j_2 k_1} + a^{j_2 j_1 k_1} \\ &-a^{k_1 j_2 j_1} - a^{j_2 k_1 j_1} \end{aligned} \tag{12.6}$$

which transforms according to the (1,1) (or adjoint) representation.

We can generalize the concept to Young tableau with more rows. The general rule is the same. Put indices in the boxes. Symmetrize in the indices in the rows. Then antisymmetrize in the indices in the columns.

In $SU(3)$, the tensors corresponding to Young tableaux with more than three boxes in any column vanish because no tensor can be completely antisymmetric in four or more indices which take on only three values. Any column with three boxes corresponds to a factor of ϵ in the three indices —

$$\epsilon^{ijk} = \begin{array}{|c|}\hline i \\ \hline j \\ \hline k \\ \hline \end{array} \tag{12.7}$$

So tableaux of the form

$$\begin{array}{|c|c|c|c|c|c|c|c|c|}\hline & \cdots & & k_1 & \cdots & k_m & i_1 & \cdots & i_n \\ \hline & \cdots & & \ell_1 & \cdots & \ell_m \\ \cline{1-6} & \cdots & \\ \cline{1-3} \end{array} \tag{12.8}$$

describe the same representation as (12.3).

12.2 Clebsch-Gordan decomposition

We can now give, without proof, an algorithm for the Clebsch-Gordan decomposition of a tensor product. To decompose the tensor product of irreducible representations α and β corresponding to tableaux A and B, you build onto A using the boxes of B in the following way. Begin by putting as in the top row of B and bs in the second row. Take the boxes from the top row of B and put them on A, building to the right and/or down, to form legal tableau (that is collections of boxes in which the numbers of boxes in the rows are not increasing as you go down, and the numbers of boxes in the columns are not increasing as you go the right), with no two as in the same column. Then take the second row and add the boxes to each of the resulting tableau to form legal tableaux with one further condition. Reading along the rows from right to left from the top row down to the bottom row, the number of as must be greater than or equal to the number of bs. This avoids double counting. The tableaux produced by this construction correspond to the irreducible representations in $\alpha \otimes \beta$.

Examples:

$$\begin{array}{|c|}\hline \\ \hline\end{array} \otimes \begin{array}{|c|}\hline a \\ \hline\end{array} = \begin{array}{|c|c|}\hline & a \\ \hline\end{array} \oplus \begin{array}{|c|}\hline \\ \hline a \\ \hline\end{array} \tag{12.9}$$

$$3 \otimes 3 = 6 \oplus \overline{3}$$

$$\begin{array}{|c|}\hline \\ \hline \\ \hline\end{array} \otimes \begin{array}{|c|}\hline a \\ \hline\end{array} = \begin{array}{|c|c|}\hline & a \\ \hline & \multicolumn{1}{c}{} \\ \cline{1-1}\end{array} \oplus \begin{array}{|c|}\hline \\ \hline \\ \hline a \\ \hline\end{array} \tag{12.10}$$

$$\overline{3} \otimes 3 = 8 \oplus 1$$

It is less trivial to do this one the other way.

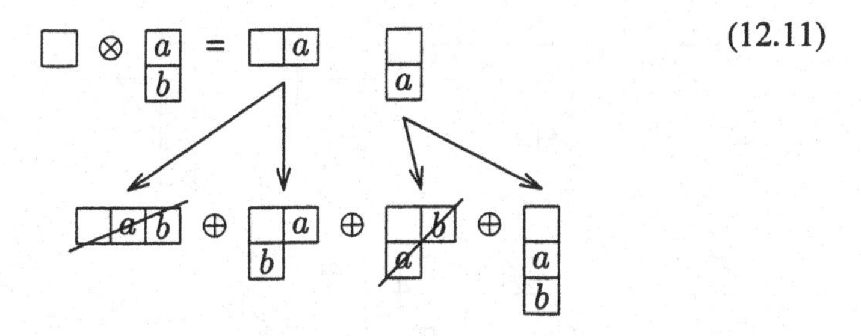

The tableaux that are crossed out do not satisfy the constraint that the number of as is greater than or equal to the number of bs. Needless to say, it is easier to do it the other way, because you have fewer boxes to move around.

Sometimes the tableaux that are produced in the first stage are useless, because there is no possible second state. Here's an example:

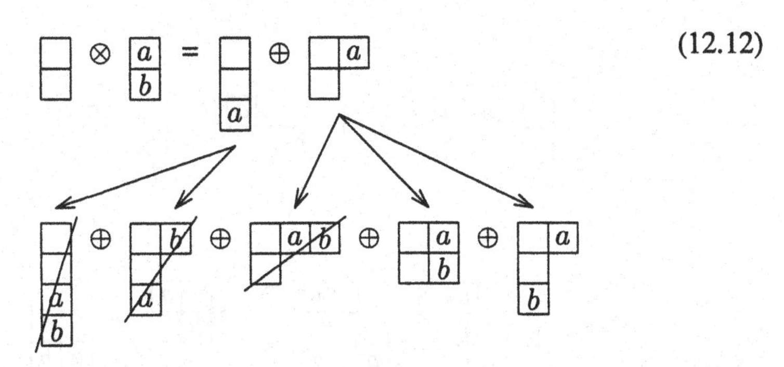

Of course we would actually never do it this way. This is the decomposition of $\overline{3} \otimes \overline{3} = \overline{6} \oplus 3$. This is just the complex conjugate of $3 \otimes 3 = 6 \oplus \overline{3}$ which we have already done, and which was much easier because it involved fewer boxes. That's a useful lesson.

Finally, here is a useful example which illustrates all of the rules — $8 \otimes 8$

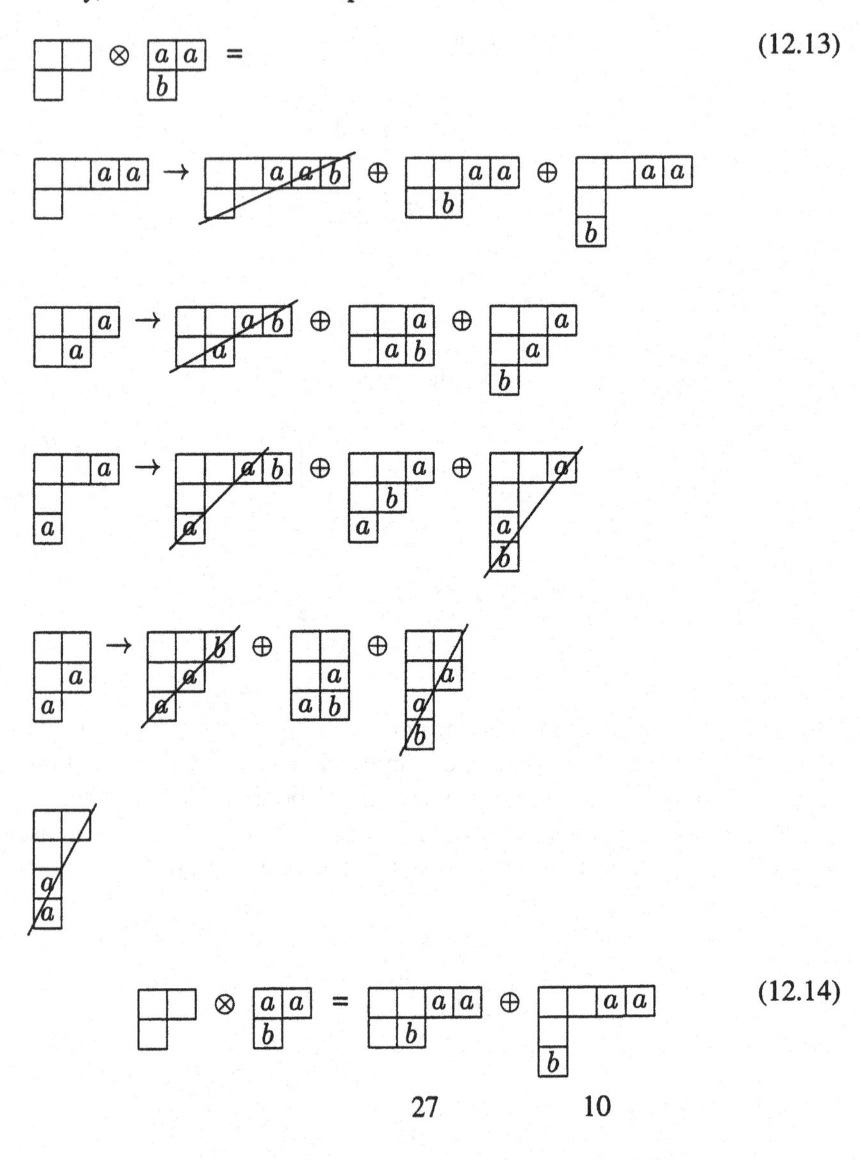

This analysis not only illustrates all the rules, it suggests one possible systematic way of ordering the calculation, in this case by putting things as far to the right as possible to start with, and then working to the left. It is important to have some such rule for your own calculations, so that you don't miss possibilities.

12.3 $SU(3) \to SU(2) \times U(1)$

Young tableaux can do many other useful things for us. One that we can already explore in $SU(3)$ is to help us understand how a representation of $SU(3)$ decomposes into representations of its subgroups. An example of this which has some phenomenological importance is the decomposition of $SU(3)$ representations into their representations of $SU(2) \times U(1)$, and particularly, of isospin and hypercharge (T_8, remember, is just $\sqrt{3}\,Y/2$).

The three components of the defining representation, the 3, decompose into a doublet with hypercharge 1/3 and a singlet with hypercharge −2/3. In general, consider a Young tableau with n boxes and look at the components in which j indices transform like doublets and $n-j$ transform like singlets. The total hypercharge will be

$$\frac{j}{3} - \frac{2(n-j)}{3} = -\frac{2n}{3} + j \tag{12.15}$$

We will denote the $n-j$ singlet components by a Young tableau of $n-j$ boxes in one row. This is the only valid Young tableau that we can build for these, because they all have the same index, and thus cannot appear in the same column.

The $SU(2)$ representations that these components transform under will be some subset of the representations that can be built out of j boxes. Now here is the point. To determine whether a given $SU(2)$ representations, α, actually appears, what we do is to compute the tensor product of α with the $n-j$ horizontal boxes. Then the number of times the original representation of $SU(3)$ appears in the tensor product is the number of times α appears in the decomposition. We will give a justification of this procedure later, when we generalize it. But now let us work out some examples. We will use a notation in which we refer to the $SU(2)$ representation by their dimension, and put a subscript to indicate the hypercharge. Thus $(2I+1)_y$ stands for an isospin I representation with hypercharge y.

First consider the 6. The result is illustrated in the following figure (where the $\bullet$ represents the trivial tableau, with no boxes).

$$\square\!\square \rightarrow \tag{12.16}$$

$$\left(\square\!\square \quad \cdot\right) \qquad 3_{2/3}$$

$$\left(\square \quad \square\right) \qquad 2_{-1/3}$$

$$\left(\cdot \quad \square\!\square\right) \qquad 1_{-4/3}$$

Each representation appears once.

Now look at the $\overline{3}$. The answer here is obvious, because this is the complex conjugate of the 3. but let us see what it looks like in this language.

$$\begin{array}{c}\square\\[-4pt]\square\end{array} \rightarrow \tag{12.17}$$

$$\left(\begin{array}{c}\square\\[-4pt]\square\end{array} \quad \cdot\right) \qquad 1_{2/3}$$

$$\left(\square \quad \square\right) \qquad 2_{-1/3}$$

Again, each representation appears once. Note that we cannot have a column of boxes on the right hand side of the ordered pairs, because this would represent an antisymmetric combination of states with only one index, which vanishes.

Next consider the adjoint representation, where we already know the answer.

$$\begin{array}{ll}\square\!\!&\!\!\square\\[-4pt]\square\!\!&\end{array} \rightarrow \tag{12.18}$$

$$\left(\begin{array}{ll}\square\!\!&\!\!\square\\[-4pt]\square\!\!&\end{array} \quad \cdot\right) \qquad 2_1$$

$$\left(\begin{array}{c}\square\\[-4pt]\square\end{array} \quad \square\right) \qquad 1_0$$

$$\left(\square\!\square \quad \square\right) \qquad 3_0$$

$$\left(\square \quad \square\!\square\right) \qquad 2_{-1}$$

Each representation appears once.

Now let's do a less trivial one, for the 27 —

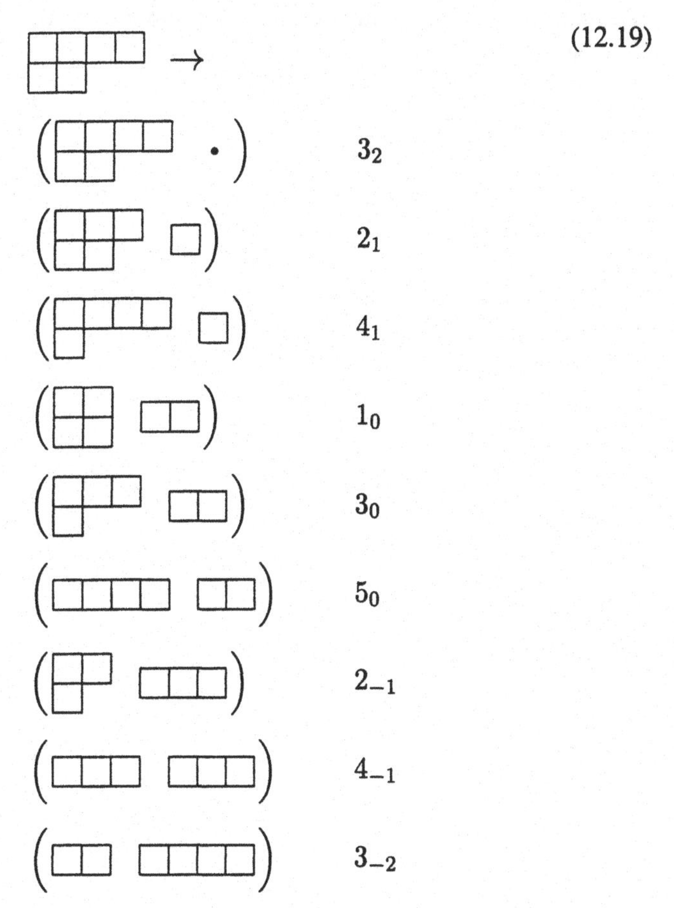

Again, each representation appears once.

Problems

12.A. Find $(2,1) \otimes (2,1)$. Can you determine which representations appear antisymmetrically in the tensor product, and which appear symmetrically.

12.B. Find $10 \otimes 8$.

12.C. For any Lie group, the tensor product of the adjoint representation with any arbitrary nontrivial representation D must contain D (think about the action of the generators on the states of D and see if you can figure out why this is so). In particular, you know that for any nontrivial $SU(3)$ representation D, $D \otimes 8$ must contain D. How can you see this using Young Tableaux?

Chapter 13

$SU(N)$

We now want to generalize the discussion of the last few chapters to $SU(N)$, the group of special unitary $N \times N$ matrices, generated by the hermitian, traceless, $N \times N$ matrices.

13.1 Generalized Gell-Mann matrices

There are a couple of different useful bases for the $SU(N)$ generators. We will start with a generalization of the Gell-Mann matrices, in which we build up from the $SU(N)$ to $SU(N+1)$ generators one step at a time. We will normalize (as in $SU(2)$ and $SU(3)$)

$$\mathrm{Tr}(T_a T_b) = \frac{1}{2}\,\delta_{ab} \tag{13.1}$$

The generators of the raising and lowering operators, we can take to have a single non-zero off-diagonal element, $1/\sqrt{2}$. The group is rank $N-1$, because there are $N-1$ independent traceless diagonal real matrices. We can choose the $N-1$ Cartan generators as follows

$$[H_m]_{ij} = \frac{1}{\sqrt{2m(m+1)}} \left(\sum_{k=1}^{m} \delta_{ik}\delta_{jk} - m\,\delta_{i,m+1}\,\delta_{j,m+1} \right) \tag{13.2}$$

 DOI: 10.1201/9780429499210-14

For example

$$H_1 = \frac{1}{2}\begin{pmatrix} 1 & 0 & \cdots \\ 0 & -1 & \cdots \\ \vdots & \vdots & \ddots \end{pmatrix} \qquad H_2 = \frac{1}{\sqrt{12}}\begin{pmatrix} 1 & 0 & 0 & \cdots \\ 0 & 1 & 0 & \cdots \\ 0 & 0 & -2 & \cdots \\ \vdots & \vdots & \vdots & \ddots \end{pmatrix}$$

$$H_3 = \frac{1}{\sqrt{24}}\begin{pmatrix} 1 & 0 & 0 & 0 & \cdots \\ 0 & 1 & 0 & 0 & \cdots \\ 0 & 0 & 1 & 0 & \cdots \\ 0 & 0 & 0 & -3 & \cdots \\ \vdots & \vdots & \vdots & \vdots & \ddots \end{pmatrix} \tag{13.3}$$

and so on.

Altogether, there are $N^2 - 1$ independent traceless hermitian matrices. These $N^2 - 1$ matrices generate the N dimensional defining representation of $SU(N)$, which we will sometimes call the N. The weights are $N - 1$ dimensional vectors,

$$[\nu^j]_m = [H_m]_{jj} = \frac{1}{\sqrt{2m(m+1)}}\left(\sum_{k=1}^{m} \delta_{jk} - m\,\delta_{j,m+1}\right) \tag{13.4}$$

These satisfy,

$$\begin{aligned}
{\nu_j}^2 &= \sum_{m=1}^{N-1} \frac{1}{2m(m+1)}\left(\sum_{k=1}^{m} \delta_{jk} - m\,\delta_{j,m+1}\right)^2 \\
&= \sum_{m=1}^{N-1} \frac{1}{2m(m+1)}\left(\sum_{k=1}^{m} \delta_{jk} + m^2\,\delta_{j,m+1}\right) \\
&= \sum_{m=j}^{N-1} \frac{1}{2m(m+1)} + \frac{(j-1)^2}{2j(j-1)} \\
&= \frac{1}{2}\sum_{m=j}^{N-1}\left(\frac{1}{m} - \frac{1}{(m+1)}\right) + \frac{(j-1)}{2j} \\
&= \frac{1}{2j} - \frac{1}{2N} + \frac{j-1}{2j} = \frac{N-1}{2N}
\end{aligned} \tag{13.5}$$

and for $i < j$,

$$
\begin{aligned}
&\nu^i \cdot \nu^j \\
&= \sum_{m=1}^{N-1} \frac{1}{2m(m+1)} \left(\sum_{k=1}^{m} \delta_{ik} - \cancel{m\,\delta_{i,m+1}} \right) \\
&\times \left(\sum_{k=1}^{m} \delta_{jk} - m\,\delta_{j,m+1} \right) \\
&= \sum_{m=j}^{N-1} \frac{1}{2m(m+1)} - \frac{(j-1)}{2j(j-1)} \\
&= \frac{1}{2j} - \frac{1}{2N} - \frac{1}{2j} = -\frac{1}{2N}
\end{aligned}
\tag{13.6}
$$

The weights all have the same length, and the angles between any two distinct weights are equal —

$$
\left|\nu^i\right|^2 = \frac{N-1}{2N} \qquad \nu^i \cdot \nu^j = -\frac{1}{2N} \;\text{ for }\; i \neq j \tag{13.7}
$$

or

$$
\nu^i \cdot \nu^j = -\frac{1}{2N} + \frac{1}{2}\,\delta_{ij} \tag{13.8}
$$

Thus the weights form a regular figure in $N-1$ dimensional space, the $N-1$ simplex.

Explicitly, the weights look like

$$
\begin{aligned}
\nu^1 &= \left(\frac{1}{2}, \frac{1}{2\sqrt{3}}, \cdots, \frac{1}{\sqrt{2m(m+1)}}, \cdots, \frac{1}{\sqrt{2(N-1)N}} \right) \\
\nu^2 &= \left(-\frac{1}{2}, \frac{1}{2\sqrt{3}}, \cdots, \frac{1}{\sqrt{2m(m+1)}}, \cdots, \frac{1}{\sqrt{2(N-1)N}} \right) \\
\nu^3 &= \left(0, -\frac{1}{\sqrt{3}}, \cdots, \frac{1}{\sqrt{2m(m+1)}}, \cdots, \frac{1}{\sqrt{2(N-1)N}} \right) \\
&\cdots \\
\nu^{m+1} &= \left(0, 0, \cdots, -\frac{m}{\sqrt{2m(m+1)}}, \cdots, \frac{1}{\sqrt{2(N-1)N}} \right) \\
&\cdots \\
\nu^N &= \left(0, 0, \cdots, 0, \cdots, -\frac{N-1}{\sqrt{2(N-1)N}} \right)
\end{aligned}
\tag{13.9}
$$

For convenience, we will choose a backwards convention for positivity —a positive weight is one in which the LAST non-zero component is positive. With this definition, the weights satisfy

$$\nu^1 > \nu^2 \cdots > \nu^{N-1} > \nu^N \tag{13.10}$$

The Es take us from one weight to another, so the roots are differences of weights, $\nu^i - \nu^j$ for $i \neq j$. The positive roots are $\nu^i - \nu^j$ for $i < j$. The simple roots are

$$\alpha^i = \nu^i - \nu^{i+1} \text{ for } i = 1 \text{ to } N-1 \tag{13.11}$$

The roots all have length 1. They satisfy using (13.8)

$$\begin{aligned} \alpha^i \cdot \alpha^j &= \frac{1}{2}\left(\delta_{ij} - \delta_{i+1,j} - \delta_{i,j+1} + \delta_{i+1,j+1}\right) \\ &= \delta_{ij} - \frac{1}{2}\,\delta_{i,j\pm 1} \end{aligned} \tag{13.12}$$

so the Dynkin diagram is

$$\bigcirc\!\!-\!\!\bigcirc \cdots \bigcirc\!\!-\!\!\bigcirc \qquad \alpha^1 \quad \alpha^2 \quad \alpha^{N-2} \; \alpha^{N-1} \tag{13.13}$$

The fundamental weights are

$$\mu^j = \sum_{k=1}^{j} \nu^k \tag{13.14}$$

It is easy to check using (13.7) that these satisfy

$$\frac{2\alpha^i \cdot \mu^j}{\alpha^{i2}} = \delta_{ij} \tag{13.15}$$

because the term proportional to $\frac{1}{2N}$ in (13.8) cancels because the simple root is a difference of fundamental weights. Thus μ^1 is the highest weight of the defining representation.

13.2 $SU(N)$ tensors

As in $SU(3)$, we can associate states with tensors,

$$|\nu^i\rangle \rightarrow |_i\rangle \tag{13.16}$$

then we can build up arbitrary representations as tensor products.

Consider, for example, the antisymmetric combination of m defining representations. The states are

$$A^{[i_1 \cdots i_m]} |i_1 \cdots i_m\rangle \tag{13.17}$$

where $A^{[i_1 \cdots i_m]}$ is completely antisymmetric. This set of states forms an irreducible representation, because of the antisymmetry. Because of the antisymmetry, no two indices can take on the same value. Thus the highest weight in this set of states arises when one of the indices is 1, another is 2, and so on. That means that the highest weight is the fundamental weight μ^m

$$\sum_{k=1}^{m} \nu^k = \mu^m \tag{13.18}$$

Thus this representation is the fundamental representation D^m.

The highest weight of any irreducible representation can be written in terms of the Dynkin coefficients, q^k, as

$$\mu = \sum_k q_k \, \mu^k \tag{13.19}$$

The q^ks are non-negative integers. The tensor associated with this representation has, for each k from 1 to $N-1$, q_k sets of k indices that are antisymmetric within each set. A simple generalization of the argument as for $SU(3)$ shows that symmetry of this tensor can be obtained from the following Young tableau, with q^k columns of k boxes

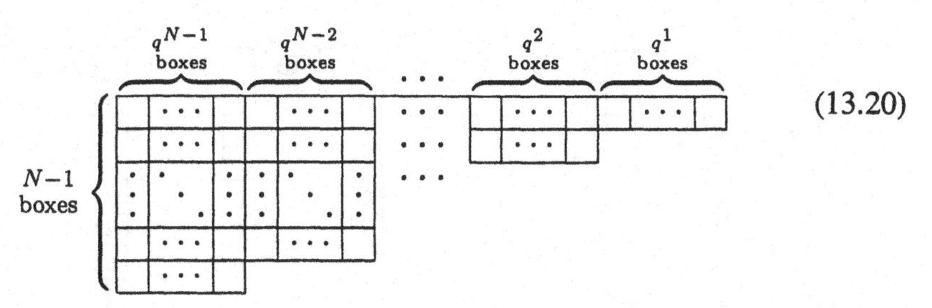

(13.20)

This gives a tensor of the right form for the same reason as it did in $SU(3)$. The highest weight state has a term in which the top row is all 1s, the second is all 2s, the third is all 3s, etc. Thus the tensor we want is obtained by first symmetrizing in the indices in the rows, and then antisymmetrizing in the indices in the columns. This is exactly the symmetrization condition that we used at the end of chapter I to construct the irreducible representations of the permutation groups. Thus the irreducible representations of $SU(N)$ with m indices are associated with the irreducible representations of S_m.

As in $SU(3)$, tableaux with more than N boxes in any column correspond to tensors which vanish identically. Any columns with N boxes contribute a factor of the completely antisymmetric tensor with N indices (which we will again call ϵ). Tableaux which are the same except for columns with N boxes correspond to the same irreducible representation.

We will sometimes denote the representation corresponding to a Young tableau by giving the number of boxes in each column of the tableau, a series of non-increasing integers, $[\ell_1, \ell_2, \cdots]$. In this notation, D^j is $[j]$.

For example the adjoint representation, in $SU(N)$ as in $SU(3)$, corresponds to a tensor with one upper and one lower index. To get it into the standard form of a tensor, we must raise the lower index with an N component ϵ. Thus the adjoint is $[N-1, 1]$.

Clebsch-Gordan decomposition works the same way as for $SU(3)$, except that now we need cs for the third row of the tableau, ds for the fourth, etc. Along with the condition that reading along the rows from right to left from the top row down to the bottom row, the number of as must be greater than or equal to the number of bs which is greater than or equal to the number of cs, etc.

For example,

$$\begin{array}{|c|}\hline \ \\ \hline \ \\ \hline\end{array} \otimes \begin{array}{|c|}\hline a \\ \hline\end{array} = \begin{array}{|c|c|}\hline \ & a \\ \hline \ & \multicolumn{1}{c}{} \\ \cline{1-1}\end{array} \oplus \begin{array}{|c|}\hline \ \\ \hline \ \\ \hline a \\ \hline\end{array} \tag{13.21}$$

$$[2] \otimes [1] = [2,1] \oplus [3]$$

or

$$\begin{array}{|c|}\hline \ \\ \hline \ \\ \hline \ \\ \hline \ \\ \hline\end{array} \otimes \begin{array}{|c|}\hline a \\ \hline b \\ \hline c \\ \hline d \\ \hline\end{array} = \begin{array}{|c|c|}\hline \ & a \\ \hline \ & b \\ \hline \ & c \\ \hline \ & d \\ \hline\end{array} \oplus \begin{array}{|c|c|}\hline \ & a \\ \hline \ & b \\ \hline \ & c \\ \hline \ \\ \cline{1-1} d \\ \cline{1-1}\end{array} \oplus \begin{array}{|c|c|}\hline \ & a \\ \hline \ & b \\ \hline \ \\ \cline{1-1} \ \\ \cline{1-1} c \\ \cline{1-1} d \\ \cline{1-1}\end{array} \oplus \begin{array}{|c|c|}\hline \ & a \\ \hline \ \\ \cline{1-1} \ \\ \cline{1-1} \ \\ \cline{1-1} b \\ \cline{1-1} c \\ \cline{1-1} d \\ \cline{1-1}\end{array} \oplus \begin{array}{|c|}\hline \ \\ \hline \ \\ \hline \ \\ \hline \ \\ \hline a \\ \hline b \\ \hline c \\ \hline d \\ \hline\end{array} \tag{13.22}$$

$$[4] \otimes [4] = [4,4] \oplus [5,3] \oplus [6,2] \oplus [7,1] \oplus [8]$$

Notice that it is the rule about number of as being greater than or equal to the number of bs and so on, that prevents us from having more than one of any of these representations. Of course, if N is less that eight, some of the tableaux in the last example will give vanishing tensors, while others will have columns of N boxes that can be eliminated without changing the representation.

Note, in general, in the Clebsch-Gordan decomposition process that the only time boxes can disappear is when columns of N boxes are removed. Thus each tableaux in the decomposition of a tableau with j boxes $\otimes$ a tableau with k boxes will have a number of boxes equal to $j+k$ modulo N. This is the analog of the conservation of **triality** in $SU(3)$. The quantity $j+k \mod N$ in $SU(N)$ is sometimes denoted by the linguistic barbarism N**-ality**.

13.3 Dimensions

For most of the small representations that we will need, it is easy to work out the dimension. For example, the dimension of $[j]$ is

$$\binom{N}{j} = \frac{N!}{j!(N-j)!} \tag{13.23}$$

The dimensions of the two-column representations can be worked out using the Clebsch-Gordan series. For example, from $[1] \otimes [2] = [2,1] \oplus [3]$, we can find the dimension of [2,1].

$$\binom{N}{2} \times \binom{N}{1} - \binom{N}{3} = \frac{N(N+1)(N-1)}{3} \tag{13.24}$$

There is, however, a simple rule for obtaining the dimensions of any representation from its tableau. It is called the **factors over hooks** rule. It is a special case of Weyl's character formula. It works as follows. Put an N in the upper left hand corner of the tableau. Then put factors in all the other boxes, by adding 1 each time you move to the right, and subtracting 1 each time you move down. The product of all these factors is F. A hook is a line passing vertically up through the bottom of some column of boxes, making a right hand turn in some box and passing out through the row of boxes. There is one hook for each box. Call the number of boxes the hook passes through h. Then if H is the product of the hs for all hooks (the same factor we used in (1.164), the dimension of the representation is

$$F/H \tag{13.25}$$

For example, for [2,1], the factors are

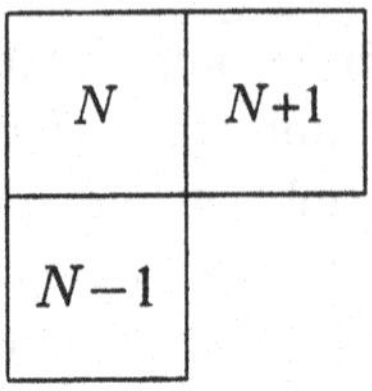

$$\begin{array}{|c|c|}\hline N & N+1 \\ \hline N-1 & \\ \cline{1-1}\end{array} \qquad F = N(N+1)(N-1) \tag{13.26}$$

The hooks are

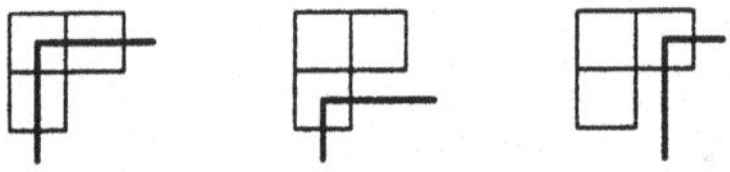

$$h = 3 \qquad h = 1 \qquad h = 1 \tag{13.27}$$

so $H = 3$. Thus

$$\frac{F}{H} = \frac{N(N+1)(N-1)}{3} \tag{13.28}$$

in agreement with (13.24).

13.4 Complex representations

Most of the representations of $SU(N)$ are complex. For example, the lowest weight of the defining representation is ν^N. But we know because the Cartan generators are traceless that

$$\sum_{j=1}^{N} \nu^j = 0 \tag{13.29}$$

and thus

$$\nu^N = -\sum_{j=1}^{N-1} \nu^j = -\mu^{N-1} \tag{13.30}$$

Thus [1] is complex, and its complex conjugate is D^{N-1} or $[N-1]$,

$$\overline{[1]} = [N-1] \tag{13.31}$$

Simlarly, the lowest weight of $[m]$ is the sum of the m smallest ν^is

$$\sum_{j=N-m+1}^{N} \nu^j = -\sum_{j=1}^{N-m} \nu^j = -\mu^{N-m} \tag{13.32}$$

and thus

$$\overline{[m]} = [N-m] \tag{13.33}$$

In general, the complex conjugate of a representation is

$$\overline{[\ell_1, \cdots, \ell_n]} = [N - \ell_n, \cdots, N - \ell_1] \tag{13.34}$$

Thus the tableau corresponding to a representation and its complex conjugate fit together into a rectangle N boxes high, as shown below:

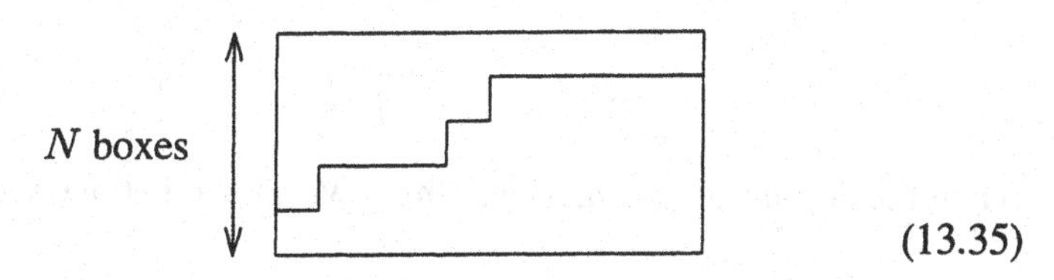

(13.35)

13.5 $SU(N) \otimes SU(M) \in SU(N+M)$

Now we can generalize the discussion of $SU(2) \times U(1) \in SU(3)$ from the last chapter. Consider the $SU(N) \otimes SU(M) \otimes U(1)$ subgroup of $SU(N+M)$ in which the $SU(N)$ acts on the first N indices and the $SU(M)$ acts on the last M indices. Both of these subgroups commute with a $U(1)$ which we can take to be M on the first N indices and $-N$ on the last M (note that it is traceless).

$$\square = (\square \;\; \cdot \;)_M \oplus (\; \cdot \;\; \square\,)_{-N} \tag{13.36}$$

We would like to determine how an arbitrary irreducible representation of $SU(N+M)$ decomposes into irreducible representations of $SU(N) \otimes SU(M) \otimes U(1)$. Suppose that we have an $SU(N+M)$ representation with $n+m$ boxes. To compute the number of times that the product of an $SU(N)$ representation A with n boxes and a $SU(M)$ representation B with m boxes appears in the $SU(N+M)$ representation when we restrict n indices to be from 1 to N and m to be from $N+1$ to $N+M$, we relax that condition and let all the indices run over all $N+M$ values, and find the tensor product $A \otimes B$ of the two representations (as representations of $SU(N+M)$). Then the number of times that the pair (A, B) appears in the decomposition is equal to the number of times the original $SU(N+M)$ representation appears in the tensor product $A \otimes B$. Note also that the $U(1)$ charge of the representation is $nM - mN$.

Notice that if we take $M = 1$, the $SU(M)$ group disappears, because there is no algebra $SU(1)$ — it has no generators. However, the construction described above still works. This is what we used in the previous chapter to do the decomposition of $SU(3)$ representations under the $SU(2) \times U(1)$ subgroup in (12.18) and (12.19).

Let us do the example $SU(3) \otimes SU(2) \otimes U(1) \in SU(5)$, where the defining representation is

$$\begin{aligned} &\square && (13.37)\\ &= (\square \quad \cdot\,)_2 \qquad (3,1)_2 \\ &\oplus (\,\cdot \quad \square)_{-3} \qquad (1,2)_{-3} \end{aligned}$$

Then the adjoint representation, which we showed above was $[N-1,1]$, is

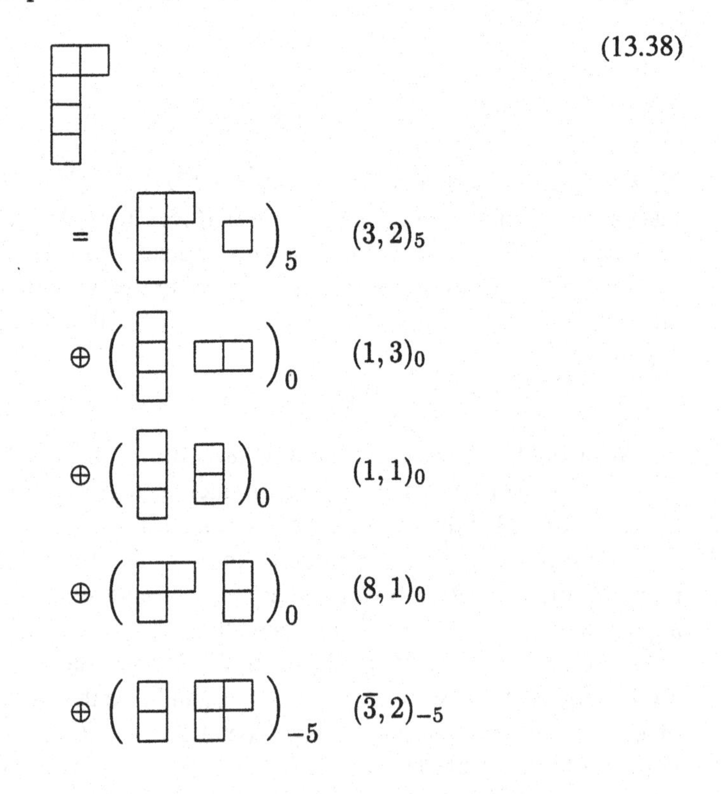

Problems

13.A. Show that the $SU(n)$ algebra has an $SU(n-1)$ subalgebra. How do the fundamental representations of $SU(n)$ decompose into $SU(n-1)$ representations?

13.B. Find $[3] \otimes [1]$ in $SU(5)$. Check that the dimensions work out.

13.C. Find [3,1]⊗[2,1]

13.D. Under the subalgebra $SU(N) \otimes SU(M) \otimes U(1) \in SU(N+M)$, where the defining representation $N+M$ transforms as $([1],[0])_M \oplus ([0],[1])_{-N}$ (in the $[\ell_1, \cdots]$ notation where the ℓs are the number of boxes in the columns of the Young tableaux, and where the first bracket indicates the $SU(N)$ representation and the second the $SU(M)$), how do the fundamental representations of $SU(N+M)$ transform? How about the adjoint representation?

13.E. Find $[2] \otimes [1,1]$ in $SU(N)$, and use the factors over hooks rule to check that the dimensions work out for arbitary N.

Chapter 14

3-D Harmonic Oscillator

This is an important chapter, but not because the three dimensional harmonic oscillator is a particularly important physical system. It is, however, a beautiful illustration of how $SU(N)$ symmetries arise in quantum mechanics.

14.1 Raising and lowering operators

The Hamiltonian (here as elsewhere we have set $\hbar = 1$) is

$$\begin{aligned} H &= \frac{1}{2m}\vec{p}^{\,2} + \frac{m\omega^2}{2}\vec{r}^{\,2} \\ &= \omega\left(a_k^\dagger a_k + 3/2\right) \end{aligned} \tag{14.1}$$

where

$$\begin{aligned} a_k &= \frac{1}{\sqrt{2m\omega}}\left(m\omega\, r_k + i\, p_k\right) \\ a_k^\dagger &= \frac{1}{\sqrt{2m\omega}}\left(m\omega\, r_k - i\, p_k\right) \end{aligned} \tag{14.2}$$

for $k = 1$ to 3. The a_k and $a_k^\dagger$ are lowering and raising operators satisfying the commutation relations of annihilation and creation operators for bosons,

$$\begin{aligned} \left[a_k, a_\ell^\dagger\right] &= \delta_{k\ell} \\ \left[a_k, a_\ell\right] &= \left[a_k^\dagger, a_\ell^\dagger\right] = 0 \\ \left[a_k^\dagger a_k, a_\ell^\dagger\right] &= a_\ell^\dagger \\ \left[a_k^\dagger a_k, a_\ell\right] &= -a_\ell \end{aligned} \tag{14.3}$$

DOI: 10.1201/9780429499210-15

If the ground state is $|0\rangle$, satisfying

$$a_k|0\rangle = 0 \tag{14.4}$$

then the energy eigenstates are

$$a^\dagger_{k_1} \cdots a^\dagger_{k_n}|0\rangle \tag{14.5}$$

with energy

$$\omega\,(n+3/2) \tag{14.6}$$

The degeneracy of these states is interesting — it is the number of symmetric combinations of the n indices, $k_1 \cdots k_n$, which is

$$\frac{(n+1)(n+2)}{2} \tag{14.7}$$

Just the dimension of the $(n,0)$ representation of $SU(3)$. This suggests that the model has an $SU(3)$ symmetry.

It is not hard to find the $SU(3)$ explicitly, because the raising and lowering operators are completely analogous mathematically to creation and annihilation operators for bosons. And we know how to form $SU(3)$ generators out of creation and annihilation operators. This guess turns out to be right. The generators of the $SU(3)$ symmetry on the Hilbert space are

$$Q_a = a^\dagger_k\,[T_a]_{k\ell}\,a_\ell \tag{14.8}$$

where $T_a = \lambda_a/2$ are the Gell-Mann matrices. You have already shown that these satisfy the commutation relations of $SU(3)$ in problem (5.B), and you can check that

$$[Q_a, H] = 0 \tag{14.9}$$

so that the energy eigenstates form representations of $SU(3)$. Also,

$$Q_a|0\rangle = 0 \tag{14.10}$$

so that the ground state is an $SU(3)$ singlet.

Just like the creation and annihilation operators, the raising and lowering operators form tensor operators under the $SU(3)$,

$$\left[Q_a, a^\dagger_k\right] = a^\dagger_\ell\,[T_a]_{\ell k} \tag{14.11}$$

thus $a^\dagger_k$ transforms like a 3. The lowering operator, a_k, transforms like a $\overline{3}$,

$$\begin{aligned}[Q_a, a_k] &= -[T_a]_{k\ell}\,a_\ell \\ &= -a_\ell\,[T_a^T]_{\ell k} = -a_\ell\,[T_a^*]_{\ell k}\end{aligned} \tag{14.12}$$

Because all the states in the Hilbert space can be obtained by the action of the raising operators on $|0\rangle$, the Hilbert space decomposes into $(n, 0)$ representations. No other representations appear.

14.2 Angular momentum

Something interesting has happened here. The $SU(3)$ symmetry is nontrivial. It implies that states with different angular momenta are exactly degenerate. Angular momentum is a subgroup of $SU(3)$ generated by (see problem 14.C)

$$L_3 = 2Q_2 \,, \quad L_1 = 2Q_7 \,, \quad L_2 = -2Q_5 \tag{14.13}$$

But for $n > 1$, the degenerate states in level n consist of more than one anglar momentum. For example, the $n = 2$ states transform like a 6 of the $SU(3)$, which under angular momentum transforms like 5+1. Thus at this level, the symmetry guarantees a degeneracy that does not follow from rotation invariance alone. But notice that we didn't have to do anything to impose this $SU(3)$ symmetry. We just imposed rotation invariance, and then the $SU(3)$ symmetry popped out, because of the linearity of the Harmonic oscillator force law. Nonlinear, anharmonic effects would spoil the $SU(3)$ symmetry. For example, a term in the potential like $(\vec{x}^2)^2$ would conserve angular momentum, but break the $SU(3)$.

This example can be extended to $SU(N)$ if you let the number of dimensions be larger or smaller. The dimensions, in this case, just give you more indices for your raising and lowering operator. For the harmonic oscillator, this looks pretty silly, since we live in three dimensions. But in a relativistic theory when we describe the Hamiltonian in terms of creation and annihilation operators, it is perfectly natural. The first approximation to the Hamiltonian of any system of N different types of particles with the same mass will be proportional to the total number operator, $a_k^\dagger a_k$ where the sum over k runs from 1 to N, just because the energy of the m particle state will be m times the energy of a one particle state. This has an $SU(N)$ symmetry. Of course, interactions and other effects that treat the particles differently may break the symmetry.

14.3 A more complicated example

It is very instructive to consider a slightly more complicated model in which all representations of $SU(3)$ appear in the Hilbert space. Consider the following two-particle Hamiltonian:

$$\begin{aligned} H &= \frac{\omega_1}{2}\left(\vec{p}^2+\vec{r}^2\right)+\frac{\omega_2}{2}\left(\vec{P}^2+\vec{R}^2\right) \\ &+\Delta\left[(\vec{r}\cdot\vec{R}-\vec{p}\cdot\vec{P})^2+(\vec{r}\cdot\vec{P}+\vec{R}\cdot\vec{p})^2\right. \\ &\left.-\vec{r}^2-\vec{p}^2-\vec{R}^2-\vec{P}^2\right] \end{aligned} \tag{14.14}$$

where the $\vec{R}$ and $\vec{P}$ are an independent set of coordinates and momenta. This describes two harmonic oscillators, in units with $m_1\omega_1 = m_2\omega_2 = 1$, coupled in what appears to be a complicated way. But in fact, it is carefully constructed to be simple in terms of raising and lowering operators. Defining

$$\begin{aligned} \vec{a}^\dagger &= (\vec{r}-i\vec{p})/\sqrt{2}, \qquad & \vec{a} &= (\vec{r}+i\vec{p})/\sqrt{2}, \\ \vec{b}^\dagger &= (\vec{R}-i\vec{P})/\sqrt{2}, \qquad & \vec{b} &= (\vec{R}+i\vec{P})/\sqrt{2} \end{aligned} \tag{14.15}$$

then the Hamiltonian becomes

$$\begin{aligned} H &= \omega_1(a_k^\dagger a_k+3/2)+\omega_2(b_k^\dagger b_k+3/2) \\ &+4\Delta\,(a_k^\dagger b_k^\dagger)\,(a_\ell b_\ell) \end{aligned} \tag{14.16}$$

The point of this is the form of the interaction term proportional to Δ, and in particular, the way the indices of the raising and lowering operators are contracted. It is constructed to commute with the operators (see problem 14.D)

$$Q_\alpha = a_k^\dagger\,[T_\alpha]_{k\ell}\,a_\ell - b_k^\dagger\,[T_\alpha^*]_{k\ell}\,b_\ell \tag{14.17}$$

These have the commutation relations of $SU(3)$, so the theory has an $SU(3)$ symmetry with the $a^\dagger$ transforming like a 3, as in the simple three dimensional harmonic oscillator, but with the $b^\dagger$ transforming as a $\overline{3}$. Because we now have raising operators that transform like $\overline{3}$s as well as 3s, we can construct states which transform like arbitrary $SU(3)$ representations.

To see what the states look like, note that H also commutes with the number operators,

$$N_a = a_k^\dagger a_k \quad \text{and} \quad N_b = b_k^\dagger b_k \tag{14.18}$$

Thus the energy eigenstates have definite numbers of $a^\dagger$s and $b^\dagger$s, and transform under irreducible representations of $SU(3)$. They look like

$$
\begin{aligned}
&|(n,m),K\rangle \\
&\equiv \left(a^\dagger_{i_1}\cdots a^\dagger_{i_n}\, b^\dagger_{j_1}\cdots b^\dagger_{j_m} - \text{traces}\right) \\
&\cdot\left(\vec{a}^\dagger\cdot\vec{b}^\dagger\right)^K|0\rangle
\end{aligned}
\tag{14.19}
$$

transforming like the (n,m) of $SU(3)$. For example, the octet states look like

$$
\left(a^\dagger_k b^\dagger_\ell - \frac{1}{3}\delta_{k\ell}\left(\vec{a}^\dagger\cdot\vec{b}^\dagger\right)\right)\left(\vec{a}^\dagger\cdot\vec{b}^\dagger\right)^K|0\rangle \tag{14.20}
$$

The indices of the $a^\dagger$s act like lower $SU(3)$ indices while the indices of the $b^\dagger$s act like upper indices. The irreducible representations must be traceless in a pair of an $a^\dagger$ index and a $b^\dagger$ index.

It is straightforward to calculate the energy eigenvalues. Because of the $SU(3)$ symmetry, we can look at any state in an irreducible representation, and in particular, at the highest weight states

$$
|(n,m),K\rangle = \left(a^\dagger_1\right)^n\left(b^\dagger_3\right)^m\left(\vec{a}^\dagger\cdot\vec{b}^\dagger\right)^K|0\rangle \tag{14.21}
$$

The first two terms in H just count $a^\dagger$s and $b^\dagger$s. The nontrivial term is the interaction term proportional to Δ. But because we know that the state is an eigenstate, we can just do it.

$$
\begin{aligned}
&(\vec{a}^\dagger\cdot\vec{b}^\dagger)\,(\vec{a}\cdot\vec{b})\,|(n,m),K\rangle \\
&= (\vec{a}^\dagger\cdot\vec{b}^\dagger)\left[\vec{a}\cdot\vec{b},(a^\dagger_1)^n\right](b^\dagger_3)^m\left(\vec{a}^\dagger\cdot\vec{b}^\dagger\right)^K|0\rangle && (a)\\
&+(\vec{a}^\dagger\cdot\vec{b}^\dagger)\,(a^\dagger_1)^n\left[\vec{a}\cdot\vec{b},(b^\dagger_3)^m\right]\left(\vec{a}^\dagger\cdot\vec{b}^\dagger\right)^K|0\rangle && (b)\\
&+(\vec{a}^\dagger\cdot\vec{b}^\dagger)\,(a^\dagger_1)^n\,(b^\dagger_3)^m\left[a_k,\left[b_k,\left(\vec{a}^\dagger\cdot\vec{b}^\dagger\right)^K\right]\right]|0\rangle && (c)
\end{aligned}
$$

(a) is

$$
\begin{aligned}
&(\vec{a}^\dagger\cdot\vec{b}^\dagger)\left[\vec{a}\cdot\vec{b},(a^\dagger_1)^n\right](b^\dagger_3)^m\left(\vec{a}^\dagger\cdot\vec{b}^\dagger\right)^K|0\rangle \\
&= n\,(\vec{a}^\dagger\cdot\vec{b}^\dagger)\,(a^\dagger_1)^{n-1}\,b_1\,(b^\dagger_3)^m\left(\vec{a}^\dagger\cdot\vec{b}^\dagger\right)^K|0\rangle \\
&= n\,(\vec{a}^\dagger\cdot\vec{b}^\dagger)\,(a^\dagger_1)^{n-1}\,(b^\dagger_3)^m\left[b_1,\left(\vec{a}^\dagger\cdot\vec{b}^\dagger\right)^K\right]|0\rangle \\
&= n\,K\,|(n,m),K\rangle
\end{aligned}
\tag{14.22}
$$

(b) works the same way, and gives

$$m\,K\,|(n,m),K\rangle \tag{14.23}$$

(c) is

$$\begin{aligned}
&(\vec{a}^\dagger\cdot\vec{b}^\dagger)\,(a_1^\dagger)^n\,(b_3^\dagger)^m\left[a_k,\left[b_k,\left(\vec{a}^\dagger\cdot\vec{b}^\dagger\right)^K\right]\right]|0\rangle \\
&=(\vec{a}^\dagger\cdot\vec{b}^\dagger)\,(a_1^\dagger)^n\,(b_3^\dagger)^m\left[a_k,K\,a_k^\dagger\left(\vec{a}^\dagger\cdot\vec{b}^\dagger\right)^{K-1}\right]|0\rangle \\
&=K\,(\vec{a}^\dagger\cdot\vec{b}^\dagger)\,(a_1^\dagger)^n\,(b_3^\dagger)^m\left(\left[a_k,a_k^\dagger\right]\left(\vec{a}^\dagger\cdot\vec{b}^\dagger\right)^{K-1}\right. \\
&\left.+a_k^\dagger\left[a_k,\left(\vec{a}^\dagger\cdot\vec{b}^\dagger\right)^{K-1}\right]\right)|0\rangle \\
&=K\,(K+2)\,|(n,m),K\rangle
\end{aligned} \tag{14.24}$$

Thus the energy eigenvalue is

$$\omega_1\left(n+K+\frac{3}{2}\right)+\omega_2\left(m+K+\frac{3}{2}\right)+4\Delta\,K\,(n+m+K+2) \tag{14.25}$$

The physics of this model is not very interesting. However, as an example of how symmetries arise, it is very instructive. There is a very useful way of thinking about what goes on here. For a moment, think of the interaction term, proportional to Δ, as a perturbation on the harmonic oscillator terms. In the absence of the Δ term, the theory has a larger symmetry — $SU(3)\times SU(3)$ (and some $U(1)$s, but we will concentrate on the non-Abelian symmetries) — separate $SU(3)$ symmetries on the $a^\dagger$ and $b^\dagger$ variables. The interaction terms couples the as and bs together and breaks the independent $SU(3)$ symmetries down the single $SU(3)$ that treats the interaction terms as a singlet, because $a_k b_k$ and $a_k^\dagger b_k^\dagger$ are singlets under the $SU(3)$ symmetry. It is often useful to organize the symmetry structure of the theory in such a structured way.

Problems

14.A. Show that the operators

$$O_{ij}^k = a_i^\dagger a_j^\dagger a_k - \frac{1}{4}\left(\delta_{ik}a_\ell^\dagger a_j^\dagger a_\ell + \delta_{jk}a_\ell^\dagger a_i^\dagger a_\ell\right)$$

transform like a tensor operator in the (2,1) representation.

14.B. Calculate the non-zero matrix elements of the operator O^3_{11} (where O^k_{ij} is defined in 14.A) between states of the form

$$a_i^\dagger \left(a_\ell^\dagger b_\ell^\dagger\right) |0\rangle$$

and

$$a_j^\dagger a_k^\dagger \left(a_\ell^\dagger b_\ell^\dagger\right) |0\rangle \,.$$

14.C. Show that (14.13) generates the standard angular momentum,

$$\vec{r} \times \vec{p} \,.$$

14.D. Show that

$$[Q_\alpha, a_k b_k] = 0 \,.$$

Chapter 15

$SU(6)$ and the Quark Model

In the spectrum of low-lying baryons, the octet with spin 1/2 and the decuplet with spin 3/2 are not very far apart in mass. Typical splittings between the octet and decuplet states are not so different from splittings within the $SU(3)$ representations. Noticing this, in the 60's, many physicists played with the idea of embedding $SU(3)$ in some larger symmetry group that would connect the two representations. Because the baryons in the 8 and 10 have different spins, the larger group cannot commute with angular momentum. You might expect this to cause problems, because it means mixing up internal symmetries and spacetime symmetries, and indeed it does. But the problems do not show up until you try to make the theory relativistic. We will discuss some of this, and possible resolutions, later.

15.1 Including the spin

The obvious way to extend the $SU(3)$ symmetry group is to include in the quark states the quark spin as well as the quark type. This suggests that we think about a 6 dimensional tensor product space in which the states have an $SU(3)$ index, and a spin index —

$$\left(|u\rangle, |d\rangle, |s\rangle\right) \left(|1/2\rangle, |-1/2\rangle\right) \tag{15.1}$$

The corresponding symmetry group is an $SU(6)$ symmetry that acts on this six dimensional tensor product space. We can write the generators as products of 3×3 matrices in the $SU(3)$ space (though not necessarily traceless) and 2×2 matrices in spin space (again, not necessarily traceless). In particular,

 DOI: 10.1201/9780429499210-16

the generators include the $SU(3)$ generators

$$\frac{1}{2}\lambda_a \tag{15.2}$$

with the identity in spin space understood, and the spin generators

$$\frac{1}{2}\sigma_j \tag{15.3}$$

with the identity in the $SU(3)$ space understood. That is, this $SU(6)$ has an $SU(3)$ subalgebra under which the 6 transforms like two 3s, and it has an $SU(2)$ subalgebra under which the 6 transforms like 3 2s. The $SU(3)$ and $SU(2)$ algebras commute with one another. We say that the $SU(6)$ has an $SU(3) \times SU(2)$ subgroup under which the 6 transforms like a $(3, 2)$. The other $SU(6)$ generators are the products

$$\frac{1}{2}\lambda_a\sigma_j \tag{15.4}$$

The total number of generators is then 8+3+24=35, which, sure enough, is the number of independent, hermitian, traceless 6×6 matrices.

The low-lying baryons, consisting of three quarks, transform like the 56 dimensional representation of $SU(6)$, the completely symmetric combination of 3 6s. Let's pause to understand how such a representation transforms.

15.2 $SU(N) \times SU(M) \in SU(NM)$

The general situation is $SU(N) \times SU(M) \in SU(NM)$. It arises only for $SU(k)$ where k is not a prime, and thus this embedding does not show up in $SU(2)$ or $SU(3)$. The idea is to always exploit the idea of a tensor product space. For $SU(NM)$, the defining representation has NM indices, and we can therefore describe it in a tensor product space, replacing the NM indices by an ordered pair of indices, i, x where i runs from 1 to N and x runs from 1 to M. Then the matrices

$$[T_a^N]_{ij}\,\delta_{xy} \qquad \text{and} \qquad \delta_{ij}\,[T_a^M]_{xy} \tag{15.5}$$

are hermitian and traceless (remember that in a tensor product space, taking the trace means contracting both types of indices) and generate $SU(N)$ and $SU(M)$, respectively.

Under the $SU(N) \times SU(M)$ subalgebra generated by these matrices, the NM transforms like (N, M) (or equivalently, ([1],[1]) — that is there are M copies of the N of $SU(N)$ which transform into one another under the $SU(M)$ — or vice versa, whichever you like). The reason that this is useful is that very often our creation operators have more than one index — like the $SU(3)$ indices and spin indices in the quark states of $SU(6)$. Then this kind of tensor product is the natural place for them, and such $SU(NM)$ symmetries arise quite naturally.

Now we can ask the decomposition question. Given a representation of $SU(NM)$, how does it transform under $SU(N) \times SU(M)$?

In particular, suppose we have a representation D, of $SU(NM)$ corresponding to some Young tableau with K boxes. In a tensor language, D is a tensor with K $SU(N)$ indices and K $SU(M)$ indices. Thus because of N-ality and M-ality, if the decomposition contains the representation (D_1, D_2) where D_1 is an $SU(N)$ representation with K_1 boxes and D_2 is an $SU(M)$ representation with K_2 boxes, then

$$K = K_1 \mod N = K_2 \mod M\,. \tag{15.6}$$

To go further, it is again easiest to argue backwards. First, add columns of N boxes to D_1 and columns of M boxes to D_2 if necessary to bring both up to a total of K boxes. Now what the decomposition means is that we can write the tensor described by a D tableau with (i, x) pairs in each box as a linear combination of products of tensors with i indices in the D_1 tableau and x indices in the D_2 tableau.

Now the idea is simple. Both the D_1 and D_2 tableaux have definite symmetry properties under permutations of the k indices. They are associated with irreducible representations of the symmetric group S_k. The question then is whether the product has any component that had the symmetry of D. If so, then (D_1, D_2) will appear in the decomposition of D.

Of course, this doesn't quite solve the problem, because we must still understand whether the tensor product of two irreducible representations of S_k contains a third. We won't try to solve this problem in general. But we will be able to tell in some special cases.

For the 56 of $SU(6)$, this analysis is simple. The Young tableau associated with the 56 is

(15.7)

which corresponds to the trivial representation of S_3. There are three possible irreducible representations of $SU(3)$ with three boxes:

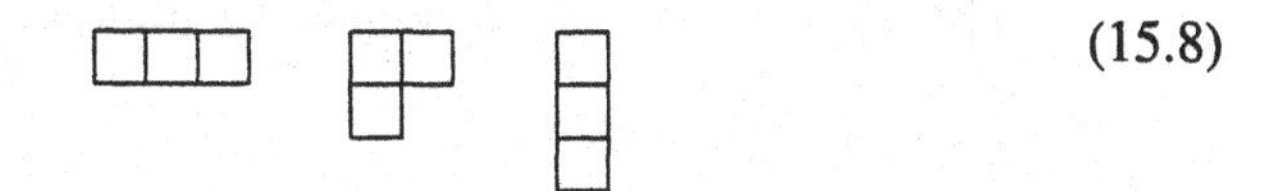

(15.8)

And there are two possible irreducible representations of $SU(2)$ with three boxes:

$$\Box\Box\Box \qquad \begin{array}{ll}\Box&\Box\\ \Box&\end{array} \tag{15.9}$$

The $SU(3) \times SU(2)$ representations in the 56 are then those ordered pairs of $SU(3)$ and $SU(2)$ representations for which the trivial representation of S_3 appears in the tensor product of the S_3 representations of their Young tableaux. For example, the representation

$$(\Box\Box\Box, \Box\Box\Box) \tag{15.10}$$

appears, because the completely symmetric representation is contained in its tensor product with itself. This corresponds to the 10 of $SU(3)$ with spin 3/2. The other possibility is

$$\left(\begin{array}{ll}\Box&\Box\\ \Box&\end{array}, \begin{array}{ll}\Box&\Box\\ \Box&\end{array}\right) \tag{15.11}$$

The tensor product of these two S_3 representations contains the trivial representation (as we will see explicitly) because any pair of $SU(3)$ and $SU(2)$ indices transforms the same way under any permutation, the product contains the representation which is unchanged by the permutation. This representation corresponds to the 8 of $SU(3)$ with spin 1/2. Thus the 56 includes precisely the low-lying baryon states.

15.3 The baryon states

Let's look explicitly at these states. The decuplet states are particularly simple because they are separately symmetric in the $SU(3)$ and $SU(2)$ indices. For example, indicating the $\pm 1/2$ spin states by $|\pm\rangle$

$$\begin{aligned}
&|\Delta^{++}, 3/2\rangle = |uuu\rangle|{+}{+}{+}\rangle \\
&|\Delta^{+}, 1/2\rangle = \frac{1}{3}\Big(|uud\rangle + |udu\rangle + |duu\rangle\Big) \\
&\cdot\Big(|{+}{+}{-}\rangle + |{+}{-}{+}\rangle + |{-}{+}{+}\rangle\Big) \\
&|\Sigma^{*0}, 1/2\rangle = \frac{\sqrt{2}}{6} \\
&\cdot\Big(|uds\rangle + |dus\rangle + |sud\rangle + |sdu\rangle + |dsu\rangle + |usd\rangle\Big) \\
&\cdot\Big(|{+}{+}{-}\rangle + |{+}{-}{+}\rangle + |{-}{+}{+}\rangle\Big)
\end{aligned} \tag{15.12}$$

The octet states are more complicated because they are antisymmetric in one pair of spins and quark labels (because of the columns in the Young tableau). To get a completely symmetric state, you can multiply the two and then add cyclic permutations. For example

$$
\begin{aligned}
|\Lambda, 1/2\rangle = & \frac{\sqrt{3}}{6} \\
& \cdot\Big\{ \big(|uds\rangle - |dus\rangle\big)\big(|{+}{-}{+}\rangle - |{-}{+}{+}\rangle\big) \\
& + \big(|sud\rangle - |sdu\rangle\big)\big(|{+}{+}{-}\rangle - |{+}{-}{+}\rangle\big) \\
& + \big(|dsu\rangle - |usd\rangle\big)\big(|{-}{+}{+}\rangle - |{+}{+}{-}\rangle\big)\Big\}
\end{aligned}
\tag{15.13}
$$

In this notation, the quark label and spin states are correlated. The first quark label goes with the first spin states, and so on. The way this Λ state works is that if you interchange both spin and quark labels of any pair, one of these three terms is unchanged, while the other two get interchanged. Note the normalization of the state. It is often more convenient to construct these states by multiplying an $SU(3)$ state that is symmetric in some pair of indices, like $|uud\rangle$ which is symmetric in the first two, by a spin state which is symmetric in the same pair, but with total spin 1/2. Then again, adding cyclic permutations gives a totally symmetric state. For example,

$$
\begin{aligned}
|P, 1/2\rangle = & \frac{\sqrt{2}}{6} \\
& \cdot\Big\{ |uud\rangle \big(2|{+}{+}{-}\rangle - |{+}{-}{+}\rangle - |{-}{+}{+}\rangle\big) \\
& + |udu\rangle \big(2|{+}{-}{+}\rangle - |{-}{+}{+}\rangle - |{+}{+}{-}\rangle\big) \\
& + |duu\rangle \big(2|{-}{+}{+}\rangle - |{+}{+}{-}\rangle - |{+}{-}{+}\rangle\big)\Big\}
\end{aligned}
\tag{15.14}
$$

In the first term, for example, the $|{+}{-}{+}\rangle$ state and the $|{-}{+}{+}\rangle$ state must have the same coefficient for symmetry under interchange of the first two quarks, but the $|{+}{+}{-}\rangle$ state is automatically symmetric, so it can have a different coefficient. We have chosen the coefficients so that the sum vanishes, which guarantees that the state will be orthogonal to the symmetric spin 3/2 state. This can never give a decuplet state. If the $SU(3)$ state is completely symmetric, the sum over cyclic permutations gives zero rather than a nontrivial state. Again, note that if you interchange both spin and quark labels of any pair, one of these three terms is unchanged, while the other two get interchanged.

15.4 Magnetic moments

Let us now see what $SU(6)$ symmetry has to say about the magnetic moments of the baryons. To determine the $SU(6)$ properties of the quarks, note that if a quark is a point particle like the electron, its magnetic moment is approximately

$$\frac{e}{2m} Q \vec{\sigma} \tag{15.15}$$

where m is the quark mass and eQ is the quark charge. This looks like an $SU(6)$ generator, so we infer that the magnetic moment operator transforms like the 35 dimensional adjoint representation.

We are interested in the matrix elements

$$\langle 56|35|56\rangle \tag{15.16}$$

But $35 \otimes 56$ is

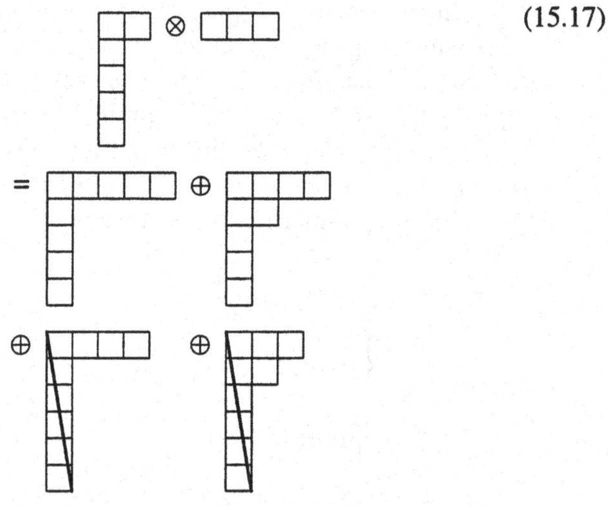

(15.17)

Because there is a unique 56 in the decomposition, we know that the matrix elements are determined by a single reduced matrix element. Therefore, we can compute the magnetic moments up to an overall constant by looking at the matrix element of any operator that transforms like the correct component of a 35. In particular, we can use the $SU(6)$ generators themselves, and conclude

$$\mu_{56} \propto \langle 56|\, Q\,\vec{\sigma}\, |56\rangle \tag{15.18}$$

We can use this relation to compute the ratio of the proton to neutron magnetic moments, which was not determined by the Coleman-Glashow analysis. If we compute the matrix element between states with the same σ_3 value, then only the $Q\sigma_3$ matrix element is not zero, so we can compute that. We do this just by applying the tensor-product rule:

$$\begin{aligned} Q\,\sigma_3\,|P,1/2\rangle &= \frac{\sqrt{2}}{6} \\ &\cdot\left\{\frac{2}{3}|uud\rangle\left(2|{+}{+}{-}\rangle - |{+}{-}{+}\rangle + |{-}{+}{+}\rangle\right)\right. \\ &+\frac{2}{3}|uud\rangle\left(2|{+}{+}{-}\rangle + |{+}{-}{+}\rangle - |{-}{+}{+}\rangle\right) \\ &-\frac{1}{3}|uud\rangle\left(-2|{+}{+}{-}\rangle - |{+}{-}{+}\rangle - |{-}{+}{+}\rangle\right) \\ &\left.+\text{cyclic permutations}\right\} \end{aligned} \tag{15.19}$$

Thus

$$\begin{aligned} &\langle P,1/2|Q\sigma_3|P,1/2\rangle \\ &= 3\,\frac{2}{36}\left(\frac{2}{3}(4+1-1)+\frac{2}{3}(4-1+1)-\frac{1}{3}(-4+1+1)\right)=1 \end{aligned} \tag{15.20}$$

The matrix element between neutron states can be obtained by simply interchanging the us and the ds in this calculation, which means interchanging the $2/3$ and $-1/3$ factors —

$$\begin{aligned} &\langle N,1/2|Q\sigma_3|N,1/2\rangle \\ &= 3\frac{2}{36}\left(-\frac{1}{3}(4+1-1)-\frac{1}{3}(4-1+1)+\frac{2}{3}(-4+1+1)\right)=-\frac{2}{3} \end{aligned} \tag{15.21}$$

So we expect

$$\frac{\mu_P}{\mu_N} = -\frac{3}{2} \tag{15.22}$$

Experimentally, in nuclear magnetons (units of $e/2m_P$)

$$\begin{gathered} \mu_P = 2.79 \qquad\qquad \mu_N = -1.91 \\ \frac{\mu_P}{\mu_N} = -1.46 \end{gathered} \tag{15.23}$$

So it works, but the logic is rather indirect. From the quark model, we extracted only the $SU(6)$ transformation law of the magnetic moment operator. We then used the Wigner-Eckart theorem to show that we can calculate the ratio by looking at the corresponding matrix elements of an $SU(6)$ generator.

We can get the result more easily by simply assuming that the quarks actually exist as nonrelativistic constituents, and that the baryons really are bound states of three quarks in an angular momentum zero state. Then the magnetic moment of the baryon is simply the sum of the magnetic moments of the quarks. The magnetic moment of a quark (by analogy with an electron) is $\frac{Qe}{2m}$ in the direction of the spin where m is the quark mass, thus it can be written

$$\frac{e}{2m} Q \vec{\sigma} \tag{15.24}$$

Therefore in the tensor product language, the sum of the quark magnetic moments is

$$\frac{e}{2m} \sum_{\text{quarks}} Q \vec{\sigma} \tag{15.25}$$

The proton magnetic moment is $\frac{e}{2m_P}\mu_P$ where m_P is the proton mass and μ_P is the magnetic moment measured in "nuclear magnetons". Thus

$$\frac{e}{2m_P}\mu_P = \frac{e}{2m}\langle P|Q\vec{\sigma}|P\rangle / \langle P|\vec{\sigma}|P\rangle = \frac{e}{2m} \tag{15.26}$$

which is exactly what we computed in the $SU(6)$ argument, but with one additional bit of information — we know the scale. If we assume that the quark mass is about 1/3 the proton mass, we have

$$\mu_P = \frac{m_P}{m} = 3 \tag{15.27}$$

This is not bad!

The quark model is thus not only simpler, it is more predictive. It also gives a reasonable account of some $SU(3)$ breaking effect in the baryon magnetic moments. For example, the Coleman-Glashow prediction for the Λ magnetic moment is

$$\mu_\Lambda = -\frac{1}{3}\mu_P = -.93 \tag{15.28}$$

You can see this directly in $SU(6)$ by computing

$$\langle \Lambda, 1/2|Q\sigma_3|\Lambda, 1/2\rangle = -\frac{1}{3} \tag{15.29}$$

But experimentally

$$\mu_\Lambda \approx -.61 \tag{15.30}$$

which is significantly different. The quark model gives

$$\frac{e}{2m_p}\mu_\Lambda = \sum_{\text{quarks}} \frac{e}{2m_{\text{quark}}} Q\sigma_3 \tag{15.31}$$

In the matrix element, only the s quark actually contributes in the Λ, (15.13), because the u and d spins are combined into a spin zero state. Thus we predict

$$\mu_\Lambda = -\frac{1}{3}\frac{m}{m_s} \tag{15.32}$$

In the quark model, we expect the s quark to be heavier than the u and d quarks to account for the larger mass of hadrons containing the s quark. For example, if we assume (very roughly) $m_\Lambda \approx m_s + 2m$ and $m_P \approx 3m$, then

$$m_s \approx \frac{3m_\Lambda - 2m_P}{3} \approx 490\,\text{MeV} \tag{15.33}$$

so

$$\mu_\Lambda \approx -.64 \tag{15.34}$$

which is a bit better.

Problems

15.A. Find the $SU(6)$ (i.e. quark model) wave functions for all the spin 1/2 baryons except P, N and Λ (which were discussed in the text).

15.B. Use the wave functions you found in (15.A) to calculate the magnetic moments,

a. in the $SU(6)$ limit, calculating the ratios to μ_P;

b. in the quark model, put in $SU(3)$ symmetry breaking by including $m_s \neq m_{u,d}$.

15.C. Show that the $|\Lambda, 1/2\rangle$ state, (15.13), is an isospin singlet.

Chapter 16

Color

There are some things wrong with the simple quark model discussed in the previous chapter. The first is that the connection between spin and statistics is wrong for these quarks. The quarks must have spin 1/2 in order to produce the spin 1/2 and 3/2 baryons. Thus we would expect them to obey Fermi-Dirac statistics, and we would expect the ground state of the three quark system to be an s-wave, completely symmetric under the exchange of the position labels of the three quarks. This would lead to an $SU(6)$ representation completely **anti**symmetric in the $SU(6)$ indices of the three quarks — which is a [3] or a 20 which transforms under $SU(3) \times SU(2)$ as $([2,1],[2,1]) \oplus ([3],[1,1,1])$ or $(8,2) \oplus (1,4)$ which is not what we want.

The second difficulty is that we need some explanation of why only the $\bar{q}q$ and qqq combinations seem to exist. This would certainly not be explained by any simple attractive force between quarks.

Finally, there is now a tremendous amount of evidence from studies of high energy QCD that the quarks carry an attribute which we call **color**. Quarks come in three colors, and the colors interact with a set of 8 gluons in an $SU(3)$ symmetric way. The $SU(3)$ associated with the color interaction is called color $SU(3)$. It has **nothing to do with Gell-Mann's $SU(3)$,** except, of course, that it provides the force that binds the light quarks into Gell-Mann's representations.

16.1 Colored quarks

The quark states transform like a 3 — they have a single color $SU(3)$ index which I will let run from 1 to 3. The quark wave functions are tensors

$$q^i \quad \text{for } i = 1 \text{ to } 3\,. \tag{16.1}$$

DOI: 10.1201/9780429499210-17

They also carry labels for Gell-Mann's $SU(3)$ (flavor), spin and position. The idea is that the color interaction binds three quarks into a color singlet baryon state, which is then described by contracting the color wave functions of the quarks with an ϵ —

$$\epsilon_{ijk}\, q^i\, q^j\, q^k \tag{16.2}$$

Then, because the ϵ is completely antisymmetric, the baryon state is symmetric under the exchange of all the other labels.

It is this binding of three quarks into an $SU(3)$ singlet baryon that suggested the name "color" for the color $SU(3)$ indices. The metaphor is based on the fact that colorless light can be produced by combining beams of the three primary colors, red, green and blue. In the same way, a **colorless** (that is color $SU(3)$ singlet) baryon state can be made out of three colored quarks. Thus the color $SU(3)$ indices are sometimes referred to as red, green and blue, rather than 1, 2 and 3.

I think that it was the use of the term color that gave rise, in turn, to the name **flavor** for the label, u, d, s, etc, that distinguishes one type of quark from another. Gell-Mann's $SU(3)$ is sometimes referred to as flavor $SU(3)$ because it transforms the light flavors into one another. Flavor is a completely independent attribute. The quarks carry both color and flavor (and also spin).

Antiquarks transform like $\overline{3}$s under color $SU(3)$, because color charges, at least those in the Cartan subgroup that can be diagonalized, are quantum numbers like electric charge and baryon number, and like electric charge and baryon number, they change sign in going from a particle to its antiparticle. If the Cartan charges change sign, that means that the weights change sign, and that means going from the 3 representation to its complex conjugate, the $\overline{3}$.

We can make a color singlet state from three antiquarks by contracting with an ϵ —

$$\epsilon^{ijk}\, \overline{q}_i\, \overline{q}_j\, \overline{q}_k \tag{16.3}$$

This describes an antibaryon. Or we can make color singlet states out of one quark and one antiquark —

$$q^i\, \overline{q}_i \tag{16.4}$$

These are the mesons. They include the pseudoscalar meson octet, the pseudoscalar singlet η' and also an octet plus singlet of spin 1 **vector mesons,**

shown below

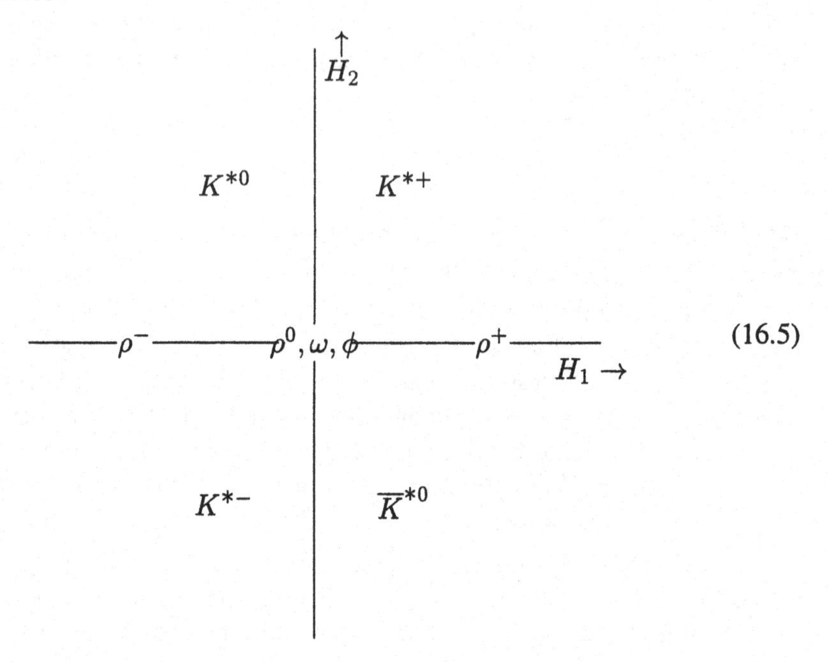

(16.5)

Why does the color force bind these states? A partial answer is that it behaves like the electromagnetic force. The gluons couple to **color charges**, that is the color $SU(3)$ generators, just as the photon, the particle of light, couples to electric charge (the generator of a $U(1)$ symmetry). The electromagnetic force between two objects is attractive when the product of their electric charges is negative. This is why the electromagnetic force tends to bind charged particles into neutral atoms and molecules. One difference with color is that there are 8 gluons, instead of one photon, and the color interaction between two colored particles is proportional to the sum of the products of their color charges. If this is negative, the force is attractive.

Specifically, consider a state of two particles, A and B, transforming according to some representations of color $SU(3)$:

$$\begin{aligned} T_a^A|r, A\rangle &= |s, A\rangle\,[T_a^A]_{sr} \\ T_a^B|x, B\rangle &= |y, B\rangle\,[T_a^B]_{yx} \end{aligned} \tag{16.6}$$

where r, s (x, y) are the color indices for the A (B) representation. The two particle state is then a tensor product

$$|v, A, B\rangle = v_{rx}|r, A\rangle|x, B\rangle \tag{16.7}$$

Now the color interaction is proportional to the product of the charges,

$$T_a^A T_a^B \tag{16.8}$$

summed over a, in the usual sense of the tensor product space. This is invariant under $SU(3)$ because the generator on the tensor product space is

$$T_a = T_a^A + T_a^B \tag{16.9}$$

which commutes with (16.8) because

$$\begin{aligned} & \left[T_a^A + T_a^B, T_b^A T_b^B\right] \\ &= \left[T_a^A, T_b^A\right] T_b^B + T_b^A \left[T_a^B, T_b^B\right] \\ &= i\, f_{abc} \left(T_c^A T_b^B + T_b^A T_c^B\right) = 0 \end{aligned} \tag{16.10}$$

The eigenstates of (16.8) are irreducible representations of color $SU(3)$ because of Schur's lemma. To compute the eigenvalues, it is convenient to write

$$T_a^A T_a^B = \frac{1}{2}(T_a^2 - T_a^{A^2} - T_a^{B^2}) \tag{16.11}$$

The object T_a^2 (summed over a) is called a **Casimir operator** (it is the analog of $\vec{J}^2$ in angular momentum $SU(2)$). It is easy to see that it commutes with the $SU(3)$ generators and is therefore a number on each irreducible representation. Note that $T_a^{A^2}$ and $T_a^{B^2}$ are fixed because they are properties of the particles involved, but T_a^2 depends on how the A and B states are combined into a state with definite color — that is, it depends on how the particular linear combination of tensor product states we are looking at transforms under $SU(3)$, and it has definite values on states that transform under an irreducible representation. Loosely speaking, T_a^2 measures the size of the color representation. The smaller T_a^2, the less "colorful" the state. Thus the color force is most attractive in the least colorful states, and therefore it tends to bind quarks and antiquarks into the least colorful states possible. This is a step in the right direction.

In a $q\bar{q}$ state, the most attractive state will be an $SU(3)$ singlet (with $T_a^2 = 0$). In a three quark state, we can consider each pair of quarks in turn. If the colors of a quark pair are combined symmetrically, the qq state is a 6. If they are combined antisymmetrically, the state is a $\overline{3}$. T_a^2 is 4/3 for the $\overline{3}$ and 10/3 for the 6, thus the antisymmetric combination is the most attractive. Thus the most favored state is one that is antisymmetric in the colors of each quark pair, which is the color singlet, completely antisymmetric state.

16.2 Quantum Chromodynamics

There is much more to color. The quantum theory of the color interaction of quarks and gluons is called **QCD, Quantum Chromodynamics**, in analogy with the quantum theory of the electromagnetic interactions of electrons and photons, **QED, Quantum Electrodynamics**. There is a dimensional parameter built into the QCD theory, $\Lambda_{\rm QCD}$, a few hundred MeV. The QCD interaction is rather weak for distances smaller than $1/\Lambda_{\rm QCD}$, however, it gets strong for distances larger than $1/\Lambda_{\rm QCD}$. We now believe that the quarks are permanently confined by the strong long-distance QCD interactions inside colorless hadrons, so that we can never completely isolate a colored quark from its hadronic surrounding. This doesn't mean that we cannot see quarks. In fact, we see quarks and gluons rather directly in scattering experiments at energies and momenta much larger than $\Lambda_{\rm QCD}$. But it does mean that quarks do not show up directly in the low energy spectrum.

With this picture of the strong interactions, we can understand **why** Gell-Mann's $SU(3)$ is a useful symmetry. The interaction of the gluons is the same for each of the quarks. The only thing that distinguishes between quarks (in the strong interactions) is their mass. A mass term in the QCD Hamiltonian looks like

$$m_u \, \overline{u} \, u + m_d \, \overline{d} \, d + m_s \, \overline{s} \, s \tag{16.12}$$

where u, d, s ($\overline{u}$, $\overline{d}$, $\overline{s}$) are annihilation (creation) operators for the u, d and s quarks. This can be rewritten

$$\begin{aligned} (m_u + m_d + m_s) \, (\overline{u} \, u + \overline{d} \, d + \overline{s} \, s)/3 && (a) \\ +(m_u - m_d) \, (\overline{u} \, u - \overline{d} \, d)/2 && (b) \\ +(2m_s - m_u - m_d) \, (2 \, \overline{s} \, s - \overline{u} \, u - \overline{d} \, d)/6 && (c) \end{aligned}$$

(a) is an $SU(3)$ invariant. (b) breaks isospin symmetry, which is one reason that we think that the u-d mass difference is very small, a few MeV. This term is a small perturbation that you can ignore unless you are interested in isospin breaking. It happens to have effects roughly the same size as the electromagnetic interactions, even though it probably has nothing to do with electromagnetism. (c) is the term that is responsible for most of the breaking of Gell-Mann's $SU(3)$. It is fairly small, because the $s - u$ and $s - d$ quark mass differences, as measured, for example by the mass splittings within $SU(3)$ representations, are not large compared to $\Lambda_{\rm QCD}$. Notice that (c) is a tensor operator, the 8 component of an octet. This is why Gell-Mann's original guess about the transformation law of the medium strong term in the Hamiltonian actually worked. We no longer have to guess because we have some understanding of the dynamics.

16.3 Heavy quarks

As you probably know, there are other quarks besides the u, d and s quarks. The c (for charm) quark, the b (for bottom or beauty) quark and the t (for top or truth) quark have all been seen in high energy collisions. They are all unstable, decaying back into the lighter quarks very quickly. But the c and b quarks last long enough to bind into hadrons just as the lighter quarks do. So there is now a rich phenomenology of particles containing c and b quarks. These particle had not been seen, and were barely even imagined when Gell-Mann first explored the consequences of $SU(3)$ symmetry.

How do these states transform under Gell-Mann's $SU(3)$? You should be able to guess what these states look like. There are meson states and baryon states. Meson states with a heavy b or c antiquark and a light u, d or s quark transform like the 3, because the light quarks are a 3. Meson states with a heavy b or c quark and a light u, d or s antiquark transform like the $\overline{3}$, because the light antiquarks are a $\overline{3}$. Meson states in which both the quark and the antiquark are heavy are $SU(3)$ singlets. They transform trivially under Gell-Mann's $SU(3)$ because the heavy quarks do not carry any $SU(3)$ properties at all. The baryons containing a single heavy quark and two light quarks transform like the 6 and the $\overline{3}$ of $SU(3)$. And so on.

16.4 Flavor $SU(4)$ is useless!

Every so often, someone gets the deceptively attractive idea of enlarging Gell-Mann's flavor $SU(3)$ symmetry to an $SU(4)$ or an $SU(5)$ including the c or the c and b quark. Alas, this seemingly obvious extension is quite useless. The trouble is that the masses of the c and b quarks are so much larger than the light quark masses, and so different from each other, that mass differences involving these quarks are **much larger than** Λ_{QCD}. The perturbative description of the breaking of $SU(3)$ that makes Gell-Mann's $SU(3)$ useful is not appropriate in the extensions to $SU(4)$ or $SU(5)$. Don't be fooled.

Problems

16.A. Find a relation between the sum of the products of the color charges in the $q\overline{q}$ state in a meson and a the qq pair in a baryon.

16.B. Suppose that a "quix", Q, a particle transforming like a 6 under color $SU(3)$ exists. What kinds of bound states would you expect with one quix and additional quarks or antiquarks? How do these states transform under Gell-Mann's $SU(3)$? **Hints:** The quix is a flavor singlet, because Gell-Mann's $SU(3)$ just transforms the light quarks. Also, you should only include states whose wave functions cannot be factored into two independent color singlets. For example, thinking just about the light quarks, you would not include a $\bar{q}qqqq$ state, because you can show that every wave function you can write down factors into a color singlet $\bar{q}q$ and a color singlet qqq. The corresponding state would presumably fall apart into a meson and a baryon.

16.C. Here is a convenient way to calculate the Casimir operators, $C(D)$ $(= T_a^2)$ for small representations, D. Note that

$$\mathrm{Tr}\, T_a^2 = \dim(D)\, C(D) = \sum_a \mathrm{Tr}(T_a T_a) = \sum_a k_D = 8k_D$$

where k_D is defined in (6.2) and $\dim(D)$ is the dimension of the representation D. $k_D = 1/2$ for $D = 3$ (or $\bar{3}$) so $C(D) = 4/3$. But k_D behaves in a simple way under $\oplus$ and $\otimes$:

$$(a) \quad k_{D_1 \oplus D_2} = k_{D_1} + k_{D_2}$$

$$(b) \quad k_{D_1 \otimes D_2} = \dim(D_1)\, k_{D_2} + \dim(D_2)\, k_{D_1}$$

Prove (a) and (b) and use them to calculate $C(8)$, $C(10)$, and $C(6)$.

Chapter 17

Constituent Quarks

The color $SU(3)$ theory of hadrons is a genuinely strongly interacting theory. There is no obvious small parameter that we can use to express QCD predictions in a perturbation series. At present, first principles QCD calculations can be done only by large-scale computer evaluations of the functional integral. While such calculations have yielded important evidence that QCD can explain the spectrum of light hadrons, the analysis is far from complete. Given this complexity, it is remarkable that a simple non-relativistic quark-model gives a reasonable qualitative picture of light hadron masses. In this section, we will go over this picture briefly.

17.1 The nonrelativistic limit

If all the quarks had masses very large compared to the QCD parameter, Λ, we could justify the nonrelativistic quark model in a simple way. In that case, the QCD interaction would behave rather like ordinary electrodynamics, except for a few peculiarities associated with the non-Abelian behavior of the Gluons, as we discussed in the previous chapter. The baryons would simply be nonrelativistic bound states. We would be able to organize a calculation of the bound-state energies in an expansion in inverse powers of the large quark masses. The leading contribution would be simply the sum of the quark masses. In next order, there would be a contribution from the color interaction depending on the space wave function but independent of the quark spins. In the limit that the u, d and s are degenerate, we would have an approximate $SU(6)$ symmetry like that discussed in Chapter 15. The leading $SU(6)$-breaking spin dependent interactions would be suppressed by inverse powers of the masses, because they are all relativistic effects of color magnetism. The

 DOI: 10.1201/9780429499210-18

most important such effect for the ground state (which should be primarily an $\ell = 0$ state) would be the color magnetic moment interactions between the constituent quark spins.

This picture works surprisingly well even though the quarks are light (though it must be supplemented to describe the pseudoscalar mesons — not a surprise, because the pion is too light to be thought of as a nonrelativistic bound state). We do not understand why it works as well as it does, but let's see how it gives a picture of the masses of the ground states in the baryon and meson sectors, the $SU(6)$ 56 of baryons (the spin 1/2 octet and spin 3/2 decuplet) and the quark-antiquark bound states in the $6 \otimes \bar{6}$ (the spin 1 and 0 octets and singlets).

What is really interesting about this picture is the way the spin-dependent interactions work. If the ground-states are primarily s-wave, there will be no contribution from a single color-magnetic moment interaction. However, there will be important contributions from gluon exchange between quarks that depend on the relative orientation of the color dipoles. Except for the color factor, this interaction looks like that between electron magnetic moments, or between current loops. If two current loops are sitting on top of one another, the state in which the dipole moments are aligned has lower energy than the state in which they are paired (antiparallel).

In both the baryon and the meson, the color force between pairs is attractive. Thus the magnetic moment interaction is like the magnetic moment interaction between an electron and a positron, with opposite charges (as explained in Chapter 16, it is the magic of color that allows each of the three pairs in the proton to behave as if they have "opposite" charges). Therefore aligned magnetic moments correspond to paired spins, and *vice versa*. Thus we expect that the state in which the spins are paired has lower energy than the state in which they are aligned. Formally, there is a term in the Hamiltonian that looks like

$$\sum_{\substack{\text{pairs} \\ i,j}} \frac{\kappa}{m_i m_j} \vec{\sigma}_i \cdot \vec{\sigma}_j \tag{17.1}$$

where the σ's are the Pauli matrices acting on the spins and κ is some function of the quark positions. The factors of $1/m_i$ are there because the color magnetic moment is inversely proportional to the quark mass. To the extent that the wave functions are independent of the type of quark (which they are in the $SU(3)$ limit) and quark spin (the limit we are considering), κ is just a constant on the whole 56 or 35.

The spin-spin interaction, (17.1), makes the spin 1 mesons, in which the spins are aligned, heavier than the spin 0 mesons in which they are paired. It also makes the baryon decuplet, in which every pair is aligned, heavier than the octet, in which some spins are paired. In particular, if we ignore the mass differences between quarks it clearly just depends on the total spin, $\vec{S}$, because we can write it as

$$\begin{aligned}&\frac{\kappa}{m^2}\sum_{\substack{\text{pairs}\\ i,j}}\vec{\sigma}_i\cdot\vec{\sigma}_j\\ &=\frac{\kappa}{2m^2}\left(\sum_{i,j}\vec{\sigma}_i\cdot\vec{\sigma}_j-\sum_j\vec{\sigma}_j\cdot\vec{\sigma}_j\right)\\ &=\frac{\kappa}{2m^2}\left(4\vec{S}^{\,2}-3n\right)\end{aligned} \tag{17.2}$$

where $n = 3$ for baryons and 2 for mesons.

The real beauty of this picture is that in the masses of the ground state particles, we see the quark mass dependence of the color magnetic interaction. For example, the splitting between Ξ^* and Ξ is smaller than the Δ-N splitting because it involves the heavier s quarks. Likewise in the meson system, the K^*-K splitting is smaller than the ρ-π splitting for the same reason. The most amusing example along these lines is the Σ-Λ splitting. These two baryons consist of the same quarks, one u, one d and one s, and they have the same total spin. The difference is the way in which the spins are put together. In the Σ, because the isospin is 1, the u and d quarks are in a symmetric flavor state, and thus the spins must be aligned, because the state must also be symmetric in spin space. In the Λ, the isospin is 0, the u and d quarks are in an antisymmetric flavor state, and thus the spins are paired. The Λ has lower energy because the color magnetic interaction between the lighter u and d quarks is more important.

Furthermore, the ratio of m_s to $m_{u,d}$ required is about the same in the mesons and the baryons. You will show this in problem 17.B.

Although this simple picture was developed in the early days of QCD, it is still not known for certain whether its success is simply fortuitous, or whether it is telling us something important about QCD, or both! Probably, both. Certainly, there are reasons to believe that the picture is more complicated, because there are other ways of estimating the quark mass ratios that give very different results. But it is also true that the constituent quark model picture described in this chapter goes over very smoothly to a sensible description of mesons and baryons containing heavy c and b quarks. Thus even though, as explained in the last chapter, we cannot use more powerful symmetry arguments to understand these states because the symmetries are badly broken, we can still understand a lot about them by using the quark model directly. The simultaneous success of the constituent quark model for both heavy and light quark states convinces me that it does capture at least some important part of the physics of QCD bound states.

Problems

17.A. Suppose the quix, Q, described in (16.B) is a heavy spin zero particle. Then in the ground state $qq\overline{Q}$ bound states of an antiquix and two light quarks (any of u, d or s), the only spin dependence should come from the light quarks (because the ground state presumably has orbital angular momentum zero). Discuss the spectrum of all the $qq\overline{Q}$ ground state particles, spin 0 and spin 1, giving their $SU(3)$ properties and their spins. **Hint:** In the ground state, we expect the space wave function of the two quarks to be symmetric under exchange.

17.B. Estimate $m_{u,d}/m_s$ by comparing the $\rho - \pi$ mass splitting with the $K^* - K$ mass splitting. Make an independent estimate of the ratio using appropriate combinations of the Σ^*-Σ and Σ-Λ. mass differences.

***17.C.** Discuss semiquantitatively the mass spectrum of baryon states with a single c quark and a pair of light quarks (various combinations of u, d and s). You will need to find some experimental information about some of these bound states to get started. You will also need to think about which of the formulas in this chapter you can trust in this context.

Chapter 18

Unified Theories and $SU(5)$

The forces of the standard model of Elementary Particle Physics are shown in the table below:

"Known" Forces

force	E&M	weak	strong
range	∞	10^{-16} cm	10^{-13} cm
strength	$\frac{1}{137}$	$\approx \frac{1}{30}$	≈ 1
particle	photon	W and Z	gluons
mass	0	$\approx 100 m_P$	0 or $\approx m_P$

gravity
∞
$\approx 10^{-38} \frac{E^2}{m_P^2}$
graviton?
0

Gravity is separate, because if we were only interested in the physics of individual particles, we wouldn't know about it at all. It is only because we have some experience with huge collections of particles put together into planets and stars that we know about gravity.

18.1 Grand unification

Each of the other forces is associated with a Lie Algebra. This suggests that it may be possible to unify all the particle interactions as different aspects of a single underlying interaction, based on a single simple Lie algebra.

We will see in this chapter that all the particle interactions fit very neatly into the simple Lie algebra $SU(5)$. We will also discuss other embeddings, based on larger algebras. These theories are called **Grand Unified theories**, where the term "grand" is added for obscure historical reasons. We will not be able to discuss the full structure of these theories without the language of

 DOI: 10.1201/9780429499210-19

quantum field theory. But we can, at least, exhibit the Lie algebraic structure of grand unified theories in some detail. I hope that this will whet your appetite for a more complete study of the physics behind the group theory.[1]

We have already discussed the color $SU(3)$ theory of the strong interactions. The $SU(2) \times U(1)$ theory of the electroweak interactions is slightly more complicated. We will give only a superficial introduction here.

18.2 Parity violation, helicity and handedness

One of the salient features of the weak interactions is that they violate parity. Parity is the symmetry in which the signs of all space coordinates are changed. It is equivalent (up to a rotation) to reflection in a mirror. Spin 1/2 particles like electrons and quarks, if they are moving, can be characterized by their **helicity**, the component of the spin in the direction of motion. For a spin 1/2 particle, the helicity is $\pm 1/2$. Particles with helicity $1/2$ are said to be **right-handed**. Those with helicity $-1/2$ are said to be **left-handed**. For reasons that are not obvious, but which follow from basic principles of quantum field theory, the antiparticle of a right-handed particle is left-handed, and the antiparticle of a left-handed particle is right-handed. Helicity (or handedness) is not invariant under a parity transformation because a mirror interchanges left and right. Thus if some interaction acts differently on the right-handed and left-handed components of a particle, the interaction is parity violating. That is what the weak interactions do. A massive particle (at least if it carries a conserved particle number) must have both left-handed and right-handed components, because the helicity of a massive particle is not relativistically invariant. It changes sign depending on the reference frame. Thus electrons and their heavier cousins, muons and taus, and all the quarks have both left- and right-handed parts. But a massless particle need not have both components. Neutrinos are known to be very light, and for our purposes, we can treat them as massless.[2] Thus far, only left-handed neutrinos and their antiparticles, right-handed antineutrinos, have been observed.

For reasons which will become clear, it is useful to describe the symmetry properties of the creations and annihilation operators, rather than those of the states. We will restrict ourselves to the interactions of the lightest particle, the u and d quarks and the electron and its neutrino. The heavier particles all seem to be copies of one of these.

---

[1]See H. Georgi and S. L. Glashow, Phys. Rev. Lett. **32** (1974) 438.

[2]There are a number of experimental indications of tiny neutrino masses. Note also that a neutrino could have a small mass that violates particle number.

The Glashow-Weinberg-Salam theory of the weak and electromagnetic interactions treats the creation and annihilation operators as tensor operators under an $SU(2) \times U(1)$ Lie algebra. We will call the $SU(2)$ generators R_a and the $U(1)$ generator S. Consider the creation operators for the right-handed particles:

$$u^\dagger\,,\quad d^\dagger\,,\quad e^\dagger\,,\quad \overline{u}^\dagger\,,\quad \overline{d}^\dagger\,,\quad \overline{e}^\dagger\,,\quad \overline{\nu}^\dagger\,, \tag{18.1}$$

which create, respectively, right-handed u quarks, d quarks, electrons, $\overline{u}$ antiquarks, $\overline{d}$ antiquarks, positrons, and antineutrinos. The color index of the quark and antiquark operators is suppressed for now because it plays no role in the weak interactions. We will come back to it later. Color $SU(3)$ commutes with the electroweak $SU(2) \times U(1)$. Under the $SU(2)$ algebra, the positron and the antineutrino form a doublet, transforming according to the spin 1/2 representation. Like the proton and neutron creation operators under isospin, these two creation operators can be regarded as the 1 and 2 components of a tensor operator:

$$\overline{\ell}_1^\dagger = \overline{e}^\dagger\,, \qquad \overline{\ell}_2^\dagger = \overline{\nu}^\dagger\,. \tag{18.2}$$

Likewise, the $\overline{d}$ and $\overline{u}$ antiquarks transform like a doublet under $SU(2)$ and we can write them as components of a tensor operator:

$$\overline{\psi}_1^\dagger = \overline{d}^\dagger\,, \qquad \overline{\psi}_2^\dagger = \overline{u}^\dagger\,. \tag{18.3}$$

The $\overline{\psi}$ field is also a tensor operator under color $SU(3)$, transforming like a $\overline{3}$, but for now, we have not written the $SU(3)$ index explicitly.

Then the commutation relations of these creation operators with the $SU(2)$ and $U(1)$ generators are the following:

$$\begin{aligned}
\left[R_a, u^\dagger\right] &= 0\,, & \left[S, u^\dagger\right] &= 2u^\dagger/3\,;\\
\left[R_a, d^\dagger\right] &= 0\,, & \left[S, d^\dagger\right] &= -d^\dagger/3\,;\\
\left[R_a, e^\dagger\right] &= 0\,, & \left[S, e^\dagger\right] &= -e^\dagger\,;\\
\left[R_a, \overline{\psi}_r^\dagger\right] &= \overline{\psi}_s^\dagger\,[\sigma^a]_{sr}/2\,, & \left[S, \overline{\psi}_r^\dagger\right] &= -\overline{\psi}_r^\dagger/6\,;\\
\left[R_a, \overline{\ell}_r^\dagger\right] &= \overline{\ell}_s^\dagger\,[\sigma^a]_{sr}/2\,, & \left[S, \overline{\ell}_r^\dagger\right] &= \overline{\ell}_r^\dagger/2\,.
\end{aligned} \tag{18.4}$$

Thus all the fields are tensor operators, with $\overline{\psi}^\dagger$ and $\overline{\ell}^\dagger$ transforming like doublets under the $SU(2)$, and the rest transforming like singlets.

The annihilation operators for the right-handed particles are just the adjoints of (18.1). They transform under the complex conjugate representation. In particular, all the S values change sign. The creation operators for the left-handed particles transform like the annihilation operators for their right-handed antiparticles. Thus, for example, the creation operator for a left-handed u quark transforms like the annihilation operator from a right-handed $\overline{u}$ antiquark.

The S values in (18.4) have been constructed so that the electric charge operator, Q, is

$$Q = R_3 + S\,. \tag{18.5}$$

You can check that

$$\begin{aligned}
&\left[Q, u^\dagger\right] = 2u^\dagger/3\,, && \left[Q, d^\dagger\right] = -d^\dagger/3\,, \\
&\left[Q, e^\dagger\right] = -e^\dagger\,, && \\
&\left[Q, \overline{u}^\dagger\right] = -2\overline{u}^\dagger/3\,, && \left[Q, \overline{d}^\dagger\right] = \overline{d}^\dagger/3\,, \\
&\left[Q, \overline{e}^\dagger\right] = \overline{e}^\dagger\,, && \left[Q, \overline{\nu}^\dagger\right] = 0\,.
\end{aligned} \tag{18.6}$$

Now the idea of the electroweak standard model is that as in QCD, each of the generators of the Lie algebra is associated with a force particle. The R_a with three W_as, and the S with X. One linear combination of W_3 and X is the photon which couples to the electric charge, Q. Thus electromagnetism is contained within this larger theory. The other particles, the $W^\pm$ (corresponding to the complex combinations $W_1 \pm iW_2$) and the Z (the combination of W_3 and X orthogonal to the photon) are responsible for the weak interactions.

18.3 Spontaneously broken symmetry

There is something peculiar going on here. The $SU(2) \times U(1)$ cannot really be a symmetry. If it were, the weak interactions would have long range, like the electromagnetic interactions. Instead, the weak interactions have very short range and their force particles, the W and Z, are massive. Furthermore, if the $SU(2) \times U(1)$ symmetry remained unbroken, the quarks and the electron would have to be massless particles, because the weak interactions treat their left-handed and right-handed helicity components differently, which is consistent with relativity only for massless particles. Some new physics gives mass to the quarks and leptons, and to all but one linear combination (the photon) of R_3 and X, without destroying the consistency of the theory. What is this physics?

The answer is the structure of the vacuum state. The $SU(2) \times U(1)$ generators commute with the Hamiltonian, but the vacuum state of the world is not an $SU(2) \times U(1)$ singlet. Hence, $SU(2) \times U(1)$ is not a good symmetry on the states of the physical Hilbert space, all of which are built on the asymmetric vacuum state. The quarks and leptons and the $W^\pm$ and Z are not degenerate with the massless photon in our vacuum.

This situation is called **spontaneous symmetry breaking**. The $SU(2) \times U(1)$ symmetry is said to be spontaneously broken down to the $U(1)$ of electromagnetism, because only the linear combination $Q = R_3 + S$ treats the vacuum state of our world as a singlet. The resulting theory gives a very good description of the weak and electromagnetic interactions.

18.4 Physics of spontaneous symmetry breaking

It would be logical to explain, at this point, what new physics it is that produces spontaneous symmetry breaking. Alas, we still do not know what this new physics is. There is, however, a model of the process that is easy to explain and worth understanding. This is the hypothesis of the Higgs field. A **field** is a quantity defined at each point in space and time, like the electric and magnetic fields of electromagnetism. The electric and magnetic fields have several components that transform nontrivially under rotations and Lorentz transformations. But imagine that there exists a **scalar** field, which is invariant under rotations and Lorentz transformations. Such a field can have a non-zero value in the vacuum state without breaking rotation symmetry or Lorentz invariance. If the field transforms nontrivially under $SU(2) \times U(1)$, a non-zero vacuum value breaks the $SU(2) \times U(1)$ symmetry spontaneously.

In the $SU(2) \times U(1)$ electroweak theory, a Higgs field transforming like a doublet of $SU(2)$ and with $S = 1/2$ does the trick. Such a field (call it ϕ) transforms as

$$\left[R_a, \phi_r^\dagger\right] = \phi_s^\dagger \, [\sigma^a]_{sr}/2\,, \qquad \left[S, \phi_r^\dagger\right] = \phi_r^\dagger/2\,. \tag{18.7}$$

If such a field exists, its self interactions can be described by a potential, $V(\phi)$ which is just the energy stored in a constant ϕ field. $V(\phi)$ is actually only a function of $\phi^\dagger\phi$ because it is invariant under $SU(2) \times U(1)$. Now the lowest energy state corresponds to the minimum value of $V(\phi)$. But it may be that ϕ is not zero at the minimum of $V(\phi)$. A simple example of a potential with a non-zero value of ϕ at its minimum is

$$V(\phi) = \lambda\left(\phi^\dagger\phi - v^2\right)^2 \tag{18.8}$$

for $\lambda > 0$. This is minimized when $\phi^\dagger\phi = v^2$. Then, for example, we can take the vacuum value of ϕ to be

$$\phi_1 = 0\,, \qquad \phi_2 = v\,. \tag{18.9}$$

Notice that (18.7) implies that the combination $R_3 + S$ acting on the vacuum value of ϕ, (18.9), gives zero, so that this particular subgroup of $SU(2) \times U(1)$, associated with electromagnetism, is not broken by the Higgs field. But any other linear combination of $SU(2) \times U(1)$ generators acting on the vacuum value of ϕ gives a non-zero result, which means that these generators are spontaneously broken — they correspond to rotations from the physical vacuum to an unphysical vacuum state. The precise form of the unbroken combination depends on the particular choice, (18.9), for the vacuum value of ϕ. However, any other choice with $\phi^\dagger\phi = v^2$ gives the same physics because it is related to (18.9) by an $SU(2) \times U(1)$ transformation. Thus it does no harm to make the choice (18.9) which leads to a simple form for the unbroken $U(1)$ generator, $Q = R_3 + S$.

The Higgs field also allows the electron and quarks to get mass. The rule is that a Higgs field can produce a mass for a spin 1/2 particle if the tensor product of the representation of the right-handed particle and the representation of the corresponding antiparticle contains the representation of the Higgs field or its complex conjugate.[3] You can check that this is the case for the electron and quark fields.

18.5 Is the Higgs real?

There is overwhelming evidence that the $SU(3) \times SU(2) \times U(1)$ model of the strong and electroweak interactions, with the $SU(2) \times U(1)$ symmetry spontaneously broken down to the $U(1)$ of electromagnetism, is an excellent description of the interactions of elementary particles down to distances of the order of 10^{-16}cm. It is important to point out, however, that we do not yet know whether the Higgs field actually exists. It is quite possible that it is simply a mathematical metaphor for some other dynamics **with the same symmetry properties**. As of this writing, we do not know whether the Higgs is physics or mathematics. But we do know that if the Higgs is not real, there is some other new physics that does the symmetry breaking. One of the most important goals for particle physicists today is to either see the Higgs field directly, or to find the physics that replaces it.

[3]In quantum field theory, this makes it possible to write an $SU(3) \times SU(2) \times U(1)$ invariant interaction terms involving the Higgs field and the particle creation and annihilation operators that becomes a mass term when the Higgs field is replaced by its vacuum value.

18.6 Unification and $SU(5)$

The $SU(2) \times U(1)$ symmetry of (18.4) is a partial unification of the weak and electromagnetic interactions. It leaves many striking features of the physics of our world unexplained. One of these is the quantization of electric charge. The hydrogen atom is known to be electrically neutral to extraordinary accuracy. This implies that there is a relation between the charges of the quarks and that charge of the electron. However, (18.5), while it can describe the charges we see, does not explain them. The problem is the S generator. The values of R_3 are quantized because of the non-Abelian nature of the $SU(2)$ algebra. However, the values of S are completely arbitrary. They are chosen to describe the quantized charges we see, but if we could embed both R_3 and S into a simple group, then the values of S, like those of R_3, would be constrained by the structure of the algebra.

This idea was a major motivating factor in the search for unified theories in the early 70s. At first, we searched for a simple group that unified only the $SU(2) \times U(1)$ electroweak theory. This search proved fruitless. It was only when color $SU(3)$ was included as well that interesting unifications could be obtained.

Thus we ask, is it possible to include the color $SU(3)$ of the strong interactions and further unify $SU(3)$ and $SU(2) \times U(1)$ into a larger algebra, G, which is spontaneously broken down to $SU(3) \times SU(2) \times U(1)$?

We will say that a set of creation operators, $a^\dagger_{xr}$, transforms according to the representation $(D, d)_s$ of $SU(3) \times SU(2) \times U(1)$ if it satisfies

$$
\begin{aligned}
\left[T_a, a^\dagger_{xr}\right] &= a^\dagger_{yr}\, [T^D_a]_{yx} \,, \\
\left[R_a, a^\dagger_{xr}\right] &= a^\dagger_{xt}\, [R^d_a]_{tr} \,, \\
\left[S, a^\dagger_{xr}\right] &= s\, a^\dagger_{xr} \,.
\end{aligned}
\tag{18.10}
$$

Thus x is a color $SU(3)$ index, associated with the $SU(3)$ representation D. The r is an $SU(2)$ index, associated with the $SU(2)$ representation d. The s is the S quantum number. (18.10) and (18.5) imply that s must be simply the average electromagnetic charge of the representation, because for each representation,

$$
\operatorname{Tr} Q = \operatorname{Tr} R_3 + \operatorname{Tr} S = \operatorname{Tr} S \tag{18.11}
$$

(the trace of R_3 always vanishes because of theorem 8.9, or equivalently, because $SU(2)$ representations are symmetrical about $R_3 = 0$).

We know the color $SU(3)$ transformation properties of all the particles, so we can read off the representations of the creation operators for the right-handed particles from (18.4):

$$\begin{array}{ccc} u^\dagger : (3,1)_{2/3} , & d^\dagger : (3,1)_{-1/3} , & e^\dagger : (1,1)_{-1} , \\ \overline{\psi}^\dagger : (\overline{3},2)_{-1/6} , & \overline{\ell}^\dagger : (1,2)_{1/2} . \end{array} \tag{18.12}$$

where we have indicated the $SU(2)$ representations by their dimensions. The full $SU(3) \times SU(2) \times U(1)$ representation of the creation operators for the right-handed particles is thus,

$$(3,1)_{2/3} \oplus (3,1)_{-1/3} \oplus (1,1)_{-1} \oplus (\overline{3},2)_{-1/6} \oplus (1,2)_{1/2} . \tag{18.13}$$

The creation operators for the left-handed fields transform like the complex conjugate of the representation (18.13):

$$(\overline{3},1)_{-2/3} \oplus (\overline{3},1)_{1/3} \oplus (1,1)_{1} \oplus (3,2)_{1/6} \oplus (1,2)_{-1/2} . \tag{18.14}$$

where we have used the fact that the $SU(2)$ representation is real, $\overline{2} = 2$. Notice that (18.13) and (18.14) are not the same, and thus the representation is complex because of the parity violating habits of the electroweak $SU(2) \times U(1)$.

(18.13) is the starting point in the search for unifying algebras. We want to find an algebra G which contains $SU(3) \times SU(2) \times U(1)$ as a subgroup, and which has a representation transforming like (18.13) under this subgroup. The rank of G must be at least four if it is to contain the four commuting generators, T_3, T_8, R_3 and S. The simplest possibility is to try the rank four algebra, $SU(5)$. We will see later that the other simple rank 4 algebras could not possibly work because they do not have complex representations, and so there is no way they could describe the complex representation, (18.13).

$SU(5)$ has a five dimensional representation of course. It actually has two, because the 5 is complex, and thus the 5 and $\overline{5}$ (or [1] and [4]) are not equivalent. Can we find an $SU(2) \times U(1)$ subgroup of $SU(5)$ such that the 5 transforms like some five dimensional subset of the creation operators? The only possible such subset is

$$(3,1)_{-1/3} \oplus (1,2)_{1/2} . \tag{18.15}$$

The five dimensional subset,

$$(3,1)_{2/3} \oplus (1,2)_{1/2} , \tag{18.16}$$

cannot work because the generator S is not traceless on it. Thus by theorem 8.9, S could not possibly be an $SU(5)$ generator for this choice.

It is straightforward to embed $SU(3)\times SU(2)\times U(1)$ in $SU(5)$ to obtain (18.15). Take the $SU(3)$ generators to be traceless matrices acting on only the first three indices in the 5,

$$\begin{pmatrix} T_a & 0 \\ 0 & 0 \end{pmatrix}, \tag{18.17}$$

and the $SU(2)$ generators to be the traceless matrices acting on the last two,

$$\begin{pmatrix} 0 & 0 \\ 0 & R_a \end{pmatrix} \tag{18.18}$$

Then S is the generator that commutes with both (18.17) and (18.18),

$$\begin{pmatrix} -I/3 & 0 \\ 0 & I/2 \end{pmatrix} \tag{18.19}$$

Thus we can put the $d^\dagger$ and $\overline{\ell}^\dagger$ creation operators into an $SU(5)$ 5 , $\lambda_j^\dagger$, as follows:

$$\begin{aligned} &\lambda_x^\dagger = d_x{}^\dagger\,, \text{ for } x = 1 \text{ to } 3\,; \\ &\lambda_4^\dagger = \overline{\ell}_1^\dagger = \overline{e}^\dagger\,, \\ &\lambda_5^\dagger = \overline{\ell}_2^\dagger = \overline{\nu}^\dagger\,, \end{aligned} \tag{18.20}$$

What about the rest of (18.13). What remains is $u^\dagger$, $e^\dagger$, and $\overline{\psi}^\dagger$ which transforms as

$$(3,1)_{2/3} \oplus (1,1)_{-1} \oplus (\overline{3},2)_{-1/6}\,. \tag{18.21}$$

This representation is 10 dimensional. $SU(5)$ has a 10 and a $\overline{10}$, the [2] and [3] representations. The 10 is an antisymmetric tensor product of two 5's, so we can determine how it transforms under the $SU(3) \times SU(2) \times U(1)$ subgroup by taking the antisymmetric product of (18.15) with itself (you can easily check that this is equivalent to using the general rule described in chapter 13). The $SU(3)$ and $SU(2)$ representations compose in the standard way. The S quantum number simply add. Thus,

$$\begin{aligned} &\left[(3,1)_{-1/3} \oplus (1,2)_{1/2}\right] \otimes \left[(3,1)_{-1/3} \oplus (1,2)_{1/2}\right]_{AS} \\ &\quad = \left[(\overline{3},1)_{-2/3} \oplus (1,1)_1 \oplus (3,2)_{1/6}\right]\,. \end{aligned} \tag{18.22}$$

This is just the complex conjugate of the representation (18.21). Thus what we actually want is the $\overline{10}$ representation, and then we can write the remaining right-handed fermion creation operators in an $SU(5)$ representation antisymmetric in two upper indices, $\xi^{jk\,\dagger} = -\xi^{kj\,\dagger}$, with

$$\begin{aligned}
\xi^{ab\,\dagger} &= \epsilon^{abc}\, u_c^\dagger \text{ for } a,b,c = 1 \text{ to } 3\,, \\
\xi^{a4\,\dagger} &= \overline{\psi}_2^{a\,\dagger} = \overline{u}^{a\,\dagger} \text{ for } a = 1 \text{ to } 3\,, \\
\xi^{a5\,\dagger} &= \overline{\psi}_1^{a\,\dagger} = \overline{d}^{a\,\dagger} \text{ for } a = 1 \text{ to } 3\,, \\
\xi^{45\,\dagger} &= e^\dagger\,.
\end{aligned} \tag{18.23}$$

This is the standard $SU(5)$ model, with the creation operators for the right- handed particles transforming like $5 \oplus \overline{10}$, or equivalently, with the creation operators for the left-handed particles transforming like $\overline{5} \oplus 10$. The most interesting thing about it is the way that everything fits.

18.7 Breaking $SU(5)$

There are two issues that must be addressed in discussing the spontaneous breaking of $SU(5)$ symmetry.

1. How is the symmetry broken down from $SU(5)$ to $SU(3) \times SU(2) \times U(1)$?

2. How do the quarks and leptons get mass?

Let us answer these in the Higgs field language.

There is a very simple solution to the problem of breaking $SU(5)$ down to $SU(3) \times SU(2) \times U(1)$, which in fact is very much analogous to Gell-Mann's solution to the problem of breaking $SU(3)$ down to the $SU(2)\times U(1)$ of isospin and hypercharge. The S generator in the adjoint representation has all the desired properties for the vacuum value of a Higgs field. Just as the hypercharge generator in the adjoint representation of $SU(3)$ commutes with isospin and hypercharge, so the $U(1)$ generator, S, in the adjoint of $SU(5)$ commutes with $SU(3) \times SU(2) \times U(1)$. Thus the 24 representation with a vacuum value in the S direction[4] is a good choice for the Higgs field that breaks $SU(5)$ down to $SU(3) \times SU(2) \times U(1)$.

Two conditions must be satisfied in order for a Higgs field to give mass directly to the fermions in an $SU(5)$ theory.[5] The Higgs representation (or its

[4]Unlike the situation in the Higgs doublet breaking $SU(2) \times U(1)$, there are physically inequivalent directions in the adjoint of $SU(5)$. One can show that the S direction is a possible vacuum value.

[5]More complicated, indirect mechanisms involving quantum mechanical "loop" effects are also possible, but we will not discuss these.

complex conjugate) must appear in the tensor product of the $SU(5)$ representations in which the fermion and its antifermion appear. And the Higgs representation must have a component that transforms under $SU(3) \times SU(2) \times U(1)$ like the Higgs field of the $SU(3) \times SU(2) \times U(1)$ model or its complex conjugate.

The right-handed positron ($\overline{e}$) and d quark fields are in the 5, while their antiparticles, the electron and $\overline{d}$ fields are in the $\overline{10}$. In $SU(5)$, $5 \otimes \overline{10}$ is $[1] \otimes [3] = [4] \oplus [3,1]$, or $\overline{5} \oplus \overline{45}$ (see problem 13.B).

From (18.15), you can check that the 5 contains a component that transforms like the $SU(3) \times SU(2) \times U(1)$ Higgs field — the $(1,2)_{1/2}$. You can check (see problem 18.E) that the 45 has the desired component as well.

Thus the electron and d quark masses can come from either of the two Higgs representations, 5 or 45.

The right-handed u and $\overline{u}$ are both in the $\overline{10}$, and

$$[3] \otimes [3] = [1] + [4,2] + [3,3] = 5 + 45 + 50 \tag{18.24}$$

We already know that both the 5 and 45 of $SU(5)$ Higgs fields have the appropriate component to act as the $SU(3) \times SU(2) \times U(1)$ Higgs field, thus they can give mass to the u quark. The 50, however, does not have any component transforming like $(1,2)_{\pm 1/2}$ under $SU(3) \times SU(2) \times U(1)$ (see problem 18.E). Thus this representation is not useful for giving mass to the u quark.

18.8 Proton decay

Another fascinating thing about the $SU(5)$ theory is that quarks, antiquarks and the electron all appear in the same irreducible representation. Because of this, some of the $SU(5)$ interactions do not conserve baryon number. Thus $SU(5)$ unification leads to proton decay. The proton is known to be extremely long-lived. But if the vacuum value of the 24 Higgs field is extremely large, then the interactions that cause proton decay are short range and the probability of two quarks inside the proton interacting to cause the proton to decay is very small. It turns out that we know approximately what the vacuum value of the 24 must be in order to explain the observed differences between the color $SU(3)$, electroweak $SU(2)$ and $U(1)$ forces.[6] Thus the rate of proton decay can actually be predicted in a given $SU(5)$ model. Since $SU(5)$ was first found theoretically, experimenters have looked for proton decay with

[6]This is not something that can be understood using group theory alone. It involves the dynamics.

more and more sensitive experiments, so far without success. In fact, the simplest version of the $SU(5)$ unified theory is fairly convincingly ruled out by these experiments. But slightly elaborated versions of the $SU(5)$ theory, particularly supersymmetric versions, are still extremely promising.

Problems

18.A. Check explicitly that mass terms for the electron and the u and d quarks are allowed in the $SU(2) \times U(1)$ model in which the symmetry is broken by the Higgs field of (18.7). **Hint:** see the discussion on page 230.

18.B. Find the symmetric tensor product of (18.15) with itself.

18.C. Do the same for (18.21).

18.D. Consider the operator

$$O = \bar{e}^{\dagger} \, \epsilon^{abc} \, u_a u_b d_c$$

where u and d are quark annihilation operators. Show that if the operator O appears in the Hamiltonian, it has the right charge and color properties to allow a proton to decay into a π^0 and a positron.

18.E. How do the 45 ([4,2]) and 50 ([3,3]) of $SU(5)$ transform under the $SU(3) \times SU(2) \times U(1)$ subgroup? Hint: it may be easier to answer the question for the complex conjugate representations and then complex conjugate.

Chapter 19

The Classical Groups

The are four infinite series of simple Lie algebras, generating what are called the classical groups. The first of these consists of the $SU(N)$ algebras that we have studied already. Their Dynkin diagrams have the form

$$\bigcirc\!\!-\!\!\bigcirc\cdots\bigcirc\!\!-\!\!\bigcirc \tag{19.1}$$

The rank n algebra $SU(n+1)$ was called A_n by Cartan, who first classified these things. In this brief section, we will go over the others very lightly, before we do the complete classification. Later, we will discuss each of them in detail.

19.1 The $SO(2n)$ algebras

The orthogonal $2n \times 2n$ matrices form the group of rotations in $2n$-dimensional real space. The group is generated by the imaginary antisymmetric $2n \times 2n$ matrices, of which $n(2n-1)$ are independent. We can choose the Cartan generators as follows:

$$[H_m]_{jk} = -i(\delta_{j,2m-1}\,\delta_{k,2m} - \delta_{k,2m-1}\,\delta_{j,2m}) \tag{19.2}$$

This amounts to breaking up the $2n$ dimensional space into n different two dimensional subspaces on which the different Cartan generators act. H_m is a little Pauli matrix, σ_2, in the appropriate 2 dimensional space, with zeros everywhere else. The eigenvectors of σ_2 are given by

$$\begin{pmatrix} 0 & -i \\ i & 0 \end{pmatrix}\begin{pmatrix} 1 \\ \pm i \end{pmatrix} = \pm 1 \begin{pmatrix} 1 \\ \pm i \end{pmatrix} \tag{19.3}$$

 DOI: 10.1201/9780429499210-20

Thus the eigenvectors of the H_m are

$$[|\pm e^k\rangle]_j = \delta_{j,2k-1} \pm i\, \delta_{j,2k} \tag{19.4}$$

satisfying

$$H_m\, |\pm e^k\rangle = \pm\delta_{km}\, |\pm e^k\rangle \tag{19.5}$$

Thus the weight vectors are $\pm$ the unit vectors e^k with components

$$[e^k]_m = \delta_{km} \tag{19.6}$$

(as usual, you must not confuse the space on which the generators act with the Cartan space in which the weight vectors live).

The roots connect one subspace with another in all possible ways, so that the roots vectors are

$$\begin{array}{ll} \text{roots} & \pm e^j \pm e^k \;\text{ for }\; j \neq k \\ \text{positive roots} & e^j \pm e^k \;\text{ for }\; j < k \\ \text{simple roots} & e^j - e^{j+1} \;\text{ for }\; j = 1 \ldots n-1 \\ & \quad \text{and }\; e^{n-1} + e^n \end{array} \tag{19.7}$$

The Dynkin diagram is

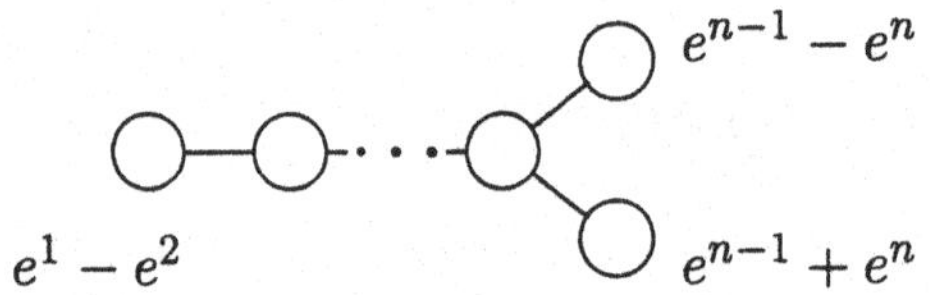

(19.8)

The algebra $SO(2n)$ was called D_n by Cartan.

19.2 The $SO(2n+1)$ algebras

For a $2n+1$ dimensional space, we can find n two dimensional subspaces, but there is a one dimensional subspace left over. Thus for the rotation group in $2n+1$ dimensional space, generated by the imaginary antisymmetric $(2n+1) \times (2n+1)$ matrices, we can choose the Cartan generators as before, and there will be weights of the form $\pm e^k$ corresponding to vectors in the kth two dimensional subspace. But there is also a 0 weight, associated with the extra dimension. Again, just as before, there will be roots connecting the various two dimensional subspaces, with the same roots as in $SO(2n)$. But now there are also roots connecting the extra one dimensional subspace with the others, with roots $\pm e^j$. Thus the roots are

$$\begin{array}{lll} \text{roots} & \pm e^j \pm e^k \text{ for } j \neq k \text{ and } \pm e^j \\ \text{positive roots} & e^j \pm e^k \text{ for } j < k \text{ and } e^j \\ \text{simple roots} & e^j - e^{j+1} \text{ for } j = 1 \ldots n-1 \\ & \text{and } e^n \end{array} \tag{19.9}$$

The root $e^{n-1} + e^n$ is not simple in $SO(2n+1)$ because it is a sum of $e^{n-1} - e^n$ and twice e^n. The Dynkin diagram is

$$\underset{e^1 - e^2}{\bigcirc} \!-\! \bigcirc \cdots \bigcirc \!-\! \bigcirc \!=\! \underset{e^n}{\bigcirc} \tag{19.10}$$

The algebra $SO(2n+1)$ was called B_n by Cartan.

19.3 The $Sp(2n)$ algebras

Finally consider the $2n \times 2n$ dimensional hermitian matrices which are tensor products of 2×2 matrices and $n \times n$ matrices, of the following form:

$$1 \otimes A\,, \qquad \sigma_1 \otimes S_1\,, \qquad \sigma_2 \otimes S_2\,, \qquad \sigma_3 \otimes S_3 \tag{19.11}$$

where the σ_a are the Pauli matrices, A is an antisymmetric $n \times n$ matrix, and S_1, S_2 and S_3 are symmetric $n \times n$ matrices. You can check that these close under commutation because of the properties of the Pauli matrices. The subset

$$1 \otimes A\,, \qquad \sigma_3 \otimes S_3 \tag{19.12}$$

for traceless S_3 generate an $SU(n)$ subalgebra of $Sp(2n)$ of the form

$$\begin{pmatrix} T_a & 0 \\ 0 & -T_a^* \end{pmatrix} \tag{19.13}$$

which generates the reducible representation $n \oplus \overline{n}$ with weights ν^j and $-\nu^j$ where the ν^j vectors are defined in (13.9). We will take the first $n-1$ elements of our Cartan subalgebra to be the Cartan subalgebra of this $SU(n)$ subgroup, $H_1, H_2 \ldots H_{n-1}$. The final element of the Cartan subalgebra is then

$$H_n = \sigma_3 \otimes I/\sqrt{2n} \tag{19.14}$$

All the generators of the $SU(n)$ subalgebra commute with H_n, thus the roots of the subalgebra have $H_n = 0$. We already know that the other components are given by the $SU(n)$ root vectors, $\nu^j - \nu^k$. The other roots correspond to the matrices $(\sigma_1 \pm i\sigma_2) \otimes S_{k\ell}$ where

$$[S_{k\ell}]_{ij} = (\delta_{ik}\,\delta_{j\ell} + \delta_{i\ell}\,\delta_{jk}) \tag{19.15}$$

with $H_n = \pm\sqrt{2/n}$ and $H_m = \pm[\nu^k + \nu^\ell]_m$.

I should say that this is a pretty stupid way of writing the roots. Later, when we talk about $Sp(2n)$ in more detail, we will introduce a better notation in which all these relations are more obvious. Anyhow, if we take ν^{n+1} to be a unit vector orthogonal to all the ν^j for $j = 1$ to n, the roots can be written as

$$\begin{array}{ll}
\text{roots} & \nu^j - \nu^k \text{ for } j \neq k \\
 & \text{and } \pm\left(\nu^j + \nu^k + \sqrt{\frac{2}{n}}\,\nu^{n+1}\right) \\
\text{positive roots} & \nu^j - \nu^k \text{ for } j < k \\
 & \text{and } \left(\nu^j + \nu^k + \sqrt{\frac{2}{n}}\,\nu^{n+1}\right) \\
\text{simple roots} & \nu^j - \nu^{j+1} \text{ for } j = 1 \ldots n-1 \\
 & \text{and } \left(2\nu^n + \sqrt{\frac{2}{n}}\,\nu^{n+1}\right)
\end{array} \tag{19.16}$$

The Dynkin diagram is

$$\bigcirc\!\!-\!\!\bigcirc\cdots\bigcirc\!\!-\!\!\bigcirc\!\!=\!\!\bigcirc \tag{19.17}$$

$\nu^1 - \nu^2$ $\qquad\qquad$ $2\nu^n + \sqrt{\frac{2}{n}}\,\nu^{n+1}$

where the $SU(n)$ roots are shorter. The algebra $Sp(2n)$ was called C_n by Cartan.

19.4 Quaternions

A useful way of thinking about the classical groups is as rotations groups in various spaces. Obviously, $SO(N)$ is the group of rotations in a real, N dimensional space. The group generators, X, in the defining representation are pure imaginary. The product iX that is exponentiated to get the representation of the group elements is completely real and antisymmetric. Similarly, $SU(N)$ is the group of "rotations" in a complex N dimensional space that preserve the norm of complex vectors. The real part of iX in the defining representation is antisymmetric, and generates an $SO(N)$ subgroup of $SU(N)$. The imaginary part of iX is symmetric and traceless.

Finally, $Sp(2N)$ can be thought of as the group of rotations in a **quaternionic** N dimensional space, where the elements are quaternions, objects of the form

$$B + \vec{j} \cdot \vec{C} \tag{19.18}$$

where the components of $\vec{j}$ are the quaternionic units, satisfying

$$j_a j_b = -\delta_{ab} + \epsilon_{abc} j_c \tag{19.19}$$

Thus a vector in the N dimensional quaternionic space looks like

$$Q \equiv \begin{pmatrix} q^1 \\ \vdots \\ q^N \end{pmatrix} = \begin{pmatrix} B^1 + \vec{j} \cdot \vec{C}^1 \\ \vdots \\ B^N + \vec{j} \cdot \vec{C}^N \end{pmatrix} \tag{19.20}$$

In this language, iX has an antisymmetric part iA, plus three independent "complex" symmetric components,

$$j_1 S_1 \,, \quad j_2 S_2 \,, \quad j_3 S_3 \,, \tag{19.21}$$

proportional to the three quaternionic imaginary units, so that the iX have the form

$$iA \,, \quad j_1 S_1 \,, \quad j_2 S_2 \,, \quad j_3 S_3 \,. \tag{19.22}$$

To see the connection of the quaternionic description of $Sp(2N)$ with (19.11), note that the quaternionic units $\vec{j}$ are equivalent to $-i$ times the Pauli matrices,

$$\vec{j} \to -i\vec{\sigma} \,, \tag{19.23}$$

so that (19.21) is related to the terms in (19.11) proportional to the Pauli matrices. To see how the $Sp(2N)$ transformations in a $2N$ dimensional space are equivalent to quaternionic rotations, consider simply building the N dimensional space of quaternions in terms of the Pauli matrices, so that the quaternionic N-vector becomes a $2N \times 2$ matrix of the form

$$Q \to \begin{pmatrix} B^1 - i\vec{\sigma} \cdot \vec{C}^1 \\ \vdots \\ B^N - i\vec{\sigma} \cdot \vec{C}^N \end{pmatrix} = \begin{pmatrix} B^1 - iC_3^1 & -C_2^1 - iC_1^1 \\ C_2^1 - iC_1^1 & B^1 + iC_3^1 \\ \vdots & \vdots \\ B^N - iC_3^N & -C_2^N - iC_1^N \\ C_2^N - iC_1^N & B^N + iC_3^N \end{pmatrix} \tag{19.24}$$

The action of (i times) the $Sp(2N)$ generators on the complex $2N$-vector defined by the first column of (19.24) is equivalent to the action of (19.22) on (19.24).

What makes this construction work is that the real numbers, the complex numbers, and the quaternions share a very important property. For all of them, an absolute value, $|z|$, can be defined with the properties that

$$|z_1\, z_2| = |z_1|\,|z_2| \qquad\qquad |z| = 0 \Rightarrow z = 0 \tag{19.25}$$

You are familiar with how this works for the real numbers and the complex numbers. For the quaternions, the absolute value is

$$\left|B + \vec{\jmath}\cdot\vec{C}\right| \equiv \sqrt{B^2 + \vec{C}^{\,2}} \tag{19.26}$$

Note that it satisfies

$$\left|B + \vec{\jmath}\cdot\vec{C}\right|^2 = \left(B - \vec{\jmath}\cdot\vec{C}\right)\left(B + \vec{\jmath}\cdot\vec{C}\right) = \left(B + \vec{\jmath}\cdot\vec{C}\right)^*\left(B + \vec{\jmath}\cdot\vec{C}\right) \tag{19.27}$$

where complex conjugation is defined as usual as changing the sign of the imaginary parts (all three of them in this case).

The existence of the absolute value with these properties means that there is a natural norm defined on the N dimensional quaternionic vector space:

$$||Q||^2 = \sum_{j=1}^{N} |q^j|^2 \tag{19.28}$$

$Sp(N)$ is the group of rotations that preserve this norm while preserving the structure of the quaternionic space.

One might wonder why there are not other possibilities for classical groups. The reason is that it is not so easy to satisfy the condition (19.25). This condition, plus the requirement that addition, subtraction, multiplication and division (except by 0) can all be consistently defined, is the defining property of a structure called a **division algebra**. There is only one other such structure, the octonians, with seven imaginary units, which we will discuss further in Chapter 27. Unfortunately, the multiplication law for octonians is not associative, so in general, rotations in an octonionic space do not form a group. However, the peculiar Lie algebras that do not fit into any of the classical groups are all related to octonions on one way or another. We have already seen the algebra G_2, and we will discuss some of these bizarre algebras further in chapter 27.

Problems

19.A. Consider the 36 matrices,

$$\sigma_a\,, \quad \tau_a\,, \quad \eta_a\,, \quad \sigma_a \tau_b \eta_c$$

where σ, τ, and η are independent sets of Pauli matricies. Show that these matrices form a Lie algebra. Find the roots, the simple roots and the Dynkin diagram. What is the algebra?

19.B. Consider the 28 matrices,

$$\sigma_a\,, \quad \tau_a\,, \quad \eta_3\,, \quad \sigma_a \eta_1\,, \quad \sigma_a \eta_2\,, \quad \tau_a \eta_1\,, \quad \tau_a \eta_2\,, \quad \sigma_a \tau_b \eta_3\,.$$

where σ, τ, and η are independent sets of Pauli matricies. Show that these matrices form a Lie algebra. Find the roots, the simple roots and the Dynkin diagram. What is the algebra?

Chapter 20

The Classification Theorem

We will now classify all of the simple Lie algebras following the argument of Dynkin. We will do this with simple analytical geometrical arguments based on the master formula, (6.36). It may be that mathematicians can visualize the meaning of the arguments in multidimensional space. I can't do that. But the analytical arguments are simple enough.

20.1 Π-systems

We know that the simple roots of any simple Lie algebra have the following properties:

A. They are linearly independent vectors.

B. If α and β are distinct simple roots, $2\alpha \cdot \beta/\alpha^2$ is a non-positive integer.

To ensure that a system of roots satisfying **A** and **B** yields a simple Lie algebra, we need one additional condition:

C. The simple root system is **indecomposable**.

A system of roots is **decomposable** if it can be split into two mutually orthogonal subsystems. A system is **indecomposable** if it is not decomposable.

It is easy to see that for decomposable simple-root systems, the simple roots in the two orthogonal subsystems commute (p and q are zero for all pairs), and the entire system of roots splits into two commuting subsets. Each subsystem, along with the Cartan generators associated with the subspace it spans, forms an invariant subalgebra. The group associated with a decomposable root system is not simple. However, it is **semi-simple**, which means

DOI: 10.1201/9780429499210-21

that is has no Abelian invariant subalgebra. Because the subalgebras commute, the groups they generate also commute. A group of this kind that is built out of two commuting subgroups, G_1 and G_2, is called a **direct product** of the two subgroups, $G = G_1 \times G_2$. An elementary example is the algebra you studied in problem 8.B, in which the Dynkin diagram consisted of two disconnected circles, associated with the group $SU(2) \times SU(2)$. All the semisimple Lie groups are direct products of simple Lie groups.

Dynkin calls a system of vectors satisfying **A**, **B** and **C** a Π-system. All we need do to classify the simple Lie algebras is to classify the possible Π-systems. This is just geometry. We know that the system of vectors can be constructed (up to a conventional normalization) from the Dynkin diagram, and we will use the term Π-system to refer interchangeably to the system of vectors or the Dynkin diagram that represents it. In terms of Dynkin diagrams, **B** is just automatic, because the diagram just determines the integers; **C** is the condition that the diagram cannot be taken apart into two diagrams without cutting any lines; and **A** is equivalent to the condition that all the simple roots are positive, from which we showed that linear independence followed if **B** is satisfied.

Let us now do some geometry.

lemma 1. The only Π-systems of three vectors are

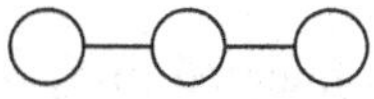 and 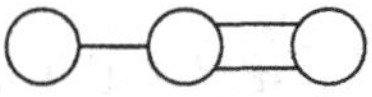 (20.1)

This follows from the simple fact that the sum of the angles between any three linearly independent vectors is less than 360°. The only possible angles in a Π-system are 90°, 120°, 135° and 150° and only one 90° degree angle is allowed by indecomposability, so these are the only possibilities. We will later see that there are three important systems that satisfy **B** and **C**, but not **A**:

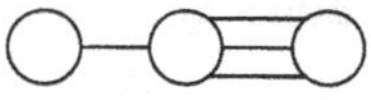 and 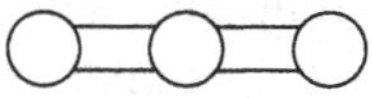 (20.2)

and 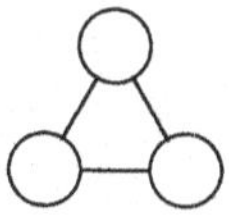

These do not satisfy **A** because the sum of the angles is 360° so the three vectors are coplanar and thus not linearly independent.

In fact, **lemma 1** is an incredibly strong constraint, because if we take from any Dynkin diagram a connected subset of circles, the result is another Dynkin diagram. This is true because the lengths and angles are unchanged and (because of **B**) the subset of vectors must also be linearly independent.

Thus any indecomposable subsystem of a Π-system is also a Π-system. Thus any three connected vectors in any Π-system must be in one of the two forms in (20.1).

A very simple corollary of this is that no triple line can appear in a Π-system of three or more vectors. Thus the only Π-system containing a triple line is

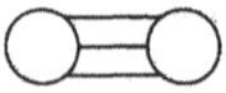

(20.3)

corresponding to the algebra G_2.

lemma 2. If a Π-system contains two vectors connected by a single line, the diagram obtained by shrinking the line away and merging the two vectors into a single circle is another Π-system.

Let α and β be the two vectors and Γ the set of all the other vectors in the Π-system. Because of **lemma 1**, we know that Γ contains no vector connected to both α and β. Thus if a vector γ is connected to α, then $\gamma \cdot \beta = 0$, and if γ is connected to β, then $\gamma \cdot \alpha = 0$. We also know that $\alpha + \beta$ has the same length as α and β. Therefore

$$\begin{aligned} &\text{if } \gamma \in \Gamma \text{ is a vector connected to } \alpha, \\ &\text{then } \gamma \cdot (\alpha + \beta) = \gamma \cdot \alpha \\ &\text{if } \gamma' \in \Gamma \text{ is a vector connected to } \beta, \\ &\text{then } \gamma' \cdot (\alpha + \beta) = \gamma' \cdot \beta \end{aligned} \tag{20.4}$$

So the set $\alpha + \beta$ and Γ is the shrunken Π-system.

Lemma 2 has two important corollaries. **a.** No Π-system has more than one double line; and **b.** no Π-system contains a closed loop. Either configuration could be shrunk into conflict with **lemma 1.**

lemma 3. If the configuration

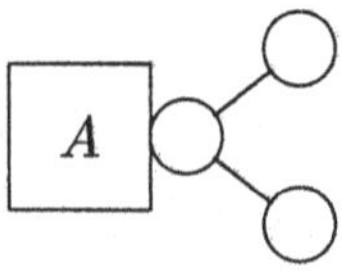

A

(20.5)

is a Π-system for some subdiagram, A, then

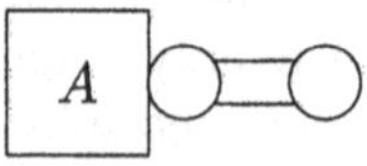

A

(20.6)

is also a Π-system.

To see this, label the vectors as follows:

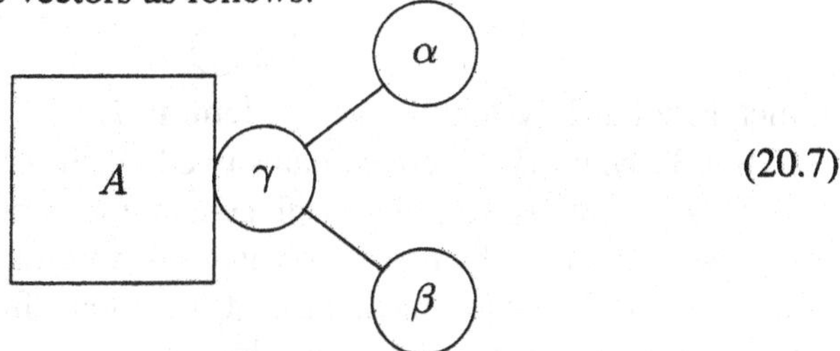

(20.7)

We know that $\alpha \cdot \beta = 0$ and

$$\begin{aligned} \frac{2\alpha \cdot \gamma}{\alpha^2} &= \frac{2\alpha \cdot \gamma}{\gamma^2} \\ = \frac{2\beta \cdot \gamma}{\beta^2} &= \frac{2\beta \cdot \gamma}{\gamma^2} = -1 \end{aligned} \tag{20.8}$$

from which it follows that

$$\begin{aligned} \frac{2\gamma \cdot (\alpha + \beta)}{\gamma^2} &= -2 \\ \frac{2\gamma \cdot (\alpha + \beta)}{(\alpha + \beta)^2} &= -1 \end{aligned} \tag{20.9}$$

Thus the following is the Dynkin diagram for the shrunken Π-system:

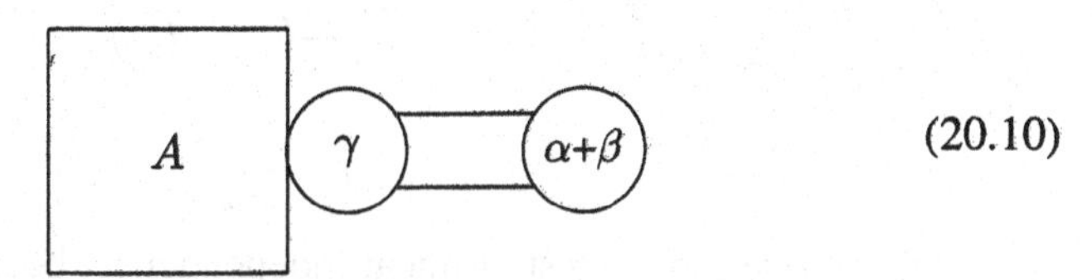

(20.10)

A corollary is that the only branches in a Π-system have the form of three single lines coming from a central circle, like

(20.11)

because anything with more branches, such as

(20.12)

or a double line such as

(20.13)

can be shrunk to

(20.14)

which is not a Π-system because of **lemma 1.**

Similarly, no Π-system contains two branches.

This is as far as we can go with general theorems. We must now consider some peculiar special cases, associated, as we will see, with what are called the exceptional groups (meaning that they don't fall into any of the infinite families discussed in the previous chapter).

lemma 4. No Π-system contains any of the following diagrams:

(a)

(b)

(c)

(d)

All of these fail the test of linear independence, because we can find a set of numbers μ_j such that, if the vectors in the diagram are α^j, the sum

$$\left(\sum_j \mu_j \, \alpha^j\right)^2 = 0 \tag{20.15}$$

The appropriate values of μ_j are indicated inside the corresponding circles below:

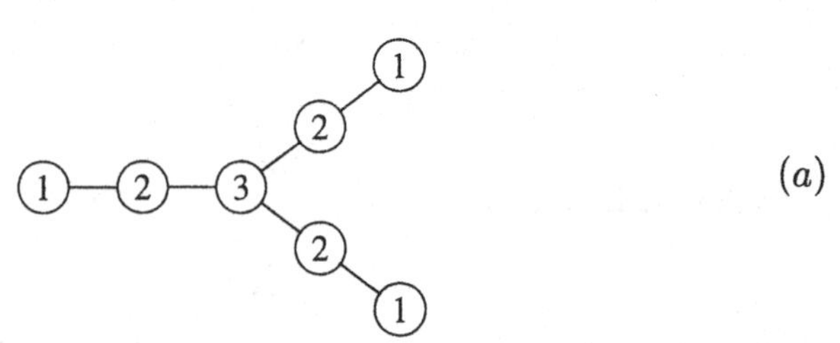

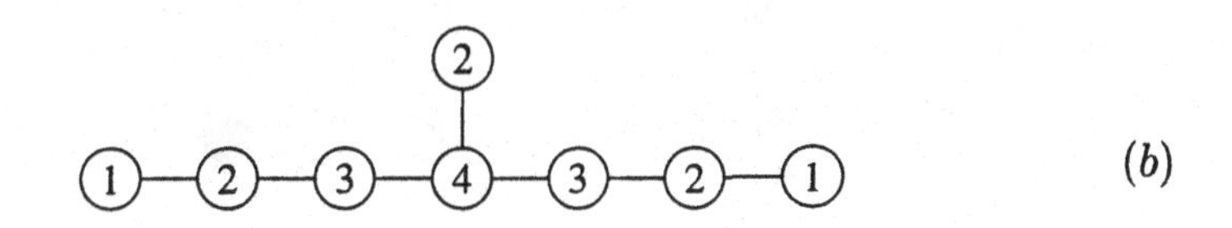

(*b*)

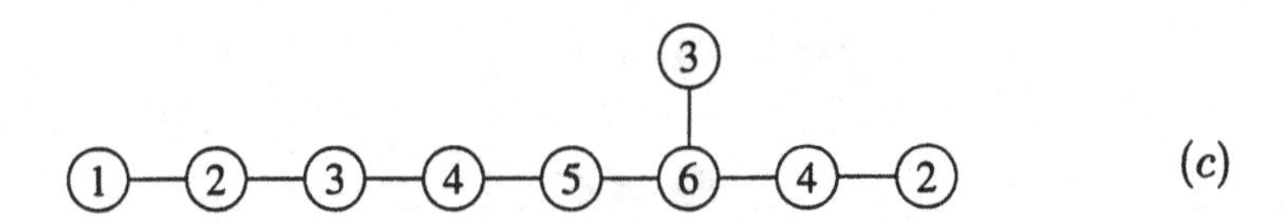

(*c*)

In (*d*), the vectors do not all have the same length. The two on the right may be either longer or shorter than the others by a factor of $\sqrt{2}$. If they are longer, use

(*d*)

If they are shorter, use

(*d*)

We will see later that all these diagrams *can* be constructed with linearly dependent vectors.

We now have enough information to complete the classification theorem. All Π systems must belong to one of 4 infinite families, or be one of 5 exceptional diagrams. All these are shown below, with shorter vectors indicated by filled circles.

A_n

B_n

C_n

D_n

G_2

(20.16)

F_4

E_6

E_7

E_8

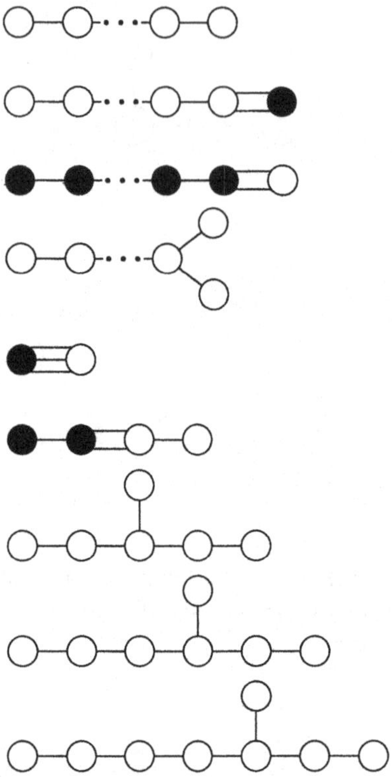

Nothing else is allowed. Everything else runs afoul of one of the geometrical constraints we have derived. There are no other infinite families. Adding more circles beyond the double line in B_n or C_n runs into (d), except for F_4. Continuing beyond the branch in D_n runs into (a), (b) or (c), except for the E_j exceptionals. And we already knew that G_2 is the unique diagram with a triple line.

Note some equivalences between the algebras. A_1, B_1 and C_1 all consist of a single circle, and thus all describe the algebra of $SU(2)$. In addition, $B_2 = C_2$. The D_n family is a bit odd. As you go down in n you keep removing circles from the left. D_4 looks like

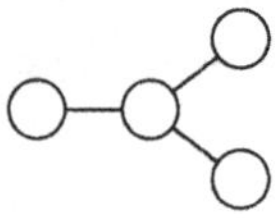

(20.17)

and D_3 degenerates into

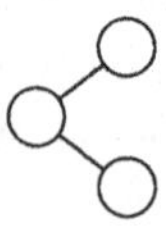

(20.18)

and thus $D_3 = A_3$. Note, further, that if we remove one more circle from D_3 to get D_2, it falls apart into two disconnected circles (the middle one must be removed to stay in the D_n family). Thus D_2 is not simple. This is the statement (an important one) that the algebra of $SO(4)$ is the same as the algebra of $SU(2) \times SU(2)$ (problem 8.B).

That is the complete list of such coincidences. All the others are distinct algebras.

20.2 Regular subalgebras

A **regular subalgebra**, R, of a simple Lie algebra, A, is a subalgebra such that the roots of R are a subset of the roots of A and the generators of the Cartan subalgebra are linear combinations of the Cartan generators of A. A regular subalgebra is called **maximal** if the rank of R is the same as the rank of A (in which case the Cartan subalgebras are identical).

If we leave a circle out of a Dynkin diagram, the result is a diagram or two disconnected diagrams. These are associated with a regular subalgebra of the original algebra, associated with a subset of the original simple roots. Thus for example, the $SU(n) \times SU(m)$ subgroup of $SU(n+m)$ is obtained in this way. We can repeat the procedure and find other regular subgroups. However, these subalgebras have lower rank than the starting group. The Cartan generators that have been left out generate $U(1)$ factors of the subalgebra. There are also non-maximal but regular subalgebras which cannot be obtained by leaving out a circle, but instead are obtained by the merging procedure that we used to establish the classification theorem.

To find the semisimple maximal regular subalgebras, we can use the following trick. Add to the system of simple roots, α^j for $j = 1$ to n, the lowest root, α^0. Because α^0 is the lowest root, $\alpha^0 - \alpha^j$ is not a root, and therefore, by the usual argument,

$$\frac{2\alpha^0 \cdot \alpha^j}{\alpha^{0\,2}} \quad \text{and} \quad \frac{2\alpha^0 \cdot \alpha^j}{\alpha^{j\,2}} \tag{20.19}$$

are non-positive integers. Therefore, this system of vectors satisfies the conditions for a Π-system except that there is one linear relation among the vectors. This is called an extended Π-system or an extended Dynkin diagram.

If we remove any vector from an extended Π-system, the remaining vectors are linearly independent. They still satisfy the master formula. They are therefore the simple roots of a regular, maximal subalgebra of the original algebra. However, the system may not be indecomposable, so the subalgebra may be semi-simple.

There is a unique extended Π-system for each Π-system, because given a Dynkin diagram, we can find the lowest root explicitly. In fact, we have already discussed all of the extended Π-systems in the discussion of the classification theorem, because they are just the systems that failed the test of linear independence, but were otherwise OK. Here they are:

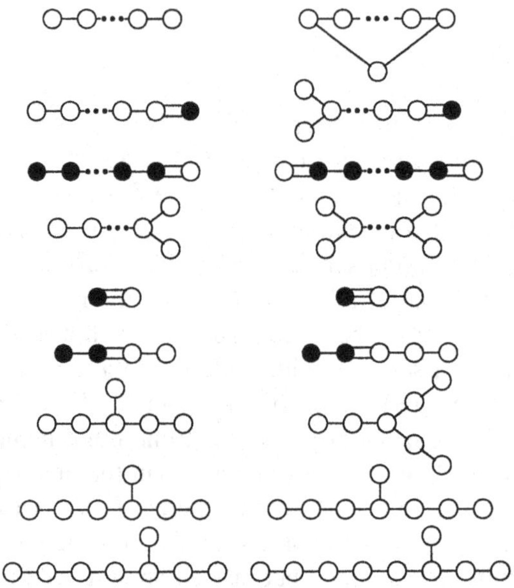

	diagram	extended diagram	
A_n			A'_n
B_n			B'_n
C_n			C'_n
D_n			D'_n
G_2			G'_2
F_4			F'_4
E_6			E'_6
E_7			E'_7
E_8			E'_8

(20.20)

There are some exceptions for small n. $A_1 = B_1 = C_1 = SU(2)$ and $D_2 = SU(2) \times SU(2)$ cannot be extended without changing notation because the highest and lowest roots of $SU(2)$ are negatives on one another, and we have not included a notation for 180° in our definition of Dynkin diagrams. It should probably be

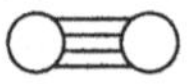

(20.21)

The other exceptions are B_2 and D_3, for which the indicated extensions do not exist because there are not enough open circles to allow for branching. The corresponding extended Π-systems are actually C'_2, and A'_3.

Let's work out some other examples.

The first thing to notice is that A_n doesn't have any semi-simple regular maximal subalgebras. Removing any circle from A'_n just takes you back to A_n again.

For the B_ns, removing an open circle from the left end of B'_n gives you back B_n again. But removing the filled circle from the right end of B'_n gives

D_n. This corresponds to the fact that $SO(2n+1)$ has an $SO(2n)$ subgroup. If instead we remove a circle from the middle of B_n' (for large enough n that the middle exists), the diagram falls apart into a B_k and a D_{n-k}, corresponding to the $SO(2k) \times SO(2n-2k+1)$ subgroup. Each of these can be further broken down, as well. In general, to enumerate all the semisimple regular maximal subalgebras, you must continue breaking things up into subgroups until you get to subalgebras that cannot be further broken down.

For the C_n, removing an open circle from either end just gives you back C_n. Removing a filled circle from the middle breaks up C_n into C_k and C_{n-k}. Finally, removing the first of last filled circle gives $A_1 \times C_{n-1}$. This is really the same thing again, because $A_1 = SU(2)$ is the same algebra as $C_1 = Sp(2)$, so this is just a degenerate special case of the same analysis.

For D_n, removing a circle from either end gives back D_n, so all we can do is to remove a circle from the middle to get $D_k \times D_{n-k}$, corresponding to $SO(2k) \times SO(2n-2k)$.

G_2 is small enough that it is easy to enumerate all the possibilities. Removing the open circle from the right of G_2' gives back G_2. Removing the filled circle from the left gives $A_2 = SU(3)$. Removing the middle circle gives $SU(2) \times SU(2)$.

F_4 has a B_4 subgroup, obtained by removing the filled circle from the left. It has an $A_1 \times A_3$ subgroup obtained by removing the second filled circle. It has an $A_2 \times A_2$ subgroup obtained by removing the left-most open circle. And it have a $C_3 \times A_1$ subgroup obtained by removing the penultimate open circle.

I'll stop here to leave some questions to ask on problem sets and exams.

20.3 Other Subalgebras

The general subject of the subalgebras of an algebra is quite complicated. But the principle is a simple one. For each algebra, there is a simplest representation, out of which all the other representations can be built. If you know how this representation transforms under a subalgebra, you can determine how any representation transforms. Conversely, each possible transformation of the simplest representation is associated with a different subalgebra.

We will come back to this notion several times in the following chapters.

Problems

20.A. Prove that decomposable Π-systems yield decomposable root systems.

20.B. Find the regular maximal subalgebras of E_6.

20.C. Find the regular maximal subalgebras of $SO(12)$. To find them all, you will have to apply the extended Dynkin diagram algorithm several times, because some of the regular maximal subalgebras themselves have nontrivial regular maximal subalgebras.

Chapter 21

$SO(2n+1)$ and Spinors

The $SO(N)$ algebras have a fascinating property which is worth exploring — spinor representations.

21.1 Fundamental weights of $SO(2n+1)$

Label the generators of $SO(2n+1)$ as

$$M_{ab} = -M_{ba} \quad \text{for } a, b = 1 \text{ to } 2n+1 \tag{21.1}$$

In the $2n+1$ dimensional defining representation

$$[M_{ab}]_{xy} = -i\left(\delta_{ax}\,\delta_{by} - \delta_{bx}\,\delta_{ay}\right) \tag{21.2}$$

It is easy to find the commutation relations

$$\begin{aligned} &[M_{ab}, M_{cd}] \\ &= -i\left(\delta_{bc}\,M_{ad} - \delta_{ac}\,M_{bd} - \delta_{bd}\,M_{ac} + \delta_{ad}\,M_{bc}\right) \end{aligned} \tag{21.3}$$

This is the standard form for rotation generators in a real vector space. The commuting generators in the Cartan sub-algebra we took to be

$$H_j = M_{2j-1,2j} \quad \text{for } j = 1 \text{ to } n \tag{21.4}$$

We can take the roots to be

$$\begin{aligned} E_{\eta e^j} &= \frac{1}{\sqrt{2}}(M_{2j-1,2n+1} + i\,\eta\,M_{2j,2n+1}) \\ E_{\eta e^j + \eta' e^k} &= \tfrac{1}{2}(M_{2j-1,2k-1} + i\,\eta\,M_{2j,2k-1} \\ &+ i\,\eta'\,M_{2j-1,2k} - \eta\,\eta'\,M_{2j,2k}) \end{aligned} \tag{21.5}$$

 DOI: 10.1201/9780429499210-22

where η, $\eta' = \pm 1$. These satisfy

$$\left[H_j, E_{\eta e^k}\right] = \eta\, [e^k]_j\, E_{\eta e^k} = \eta\, \delta_{jk}\, E_{\eta e^k} \tag{21.6}$$

so that the simple roots are

$$\begin{aligned} &\alpha^j = e^j - e^{j+1} \quad \text{for } j = 1 \text{ to } n-1 \\ &\alpha^n = e^n \end{aligned} \tag{21.7}$$

corresponding to the diagram

$$\bigcirc\!\!-\!\!\bigcirc\cdots\bigcirc\!\!-\!\!\bigcirc\!\!=\!\!\bigcirc \tag{21.8}$$

$$\alpha^1 \quad \alpha^2 \quad \cdots \qquad \alpha^{n-1} \quad \alpha^n$$

The fundamental weights are

$$\begin{aligned} &\mu^j = \sum_{k=1}^{j} e^k \quad \text{for } j = 1 \text{ to } n-1 \\ &\mu^n = \frac{1}{2} \sum_{k=1}^{n} e^k \end{aligned} \tag{21.9}$$

The last one is different because of the different normalization of α^n. This last representation is the **spinor**. By Weyl reflections in the roots e^j, we get from μ^n the set of weights

$$\frac{1}{2}(\pm e^1 \pm e^2 \cdots \pm e^n) \tag{21.10}$$

All of these are unique, because they are equivalent by reflection to the highest weight state. Furthermore, each could be the highest weight for some other definition of positivity, thus there are no other weights and the representation is 2^n dimensional. It is convenient to treat the 2^n dimensional space as a tensor product of n two dimensional spaces. Then an arbitrary matrix in the space can be built as a tensor product of Pauli matrices. Call the Pauli matrices for the jth space σ_a^j, so that

$$\begin{aligned} &|\pm e^1/2 \pm e^2/2 \cdots \pm e^n/2\rangle \\ &\equiv |\pm e^1/2\rangle \otimes |\pm e^2/2\rangle \otimes \cdots \otimes |\pm e^n/2\rangle \\ &\sigma_a^j\, |x\, e^j\rangle = |x'\, e^j\rangle\, [\sigma_a]_{x'x} \end{aligned} \tag{21.11}$$

where $x, x' = \pm 1/2$.

In this notation, the Cartan generators are

$$H_j = \frac{1}{2}\sigma_3^j \tag{21.12}$$

These all satisfy

$$H_j^2 = \frac{1}{4} \tag{21.13}$$

We could have chosen any M_{ab} to be a Cartan generator, so it is clear that

$$M_{ab}{}^2 = \frac{1}{4} \tag{21.14}$$

in this representation for any $a \neq b$. Now consider the roots

$$E_{e^j} = \frac{1}{\sqrt{2}}\left(M_{2j-1,2n+1} + iM_{2j,2n+1}\right) \tag{21.15}$$

Because we can only raise a state in the representation at most once, we must have

$$(E_{e^j})^2 = 0 \tag{21.16}$$

Expanding out $(E_{e^j})^2$, you can see that this implies

$$\{M_{2j-1,2n+1}, M_{2j,2n+1}\} = 0 \tag{21.17}$$

Again, the particular choice of axes is arbitrary, so we must have

$$\{M_{j\ell}, M_{k\ell}\} = 0 \quad \text{for} \quad j \neq k \neq \ell \neq j \tag{21.18}$$

Let us now construct the Es explicitly. Because E_{e^1} is a raising operator, we know that

$$\begin{aligned} &E_{e^1}|-e^1/2 + x_2 e^2 + \cdots x_n e^n\rangle \\ &= f(x_2, \cdots, x_n)\,|e^1/2 + x_2 e^2 + \cdots x_n e^n\rangle \\ &E_{-e^1}|-e^1/2 + x_2 e^2 + \cdots x_n e^n\rangle = 0 \end{aligned} \tag{21.19}$$

for some function, f. We can compute the norm of f using standard tricks.

$$\begin{aligned} &|f(x_2, \cdots, x_n)|^2 = \langle -e^1/2 + x_2 e^2 + \cdots x_n e^n| \\ &E_{-e^1}E_{e^1}|-e^1/2 + x_2 e^2 + \cdots x_n e^n\rangle \\ &= \langle -e^1/2 + \cdots|\,\{E_{-e^1}, E_{e^1}\}\,|-e^1/2 + \cdots\rangle \end{aligned} \tag{21.20}$$

but using (21.14) and (21.17) we see that[1]

$$\{E_{-e^1}, E_{e^1}\} = \frac{1}{2} \tag{21.21}$$

and thus

$$|f(x_2, \cdots, x_n)|^2 = \frac{1}{2} \tag{21.22}$$

Now for each $x_2, \cdots, x_n$, we can choose the relative phase of the states so that f is positive. We can do this by adjusting the phases of the states with $x_1 = -1/2$. Then

$$f(x_2, \cdots, x_n) = \frac{1}{\sqrt{2}} \tag{21.23}$$

independent of x. Comparing with our definition of σ_a^1, we can write

$$E_{\pm e^1} = \frac{1}{2}\sigma_\pm^1 \tag{21.24}$$

Let us now apply the same argument to $E_{\pm e^2}$.

$$\begin{aligned}
&E_{e^2}|x_1 e^1 - e^2/2 + \cdots x_n e^n\rangle \\
&= f_2(x_1, x_3, \cdots, x_n)\, |x_1 e^1 + e^2/2 + \cdots x_n e^n\rangle \\
&\text{where} \quad |f_2(x_1, x_3, \cdots, x_n)|^2 = \frac{1}{2}
\end{aligned} \tag{21.25}$$

for some function, f_2. For $x_1 = 1/2$, we can choose the relative phases of the states to make $f_2 = 1/\sqrt{2}$, as before so that

$$E_{e^2}|e^1/2 - e^2/2 + \cdots\rangle = \frac{1}{\sqrt{2}}|e^1/2 + e^2/2 + \cdots\rangle \tag{21.26}$$

But now we cannot change the phases of the $x_1 = -1/2$ states, because we have already fixed them in the previous argument. In particular, we know that $\{E_{\pm e^1}, E_{\pm e^2}\} = 0$, and therefore we can do the following computation

$$\begin{aligned}
&E_{e^2}|-e^1/2 - e^2/2 + \cdots\rangle \\
&= \sqrt{2}\, E_{e^2}\, E_{-e^1}|e^1/2 - e^2/2 + \cdots\rangle \\
&= -\sqrt{2}\, E_{-e^1}\, E_{e^2}|e^1/2 - e^2/2 + \cdots\rangle \\
&= -E_{-e^1}|e^1/2 + e^2/2 + \cdots\rangle \\
&= -\frac{1}{\sqrt{2}}|-e^1/2 + e^2/2 + \cdots\rangle
\end{aligned} \tag{21.27}$$

[1] We could equally well use the fact that $[E_{\alpha^1}, E_{-\alpha^1}] = \vec{\alpha}^1 \cdot \vec{H}$.

That is we need an extra minus sign for the $-e^1/2$ state. Or in terms of σ_a^j matrices,

$$E_{\pm e^2} = \frac{1}{2}\sigma_3^1\,\sigma_\pm^2 \tag{21.28}$$

where the tensor product is understood, as usual.

This is easy to understand. The σ_3^1 is there to ensure that $E_{\pm e^1}$ and $E_{\pm e^2}$ anticommute. Since $E_{\pm e^1}$ is built out of σ_1^1 and σ_2^1, $E_{\pm e^2}$ must be proportional to σ_3^1.

Continuing in exactly the same way, we find

$$\begin{aligned} E_{\pm e^3} &= \frac{1}{2}\sigma_3^1\,\sigma_3^2\,\sigma_\pm^3 \\ &\cdots \\ E_{\pm e^j} &= \frac{1}{2}\sigma_3^1\,\cdots\,\sigma_3^{j-1}\,\sigma_\pm^j \end{aligned} \tag{21.29}$$

For the hermitian generators, we can then write

$$\begin{aligned} M_{2j-1,2n+1} &= \frac{1}{2}\sigma_3^1\,\cdots\,\sigma_3^{j-1}\,\sigma_1^j \\ M_{2j,2n+1} &= \frac{1}{2}\sigma_3^1\,\cdots\,\sigma_3^{j-1}\,\sigma_2^j \end{aligned} \tag{21.30}$$

Now we know everything, because we can constuct all the other generators by commutation:

$$M_{ab} = -i\,[M_{a,2n+1}, M_{b,2n+1}] \tag{21.31}$$

for $a, b \neq 2n+1$. Each M_{ab} is just $\pm 1/2$ times a product of Pauli matrices.

This representation is called the **spinor** representation of $SO(2n+1)$. As you see, it is a generalization of the spin 1/2 representation of $SO(3)$ (or $SU(2)$).

21.2 Real and pseudo-real

From (21.10), we can see that the spinor representation, μ^n, is equivalent to its complex conjugate. Nevertheless, the spinor representations have interesting properties under complex conjugation. Before discussing this specifically, we will address the problem more generally.

Suppose that T_a generates a real irreducible unitary representation of a simple Lie algebra, so that

$$T_a = -R T_a^* R^{-1} = -R T_a^T R^{-1} \tag{21.32}$$

We can prove that R is unique up to a trivial scale factor. For suppose that there is another non-singular matrix, Q, that also satisfies

$$T_a = -Q\,T_a^*\,Q^{-1} \tag{21.33}$$

Then

$$T_a^* = -Q^{-1}\,T_a\,Q \tag{21.34}$$

and thus

$$T_a = R\,Q^{-1}\,T_a\,Q\,R^{-1} \tag{21.35}$$

or

$$\left[T_a, R\,Q^{-1}\right] = 0 \tag{21.36}$$

for all a. But if a matrix commutes with all the generators of an irreducible representation, Schur's Lemma implies that it is a multiple of the identity, and thus

$$R\,Q^{-1} = \lambda\,I \qquad R = \lambda\,Q \tag{21.37}$$

and thus R is essentially unique.

Now let us use this result to show that R must be either symmetric or antisymmetric. Because the T_a are hermitian, we can write

$$\begin{aligned}
&T_a = -R\,T_a^T\,R^{-1}\,, \qquad T_a^T = -{R^{-1}}^T\,T_a\,R^T \\
&\Rightarrow T_a = R\,{R^{-1}}^T\,T_a\,R^T\,R^{-1} \\
&\Rightarrow \left[T_a, R^T\,R^{-1}\right] = 0
\end{aligned} \tag{21.38}$$

Therefore, using Schur's Lemma again, we conclude that $R^T\,R^{-1} = \lambda I$ or

$$R^T = \lambda\,R \tag{21.39}$$

But since transposing twice gets us back to where we started, we must have $\lambda^2 = 1$, or $\lambda = \pm 1$, which is the desired result. Either

$$R = A = -A^T \qquad \text{or} \qquad R = S = S^T \tag{21.40}$$

Thus there can be two kinds of real representations. If $R = S$, the representation is called **real-positive**, or just **real**. If $R = A$, the representation is called **real-negative**, or **pseudo-real**.

21.3 Real representations

To see what this difference means in practice, suppose that T_a is equivalent to a representation consisting of purely imaginary antisymmetric matrices,

$$T'_a = U^{-1} T_a U , \qquad T'_a = -T'^*_a \tag{21.41}$$

The corresponding group representation is completely real, since it has the form

$$e^{i\epsilon_a T'_a} \tag{21.42}$$

In this case, we can compute the matrix R —

$$\begin{aligned} T'_a &= -T'^T_a = -U^T T_a^T U^{-1T} = U^{-1} T_a U \\ \Rightarrow T_a &= -U U^T T_a^* (U U^T)^{-1} \end{aligned} \tag{21.43}$$

Thus the matrix R is the symmetric matrix $U U^T$.

The converse is also true. Any irreducible representation of a simple Lie algebra with a symmetric R is equivalent to a representation that is pure imaginary. Let's see how this argument goes. All the irreducible representations are equivalent to unitary representations, so we will assume that the generators are hermitian. Then it is easy to see that R is proportional to a unitary matrix. We can write the equivalence condition as

$$-T_a^T = R^{-1} T_a R = R^\dagger T_a R^{-1\dagger} \tag{21.44}$$

where the second equality follows from Hermitivity. But then

$$R R^\dagger T_a = T_a R R^\dagger \tag{21.45}$$

and $R R^\dagger \propto I$ by Schur's Lemma. The overall constant just cancels out in the similarity transformation, so we may as well take R to be unitary. Now suppose that it is also symmetric. It is a linear algebra fact that we can write a symmetric unitary matrix as a product of unitary matrix and its transpose[2]

$$R = V V^T \tag{21.46}$$

But then the generators

$$T'_a = V^{-1} T_a V \tag{21.47}$$

are antisymmetric — for example, we can multiply the reality condition (21.44) on the left by V^T and on the right by V^{-1T} to obtain

$$-V^T T_a^T V^{-1T} = V^{-1} T_a V . \tag{21.48}$$

[2]If R is symmetric and unitary, then $R R^* = R^* R = I$, and therefore the real and imaginary parts of R commute. Thus they can be simultaneously diagonalized by an orthogonal transformation. The reader should be able to take it from there.

21.4 Pseudo-real representations

The pseudo-real representations, in contrast, cannot be transformed into purely imaginary matrices, even though they are equivalent to their complex conjugates. The simplest example of a pseudo-real representation is the spin 1/2 representation of $SU(2)$ generated by the Pauli matrices, $T_a = \sigma_a/2$. Obviously, there is no way to generate this representation with antisymmetric imaginary 2×2 matrices, because there is only one such matrix and there are three generators. But still, the representation is equivalent to its complex conjugate because

$$\sigma_a = -\sigma_2\, \sigma_a^*\, \sigma_2 \tag{21.49}$$

And indeed, the transformation matrix, σ_2, is antisymmetric, so the representation is pseudo-real.

21.5 R is an invariant tensor

There is another very useful way of thinking about the matrix R. It is an invariant tensor. To see this, note that if the representation is unitary, as we always assume, we can rewrite (21.32) as

$$T_a\, R = -R T_a^* \Rightarrow T_a\, R + R T_a^T = 0\,. \tag{21.50}$$

In terms of explicit indices, this can be written

$$[T_a]_z^x\, R^{zy} + [T_a]_z^y\, R^{xz} = 0 \tag{21.51}$$

which implies that R is an invariant tensor in the tensor product space $D \otimes D$ where D is generated by the T_a. For any real representation, $D \otimes D$ contains the trivial representation, 1, only once because R is unique. If D is real-positive, the coefficient of the representation 1 in $D \otimes D$, which is precisely the invariant tensor R, is symmetric in the exchange of the two equivalent Ds. If D is real-negative (pseudo-real), the coefficient R is antisymmetric.

21.6 The explicit form for R

To find the matrix R for the spinor representation of $SO(2n+1)$, it is enough to satisfy (21.32) for the generators

$$\begin{aligned} M_{2j-1,2n+1} &= \frac{1}{2}\,\sigma_3^1 \cdots \sigma_3^{j-1}\,\sigma_1^j \\ M_{2j,2n+1} &= \frac{1}{2}\,\sigma_3^1 \cdots \sigma_3^{j-1}\,\sigma_2^j \end{aligned} \tag{21.52}$$

for all j, because the rest can be obtained by commutation of these, and if (21.32) is satisfied for the generators (21.52), it is satisfied for any commutator of two of them. We can now build up R as a tensor product of 2×2 matrices in the various subspaces,

$$R \equiv \prod_j r^j \tag{21.53}$$

We must have

$$\begin{aligned}
-r^1 \, \sigma_{1,2}^{1\,*} \, r^{1^{-1}} &= \sigma_{1,2}^1 \\
-r^1 \, \sigma_3^{1*} \, r^{1^{-1}} \cdot r^2 \, \sigma_{1,2}^{2\,*} \, r^{2^{-1}} &= \sigma_3^1 \, \sigma_{1,2}^2 \\
-\left(\prod_{k=1}^{j-1} r^k \, \sigma_3^{k*} \, r^{k^{-1}} \right) \cdot r^j \, \sigma_{1,2}^{j\,*} \, r^{j^{-1}} & \\
= \left(\prod_{k=1}^{j-1} \sigma_3^k \right) \sigma_{1,2}^j &
\end{aligned} \tag{21.54}$$

Now we can determine the r^j one at a time. r^1 must anticommute with σ_1^1 and commute with σ_2^1. Thus we can take $r^1 = \sigma_2^1$. Then from the next condition, r^2 must commute with σ_1^2 and anticommute with σ_2^2. Thus we can take $r^2 = \sigma_1^2$. And so on — the σ_2s and σ_1s alternate.

Thus we can take the matrix R to be

$$R = \prod_{\substack{j \\ \text{odd}}} \sigma_2^j \prod_{\substack{k \\ \text{even}}} \sigma_1^k \tag{21.55}$$

Note that

$$R^2 = 1 \,, \quad \text{or} \quad R = R^{-1} \qquad \text{and} \qquad R^T = (-1)^{n(n+1)/2} \, R \tag{21.56}$$

So for $n = 1$ and 2, R is antisymmetric; for $n = 3$ and 4, R is symmetric; for $n = 5$ and 6, R is antisymmetric; etc. Thus

Algebra	Spinors
$SO(8k+3)$	pseudo-real
$SO(8k+5)$	pseudo-real
$SO(8k+7)$	real
$SO(8k+1)$	real

(21.57)

Problems

21.A. Show that the set of 10 matrices, $\frac{1}{2}\vec{\sigma}$, $\frac{1}{2}\vec{\sigma}\tau_1$, $\frac{1}{2}\vec{\sigma}\tau_3$, and $\frac{1}{2}\tau_2$ generate the spinor representation of $SO(5)$. Find the matrix $R = R^{-1}$ such that $T_a = -RT_a^* R$ for this representation.

21.B. Identify any convenient $SO(2n-1)$ subalgebra of $SO(2n+1)$ and determine how the spinor representation of $SO(2n+1)$ transforms under the subalgebra.

Chapter 22

$SO(2n+2)$ Spinors

We can use the results for $SO(2n+1)$ to work out the spinor representations for $SO(2n+2)$. We will use the same notation as in the previous discussions.

22.1 Fundamental weights of $SO(2n+2)$

The Dynkin diagram for $SO(2n+2)$ is

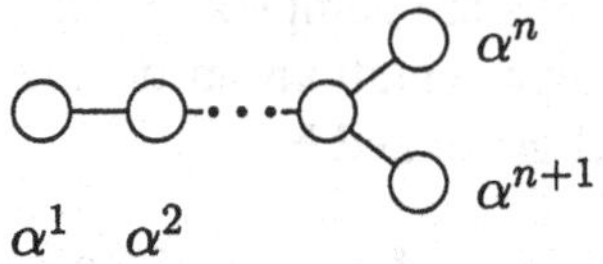

$$\alpha^1 \quad \alpha^2 \quad \cdots \quad \alpha^n, \ \alpha^{n+1} \tag{22.1}$$

where

$$\alpha^j = e^j - e^{j+1} \text{ for } j = 1 \text{ to } n\,, \quad \alpha^{n+1} = e^n + e^{n+1} \tag{22.2}$$

All the roots have the form

$$\pm e^j \pm e^k\,, \quad j \neq k \tag{22.3}$$

We are interested in the two special representations corresponding to the last two fundamental weights,

$$\begin{aligned} \mu^n &= \frac{1}{2}\Big(e^1 + \cdots + e^n - e^{n+1}\Big) \\ \mu^{n+1} &= \frac{1}{2}\Big(e^1 + \cdots + e^n + e^{n+1}\Big) \end{aligned} \tag{22.4}$$

Call the corresponding representations D^n and D^{n+1}. The weights of D^n are of the form

$$\frac{1}{2}\sum_{j=1}^{n+1} \eta_j\, e^j \tag{22.5}$$

 DOI: 10.1201/9780429499210-23

where

$$\eta_j = \pm 1 \qquad \text{and} \qquad \prod_{j=1}^{n+1} \eta_j = -1 \tag{22.6}$$

This is because the roots always change the sign of two of the ηs. The weights of D^{n+1} have the same form but with

$$\prod_{j=1}^{n+1} \eta_j = 1 \tag{22.7}$$

Now that we know the weights, we can determine the reality properties of the representations. If n is odd, so that the algebra has the form $SO(4m)$ for integer m, then there are an even number of ηs. Thus $-\mu^n$ is a weight in D^n, because it still satisfies $\prod_{j=1}^{n+1} \eta_j = -1$. Evidently, it is the lowest weight in D^n. Similarly, $-\mu^{n+1}$ is the lowest weight in D^{n+1}. Thus, for n odd, the spinor representations of $SO(2n+2)$ are real (or pseudo-real).

But for even n, $-\mu^n$ is the lowest weight in D^{n+1}, because it satisfies $\prod_{j=1}^{n+1} \eta_j = 1$. Likewise, $-\mu^{n+1}$ is the lowest weight of D^n. Thus for even n, the spinor representations are complex. $\overline{D^n}$ is D^{n+1} and $\overline{D^{n+1}}$ is D^n.

To construct these representations in detail, consider the $SO(2n+1)$ subgroup of $SO(2n+2)$ generated by

$$M_{jk} \quad \text{for} \quad j,k \leq 2n+1 \tag{22.8}$$

This eliminates the Cartan generator

$$H_{n+1} = M_{2n+1,2n+2} \tag{22.9}$$

The $4n$ generators with weights

$$\pm e^j \pm e^{n+1} \quad \text{for} \quad j \leq n \tag{22.10}$$

which are linear combinations of

$$M_{j,2n+1} \quad \text{and} \quad M_{j,2n+2} \quad \text{for} \quad j \leq 2n \tag{22.11}$$

collapse to the $2n$ generators $M_{j,2n+1}$ with weights

$$\pm e^j \quad \text{for} \quad j \leq n \tag{22.12}$$

Under this $SO(2n+1)$ subgroup both D^n and D^{n+1} transform like the spinor representation we just analyzed. Because H_{n+1} is not a generator of this subgroup, we can take over the previous analysis without change if we just ignore the $n+1$ component of all the weight vectors. When we do this, the weight vectors are the same for both D^n and D^{n+1}:

$$\frac{1}{2}\sum_{j=1}^{n} \eta_j\, e^j \quad \text{for} \quad \eta_j = \pm 1 \tag{22.13}$$

In the tensor product notation, $\eta_j = \sigma_3^j$ and the generators are given by

$$\begin{aligned} M_{2j-1,2n+1} &= \frac{1}{2}\sigma_3^1 \cdots \sigma_3^{j-1}\,\sigma_1^j \\ M_{2j,2n+1} &= \frac{1}{2}\sigma_3^1 \cdots \sigma_3^{j-1}\,\sigma_2^j \end{aligned} \tag{22.14}$$

Now going back to the full $SO(2n+2)$ algebra, in the representation D^n, we know that

$$\eta_{n+1} = -\prod_{j=1}^{n} \eta_j = -\sigma_3^1 \cdots \sigma_3^n \tag{22.15}$$

Therefore

$$H_{n+1} = M_{2n+1,2n+2} = -\frac{1}{2}\sigma_3^1 \cdots \sigma_3^n \tag{22.16}$$

Now all the generators can by found using the commutation relations.

In D^{n+1}

$$\eta_{n+1} = \prod_{j=1}^{n} \eta_j = \sigma_3^1 \cdots \sigma_3^n \tag{22.17}$$

and

$$H_{n+1} = M_{2n+1,2n+2} = \frac{1}{2}\sigma_3^1 \cdots \sigma_3^n \tag{22.18}$$

Now that we have explicit forms for the representations, we can examine the reality properties we discussed earlier in more detail. As in $SO(2n+1)$, we define the matrix

$$R = R^{-1} = \prod_{\substack{j \\ \text{odd}}} \sigma_2^j \prod_{\substack{k \\ \text{even}}} \sigma_1^k \tag{22.19}$$

and then look at the complex conjugate representations

$$-R T_a^* R^{-1} \tag{22.20}$$

for D^n and D^{n+1}. We already know that

$$-R M_{jk}^* R^{-1} = M_{jk} \tag{22.21}$$

for the $SO(2n+1)$ subgroup, $j,k \leq 2n+1$. It is then enough to see what happen to $M_{2n+1,2n+2}$ under complex conjugation. If n is odd,

$$-R\,M^*_{2n+1,2n+2}\,R^{-1} = M_{2n+1,2n+2} \tag{22.22}$$

and then the equivalence can be established for all the generators using the commutation relations. Thus, as we already knew, the representation is real. But now we know R, so we can determine whether it is real-positive or pseudo-real. For $n = 4k+1$, R is antisymmetric, so D^n and D^{n+1} are pseudo-real. For $n = 4k+3$, R is symmetric, so D^n and D^{n+1} are real.

For even n,

$$-R\,M^*_{2n+1,2n+2}\,R^{-1} = -M_{2n+1,2n+2} \tag{22.23}$$

so the representations D^n and D^{n+1} get interchanged and thus they are complex.

The full story of the reality properties of the $SO(N)$ spinors is then as follows:

(22.24)

Algebra	Spinors
$SO(8k+3)$	pseudo-real
$SO(8k+4)$	pseudo-real
$SO(8k+5)$	pseudo-real
$SO(8k+6)$	complex
$SO(8k+7)$	real
$SO(8k)$	real
$SO(8k+1)$	real
$SO(8k+2)$	complex

Problems

22.A. Show that $SO(2n)$ has a regular maximal subalgebra, $SO(2m) \times SO(2n-2m)$. How do the spinor representations of $SO(2n)$ transform under the subgroup?

22.B. Show that $SO(2n+1)$ has a regular maximal subalgebra, $SO(2m) \times SO(2n-2m+1)$. How do the spinor representations of $SO(2n+1)$ transform under the subgroup?

22.C. Show that $SO(4)$ has the same algebra as $SU(2) \times SU(2)$. Thus, it is not simple. Nevertheless, the arguments in the chapter apply. Explain how.

22.D. Show that the $SO(6)$ algebra and the $SU(4)$ algebra are equivalent, with the 4 of $SU(4)$ corresponding to a spinor representation of $SO(6)$. In $SU(4)$, $4 \otimes 4 = 6 \oplus 10$. The 6 is the vector representation of $SO(6)$. What is the 10, in the $SO(6)$ language?

Chapter 23

$SU(n) \subset SO(2n)$

In order to discuss the embedding of $SU(n)$ in $SO(2n)$ efficiently, we will introduce another language for talking about the spinor representations — Clifford algebras.

23.1 Clifford algebras

A Clifford algebra is a set of operators satisfying the following anticommutation relations:

$$\{\Gamma_j, \Gamma_k\} = 2\,\delta_{jk}\,, \qquad \text{for } j,k = 1 \text{ to } N \tag{23.1}$$

If you are given such a set of operators, you can immediately construct a representation of $SO(N)$ as follows:

$$M_{jk} = \frac{1}{4i}\,[\Gamma_j, \Gamma_k] \tag{23.2}$$

It is easy to check, using the Jacobi identity, that the M_{jk} have the commutation relations of $SO(N)$, and furthermore that the Γs are a set of tensor operators transforming according to the N dimensional representation of $SO(N)$, satisfying

$$[M_{jk}, \Gamma_\ell] = i\left(\delta_{j\ell}\,\Gamma_k - \delta_{k\ell}\,\Gamma_j\right) = \Gamma_m\,[M_{jk}^{D^1}]_{m\ell} \tag{23.3}$$

where

$$[M_{jk}^{D^1}]_{\ell m} = -i\left(\delta_{j\ell}\,\delta_{km} - \delta_{jm}\,\delta_{k\ell}\right) \tag{23.4}$$

generate the vector representation, D^1, with highest weight μ^1.

DOI: 10.1201/9780429499210-24

For $SO(2n+1)$ it is straightforward to find a Clifford algebra that gives the spinor representation we constructed. It is

$$
\begin{array}{ll}
\Gamma_1 = \sigma_2^1 \sigma_3^2 \cdots \sigma_3^n & \Gamma_2 = -\sigma_1^1 \sigma_3^2 \cdots \sigma_3^n \\
\Gamma_3 = \sigma_2^2 \cdots \sigma_3^n & \Gamma_4 = -\sigma_1^2 \cdots \sigma_3^n \\
\cdots & \\
\Gamma_{2n-1} = \sigma_2^n & \Gamma_{2n} = -\sigma_1^n \\
\multicolumn{2}{c}{\Gamma_{2n+1} = \sigma_3^1 \sigma_3^2 \cdots \sigma_3^n}
\end{array}
\tag{23.5}
$$

Notice that the product

$$\Gamma_1 \Gamma_2 \cdots \Gamma_{2n+1} = i^n \tag{23.6}$$

is proportional to the identity element.

There is a connection of this with $SO(2n+2)$, even though we don't have enough elements in the Clifford algebra to generate it. These elements of the Clifford algebra are actually proportional to $M_{j,2n+2}$ in the spinor representation that we constructed. These generators transform like the components of a vector under the $SO(2n+1)$ subgroup of $SO(2n+2)$.

While we cannot construct a representation of the $SO(2n+2)$ algebra from this Clifford algebra (at least, we cannot use (23.2) to construct the generators because we do not have enough Γs), we can construct the $SO(2n)$ algebra just by leaving out Γ_{2n+1}. But the resulting representation is reducible, because there is a nontrivial matrix that commutes with all of the generators — namely Γ_{2n+1} itself —

$$\Gamma_{2n+1} = (-i)^n \, \Gamma_1 \Gamma_2 \cdots \Gamma_{2n} = \prod_{j=1}^{n} \sigma_3^j \tag{23.7}$$

We can use this to construct projection operators onto the two invariant subspaces that transform according to the irreducible representations, D^{n-1} and D^n.

$$
\begin{array}{ll}
\frac{1}{2}\left(1 - \Gamma_{2n+1}\right) & \text{projects onto} \quad D^{n-1} \\
\frac{1}{2}\left(1 + \Gamma_{2n+1}\right) & \text{projects onto} \quad D^{n}
\end{array}
\tag{23.8}
$$

because $\prod_{j=1}^{n} \sigma_3^j$ is -1 on the first subspace and $+1$ on the second.

23.2 Γ_m and R as invariant tensors

The main reason that the Clifford algebra construction is useful is that Γs are a set of invariant tensors (see (23.3). For that reason, we can use it to help with the process of Clebsch-Gordan decomposition of products of spinors. Let us return to $SO(2n+1)$ and write out the commutation relation (23.3) in terms of components — it looks like

$$\begin{gathered} [M_{jk}]_{xz}\,[\Gamma_\ell]_{zy} - [M_{jk}^*]_{yz}\,[\Gamma_\ell]_{xz} \\ +[M_{jk}^{D^1}]_{\ell m}\,[\Gamma_m]_{xy} = 0 \end{gathered} \tag{23.9}$$

Thus $[\Gamma_\ell]_{xy}$ is an invariant tensor in which the x index transforms like the spinor representation, D^n, the y index like the conjugate representation, $\overline{D}^n$, and the ℓ like the vector representation, D^1. Of course, we know that in $SO(2n+1)$ the representation $\overline{D}^n$ is equivalent to D^n by a similarity transformation involving the matrix R, but in this form for the Γs, we have not done the similarity transformation, it is the $\overline{D}^n$ matrices that appear explicitly. We could put in the transformation explicitly. Suppose we do that using

$$-M_{jk}^* = -M_{jk}^T = R^{-1}\,M_{jk}\,R \tag{23.10}$$

It is useful to write this in terms of upper and lower indices. If the matrix elements of the generators of D^n (say) are written as

$$[M_{jk}]^x_y \tag{23.11}$$

where x refers to the row (the first index of the matrix) and y to the column (the second index), then R behaves as an invariant tensor with two upper indices. We showed this in chapter 21, but let's do it again. If we multiply on the left by R and put the indices in, (23.10) becomes

$$-R^{xz}\,[M_{jk}]^y_z = [M_{jk}]^x_z\,R^{zy} \tag{23.12}$$

or in prettier form

$$[M_{jk}]^x_z\,R^{zy} + [M_{jk}]^y_z\,R^{xz} = 0 \tag{23.13}$$

which is the statement that R^{xy} is an invariant tensor

The elements of the Clifford algebra have one upper and one lower index, like the M_{jk}s

$$[\Gamma_j]^x_y \tag{23.14}$$

These do not have a definite symmetry property (or more precisely, Γ_j has different symmetry for different values of j and these properties depend on the basis) because the upper and lower indices are different kinds of objects so it doesn't make any sense to interchange them. But by multiplying by R, we raise the lower index

$$[\Gamma_j\, R]^{xy} = [\Gamma_j]^x_z\, R^{zy} \tag{23.15}$$

This object can and does have definite symmetry.

Now let us rewrite the commutation relation of M_{jk} with Γ_ℓ with upper and lower indices

$$\begin{aligned}[M_{jk}]^x_z\,[\Gamma_\ell]^z_y + [R^{-1}\,M_{jk}\,R]^y_z\,[\Gamma_\ell]^x_z \\ +[M^{D^1}_{jk}]_{\ell m}\,[\Gamma_m]^x_y = 0\end{aligned} \tag{23.16}$$

which we can rewrite by multiplying by R^{yw} and summing over y

$$\begin{aligned}&[M_{jk}]^x_z\,[\Gamma_\ell\,R]^{zw} + [R^T\,R^{-1}\,M_{jk}\,R]^w_z\,[\Gamma_\ell]^{xz} \\ &\qquad +[M^{D^1}_{jk}]_{\ell m}\,[\Gamma_m\,R]^{xw} \\ &= [M_{jk}]^x_z\,[\Gamma_\ell\,R]^{zw} + [R^T\,R^{-1}\,M_{jk}]^w_z\,[\Gamma_\ell\,R^T]^{xz} \\ &\qquad +[M^{D^1}_{jk}]_{\ell m}\,[\Gamma_m\,R]^{xw} \\ &= [M_{jk}]^x_z\,[\Gamma_\ell\,R]^{zw} + [M_{jk}]^w_z\,[\Gamma_\ell\,R]^{xz} \\ &\qquad +[M^{D^1}_{jk}]_{\ell m}\,[\Gamma_m\,R]^{xw} = 0\end{aligned} \tag{23.17}$$

In the last step we have used the fact that R is either symmetric or antisymmetric, so I can replace the two R^Ts by two Rs. Thus it is the product $\Gamma_\ell\,R$ that is an invariant tensor in the space $D^n \otimes D^n \otimes D^1$. These products must be either symmetric or antisymmetric in exchange of the identical D^n labels, unlike the Γs themselves. This will seem familiar to those of you who have studied the Dirac equation, where something similar happens with the Dirac γ matrices.

To see whether $\Gamma_j\,R$ is symmetric or antisymmetric, we can check any Γ_j. The easiest is Γ_{2n+1} which is just a product of σ_3s in all the spaces. When we multiply by R, the result is a kind of mirror image of R, where, up to factors of i, the σ_1s and σ_2s are simply interchanged. If we define

$$R^T = \eta_R\,R \qquad \text{and} \qquad (\Gamma_j R)^T = \eta_{\Gamma R}\,\Gamma_j R \tag{23.18}$$

then

$$\eta_R \cdot \eta_{\Gamma R} = (-1)^n \tag{23.19}$$

because in each two dimensional subspace, there is a σ_1 in either R or $\Gamma_{2n+1}R$ and a σ_2 in the other. Since we know that

$$\eta_R = (-1)^{n(n+1)/2} \tag{23.20}$$

we find that

$$\eta_{\Gamma R} = (-1)^{n(n-1)/2} \tag{23.21}$$

23.3 Products of Γs

The reason that it is sometimes better to stick to the Γs rather than using the symmetric Γ Rs, is that matrix products of the Γs also yield invariant tensors. These products have one upper and one lower index, so like Γ_j they live in the space $D^n \otimes \overline{D}^n$, and we can use them to complete the process of Clebsch-Gordan decomposition. For example, consider the commutator, $[\Gamma_j, \Gamma_k]$. We already know that this is proportional to M_{jk}, which is an antisymmetric tensor operator. But we could also conclude this using the Jacobi identity:

$$\begin{aligned} &[M_{jk}, [\Gamma_\ell, \Gamma_m]] = [[M_{jk}, \Gamma_\ell], \Gamma_m] + [\Gamma_\ell, [M_{jk}, \Gamma_m]] \\ &= [\Gamma_{m'}, \Gamma_m]\, [M_{jk}^{D^1}]_{m'\ell} + [\Gamma_\ell, \Gamma'_m]\, [M_{jk}^{D^1}]_{m'm} \end{aligned} \tag{23.22}$$

which means that $[\Gamma_j, \Gamma_k]$ transforms like the antisymmetric tensor product of two vectors.

More generally the interesting combinations are the products that are antisymmetric in the vector indices, defined as a sum over permutations of the indices. For $m \leq n$, define

$$\Gamma_{[j_1}\Gamma_{j_2}\cdots\Gamma_{j_m]} \equiv \frac{1}{m!} \sum_{P\binom{k_1\cdots k_m}{j_1\cdots j_m}} \pm \Gamma_{k_1}\Gamma_{k_2}\cdots\Gamma_{k_m} \tag{23.23}$$

with the − sign for the odd permutations. The other combinations are not interesting because the anticommutator of two Γs can be eliminated using the Clifford algebra. The Jacobi identity can be used in exactly the same way to show that this transforms like an antisymmetric tensor with m vector indices. Of course, we don't really have to do this complicated sum because the Clifford algebra guarantees that the Γ_j anticommute for different values of j, so really all the sum is doing is ensuring that no two indices are the same.

We do not get any new matrices for $m > n$, because of the fact that the product of all $2n$+1 Γs is proportional to the identity, so the antisymmetric product of n+κ Γs is related to the antisymmetric product of n−κ+1 Γs.

The product $\Gamma_{[j_1}\Gamma_{j_2}\cdots\Gamma_{j_\mu]}$ tells us explicitly how the antisymmetric μ index tensor appears in the tensor product $D^n \otimes \overline{D}^n$. The tensor product of two spinors in $SO(2n+1)$ can be decomposed into antisymmetric tensors of rank 0 (the trivial representation), 1 (the vector), 2 (the adjoint for $n > 1$), and so on up to rank n. You can see that the dimensions work out because of the binomial theorem:

$$\begin{aligned} 2^{2n+1} &= (1+1)^{2n+1} = \sum_{k=0}^{2n+1} \binom{2n+1}{k} \\ \Rightarrow 2^n \times 2^n &= \sum_{k=0}^{n} \binom{2n+1}{k} \end{aligned} \tag{23.24}$$

where the second line follows because

$$\binom{2n+1}{2n+1-k} = \binom{2n+1}{k} \tag{23.25}$$

so that the sum from $k = 0$ to n is the same as the sum from n+1 to $2n$+1.

Somewhat more bizarre is the breakup into symmetric versus antisymmetric products. All the products have one upper and one lower index, so again, to get something that is either symmetric or antisymmetric, we must multiply by R. The result of multiplying by R the product of k distinct Γ_j matrices (which is the antisymmetric product, because the different indices anticommute) can be written as

$$\begin{aligned} &\Gamma_{j_1}\Gamma_{j_2}\cdots\Gamma_{j_k}R \\ &= (\Gamma_{j_1}R)R^{-1}(\Gamma_{j_2}R)R^{-1}\cdots(\Gamma_{j_k}R) \end{aligned} \tag{23.26}$$

The transpose is

$$\begin{aligned} &(\Gamma_{j_k}R)^T R^{-1T}\cdots(\Gamma_{j_2}R)^T R^{-1T}(\Gamma_{j_1}R)^T \\ &= \eta_R^{k-1}\eta_{\Gamma R}^k(\Gamma_{j_k}R)R^{-1}\cdots(\Gamma_{j_2}R)R^{-1}(\Gamma_{j_1}R) \\ &= \eta_R^{k-1}\eta_{\Gamma R}^k\Gamma_{j_k}\cdots\Gamma_{j_2}\Gamma_{j_1}R \end{aligned} \tag{23.27}$$

where we have used the fact that R^{-1} (which has two lower indices) has the same symmetry properties as R itself. To get the Γ_j factors back into the original order we have to make

$$(k-1)+(k-2)+\cdots+1 = k(k-1)/2 \tag{23.28}$$

transpositions, each of which gives a $-$ sign because of the Clifford algebra. Thus we have finally

$$\begin{aligned}&\left(\Gamma_{[j_1}\Gamma_{j_2}\cdots\Gamma_{j_k]}R\right)^T \\ &= (-1)^{k(k-1)/2}\eta_R^{k-1}\eta_{\Gamma R}^k\Gamma_{[j_1}\Gamma_{j_2}\cdots\Gamma_{j_k]}R\end{aligned} \tag{23.29}$$

We can write the factor in front somewhat more elegantly as follows

$$\begin{aligned}&(-1)^{k(k-1)/2}\eta_R^{k-1}\eta_{\Gamma R}^k \\ &= (-1)^{k(k-1)/2}\eta_R\left(\eta_R\eta_{\Gamma R}\right)^k \\ &= (-1)^{k(k-1)/2}(-1)^{n(n+1)/2}(-1)^{nk} \\ &= (-1)^{k(k-1)/2}(-1)^{n(n+1)/2}(-1)^{-nk} \\ &= (-1)^{(n-k)(n-k+1)/2}\end{aligned} \tag{23.30}$$

Thus in $SO(2n+1)$, the symmetry properties of the k index antisymmetric tensor in the tensor product of $D^n \otimes D^n$ repeat with period 4, and look like this:

$k \diagdown n$	1	2	3	4
0	$-$	$-$	$+$	$+$
1	$+$	$-$	$-$	$+$
2	$+$	$+$	$-$	$-$
3	$-$	$+$	$+$	$-$

(23.31)

Thus for $SO(3)$ the 1 is antisymmetric and D^1 is symmetric. For $SO(5)$ the 1 and D^1 are antisymmetric while D^2 is symmetric. For $SO(7)$ the D^1 and D^2 are antisymmetric while 1 and D^3 are symmetric. For $SO(9)$ the D^2 and D^3 are antisymmetric while 1, D^1 and D^4 are symmetric. And so on!

For $SO(2n)$, the analysis is more complicated because the 2^n dimensional space on which the Clifford algebra acts is not irreducible. We will use the projection operators onto the irreducible subspaces

$$P_\pm = \frac{1}{2}\left(1 \pm \Gamma_{2n+1}\right) \tag{23.32}$$

P_+ projects onto D^n while P_- projects onto D^{n-1}. Now a general $2^n \times 2^n$ transforms under commutation with the generators like the tensor product

$$\left(D^n \oplus D^{n-1}\right)\left(\overline{D}^n \oplus \overline{D}^{n-1}\right) \tag{23.33}$$

Then we can project out all four possible transformation laws with the projection operators. If K is an arbitrary $2^n \times 2^n$ matrix, then:

$$\begin{array}{lll} P^+ \, K \, P^+ & \text{transforms like} & D^n \otimes \overline{D}^n \\ P^- \, K \, P^- & \text{transforms like} & D^{n-1} \otimes \overline{D}^{n-1} \\ P^+ \, K \, P^- & \text{transforms like} & D^n \otimes \overline{D}^{n-1} \\ P^- \, K \, P^+ & \text{transforms like} & D^{n-1} \otimes \overline{D}^n \end{array} \tag{23.34}$$

As before, we can construct the K matrices out of antisymmetric products of the Γs. We need not include Γ_{2n+1} because we are interested only in $SO(2n)$ (and the projection operators make it ± 1 anyway). Furthermore, once we leave out Γ_{2n+1}, the projection operators pick out either the even or odd products, because the other Γs anticommute with Γ_{2n+1}, and therefore

$$\Gamma_j \, P_+ = P_- \, \Gamma_j \quad \text{for } j = 1 \text{ to } 2n \tag{23.35}$$

Thus only an odd number of Γs can appear between P_+ and P_- (in either order) and only an even number of Γs can appear between two P_+ or two P_-. As before, we can ignore tensors with rank $m > n$.

23.4 Self-duality

There is an additional subtlety that occurs for tensors of rank n. Not all the components of the rank n antisymmetric tensor in $D^n \otimes \overline{D}^n$ or $D^{n-1} \otimes \overline{D}^{n-1}$ are independent. For example in $D^n \otimes \overline{D}^n$

$$\begin{aligned} & P_+ \, \Gamma_1 \, \Gamma_2 \cdots \Gamma_n \, P_+ = P_+ \, \Gamma_1 \, \Gamma_2 \cdots \Gamma_n \, \Gamma_{2n+1} \, P_+ \\ & = (-i)^n \, P_+ \, \Gamma_1 \cdots \Gamma_n \, \Gamma_1 \cdots \Gamma_{2n} \, P_+ \\ & = (-i)^n \, (-1)^{n(n-1)/2} \, P_+ \, \Gamma_{n+1} \, \Gamma_{n+2} \cdots \Gamma_{2n} \, P_+ \\ & = (-i)^{n^2} \, \frac{1}{n!} \, \epsilon_{12\cdots n \, j_1 \cdots j_n} \, P_+ \, \Gamma_{j_1} \cdots \Gamma_{j_n} \, P_+ \end{aligned} \tag{23.36}$$

Evidently, the rank n tensors satisfy a self-duality condition, the nature of which depends on n. If n is even, the relation is real, and the tensor is either **self-dual** (for $D^n \otimes \overline{D}^n$), satisfying

$$A_{j_1 \cdots j_n} = \frac{1}{n!} \, \epsilon_{j_1 \cdots j_n \, k_1 \cdots k_n} \, A_{k_1 \cdots k_n} \tag{23.37}$$

or **anti-self-dual** (for $D^{n-1} \otimes \overline{D}^{n-1}$), satisfying

$$A_{j_1 \cdots j_n} = -\frac{1}{n!} \epsilon_{j_1 \cdots j_n k_1 \cdots k_n} A_{k_1 \cdots k_n} \tag{23.38}$$

If n is odd, on the other hand, the relation is complex:

$$A_{j_1 \cdots j_n} = \pm i \frac{1}{n!} \epsilon_{j_1 \cdots j_n k_1 \cdots k_n} A_{k_1 \cdots k_n} \tag{23.39}$$

and thus the representations are complex. It is through this complex self-duality condition that the complexity of the $SO(4n+2)$ spinor representations is manifested in the ordinary tensor representations.

We summarize all of this below, incorporating the fact that for $SO(2n)$ with even n, $\overline{D}^n = D^n$ and $\overline{D}^{n-1} = D^{n-1}$, while for odd n, $\overline{D}^n = D^{n-1}$ and $\overline{D}^{n-1} = D^n$. We indicate the rank m antisymmetric tensor representation by (m).

$$\begin{aligned}
SO(2n+1): \; D^n \otimes D^n &= \sum_{k=0}^{n} (k) \\
SO(4m): \; D^{2m} \otimes D^{2m-1} &= \sum_{k=0}^{m-1} (2k+1) \\
D^{2m} \otimes D^{2m} &= \sum_{k=0}^{m-1} (2k) + (2m)_+ \\
D^{2m-1} \otimes D^{2m-1} &= \sum_{k=0}^{m-1} (2k) + (2m)_- \\
SO(4m+2): \; D^{2m+1} \otimes D^{2m} &= \sum_{k=0}^{m} (2k) \\
D^{2m+1} \otimes D^{2m+1} &= \sum_{k=0}^{m-1} (2k+1) + (2m+1)_1 \\
D^{2m} \otimes D^{2m} &= \sum_{k=0}^{m-1} (2k+1) + (2m+1)_2
\end{aligned} \tag{23.40}$$

where $(2m)^{\pm}$ are self-dual and anti-self-dual respectively, and $(2m+1)_1$ and $(2m+1)_2$ satisfy complex self-duality conditions.

23.5 Example: $SO(10)$

For example, consider the 16 dimensional D^5 representation of $SO(10)$. We see that

$$16 \otimes 16 = (1) \oplus (3) \oplus (5)_1 \tag{23.41}$$

The (1) is the 10 dimensional vector. The (3) is the $10 \cdot 9 \cdot 8/6 = 120$ dimensional three index antisymmetric tensor. The (5) is the 5 index antisymmetric tensor with a complex self-duality condition, with dimension

$$\frac{1}{2}\,\frac{10 \cdot 9 \cdot 8 \cdot 7 \cdot 6}{5 \cdot 4 \cdot 3 \cdot 2} = 126 \tag{23.42}$$

The symmetry of these under exchange of the identical 16's can be found by looking at the discussion of the symmetry of the $\Gamma_j R$s for $SU(2n+1)$ — in this case we want $n = 5$ to focus on the 2^5 dimensional space of the Clifford algebra. The symmetry factor for (k) is $(-1)^{(n-k)(n-k-1)/2}$, so the (1) and (5) are symmetric and the (3) is antisymmetric.

Likewise

$$16 \otimes \overline{16} = (0) \oplus (2) \oplus (4) = 1 \oplus 45 \oplus 210 \tag{23.43}$$

23.6 The $SU(n)$ subalgebra

Now we can identify an $SU(n)$ subgroup of $SO(2n)$. From the Clifford algebra, we can construct the objects

$$\begin{aligned} A_j &= \frac{1}{2}\Big(\Gamma_{2j-1} - i\,\Gamma_{2j}\Big) \\ A_j^\dagger &= \frac{1}{2}\Big(\Gamma_{2j-1} + i\,\Gamma_{2j}\Big) \end{aligned} \tag{23.44}$$

Because of the Clifford algebra, these satisfy

$$\begin{aligned} &\{A_j, A_k\} = \left\{A_j^\dagger, A_k^\dagger\right\} = 0 \\ &\left\{A_j, A_k^\dagger\right\} = \delta_{jk} \end{aligned} \tag{23.45}$$

This is, therefore, a set of creation and annihilation operators, from which we already know how to construct an $SU(n)$ algebra. Using the matrix elements, $[T_a]_{jk}$, of the defining representation of $SU(n)$, we define

$$T_a = \sum_{j,k} A_j^\dagger\,[T_a]_{jk}\,A_k \tag{23.46}$$

We know that the T_a generate $SU(n)$ on the 2^n dimensional space of the Clifford algebra, but we must also show that it is a subalgebra of $SO(2n)$. To see this we can write

$$\begin{aligned} A_j^\dagger A_k &= \frac{1}{2}\left\{A_j^\dagger, A_k\right\} + \frac{1}{2}\left[A_j^\dagger, A_k\right] = \frac{1}{2}\delta_{jk} \\ &+\frac{i}{2}M_{2j-1,2k-1} + \frac{1}{2}M_{2j-1,2k} - \frac{1}{2}M_{2j,2k-1} + \frac{i}{2}M_{2j,2k} \end{aligned} \tag{23.47}$$

The δ_{jk} term does not contribute to T_a because $[T_a]_{jk}$ is traceless. The other terms are $SO(2n)$ generators. So, the T_a are linear combinations of the $SO(2n)$ generators, thus they generate a subalgebra.

Denote by $|0\rangle$ the state that is annihilated by all the A_j. In our representation of the Clifford algebra, all the A_js are proportional to lowering operators, σ_-^j, so $|0\rangle$ is the state for which all the $\sigma_3^j = -1$. It is in the representation D^n for n even and D^{n-1} for n odd.

As always, the creation operators, $A_j^\dagger$, are tensor operators:

$$\left[T_a, A_j^\dagger\right] = \sum_k A_k^\dagger \, [T_a]_{kj} \tag{23.48}$$

They transform according to the n dimensional defining representation of $SU(n)$.

When we act on $|0\rangle$ with some number of $A^\dagger$s (say m), we therefore get a set of states transforming like the antisymmetric tensor product of m ns because the $A^\dagger$s anticommute. This is the fundamental representation $[m]$. Otherwise, this construction is similar to what we did with the 3-dimensional harmonic oscillator.

Note also that if m is even, the states formed by acting on $|0\rangle$ with m $A^\dagger$s are in the same irreducible representation of $SO(2n)$ as the $|0\rangle$ state, because Γ_{2n+1} anticommutes with each $A^\dagger$, so it commutes with the product of an even number of them. But for odd m, the states are in the other irreducible $SO(2n)$ representation.

Putting this together, we can summarize the embedding of $SU(n)$ in $SO(2n)$ as follows. For $SO(4n+2)$,

$$D^{2n+1} = \sum_{j=0}^{n} [2j+1] \qquad D^{2n} = \sum_{j=0}^{n} [2j] \tag{23.49}$$

This is consistent with the fact that $\overline{D^{2n}} = D^{2n+1}$ because $\overline{[j]} = [2n+1-j]$.

For $SO(4n)$,

$$D^{2n} = \sum_{j=0}^{n} [2j] \qquad D^{2n-1} = \sum_{j=0}^{n-1} [2j+1] \tag{23.50}$$

There is an $SO(2n)$ generator which commutes with all of the generators of the $SU(n)$ subgroup we have constructed. It is

$$S = \sum_{j=1}^{n} M_{2j-1,2j} = \sum_{j=1}^{n} A_j^\dagger A_j - \frac{n}{2} \tag{23.51}$$

S generates a $U(1)$ subalgebra. In the space of the Clifford Algebra

$$S = \frac{1}{2} \sum_{j=1}^{n} \sigma_3^j \tag{23.52}$$

Thus $S|0\rangle = -\frac{n}{2}|0\rangle$. Then, since $\left[S, A_j^\dagger\right] = A_j^\dagger$, the creation operators raise S. Thus the representation $[m]$ in the spinor representations of $SO(2n)$ has $S = m - n/2$.

Problems

23.A. Check that the dimensions work out in (23.49) and (23.50) by using the binomial theorem.

23.B. Use (23.9) and (23.44) to determine how the vector representation, D^1, of $SO(2n)$ transforms under $SU(n)$. Explain why this result is obvious.

23.C. Let $u^{jk\ell}$ be a completely antisymmetric tensor in $SO(6)$. A self-duality condition has the form

$$u^{jk\ell} = \lambda\, \epsilon^{jk\ell abc}\, u^{abc}\,.$$

What are the possible values of λ?

23.D. How do the spinor representations of $SO(14)$ transform under the following subgroups:

a. $SU(7)$?

b. $SO(4) \times SU(5)$?

Chapter 24

$SO(10)$

$SU(5)$ is the only choice for a unifying algebra that makes use only of the matter particles that we actually see. However, it is not obvious that this is a necessary property. We know that the unifying symmetry, whatever it is (and assuming that it exists at all!) must be spontaneously broken to account for the differences in the interactions that we see. It could be that the process of spontaneous breaking gives a large mass to some of the matter particles of a representation, thus effectively removing them. The next simplest unification, based on the algebra $SO(10)$, adds just one such matter particle - a right-handed neutrino.

24.1 $SO(10)$ and $SU(4) \times SU(2) \times SU(2)$

Let us begin the discussion of $SO(10)$ unification by considering (23.49) for $n = 2$. This shows that the $SU(5)$ content of the spinor representations of $SO(10)$ is as follows:

$$\begin{aligned} D^5 &= [1] + [3] + [5] = 5 + \overline{10} + 1\,, \\ D^4 &= [0] + [2] + [4] = 1 + 10 + \overline{5}\,. \end{aligned} \tag{24.1}$$

Thus D^5 has the right $SU(5)$ content to describe the right-handed creation operator of a family of quarks and leptons, with the addition of one $SU(5)$ singlet. Likewise, D^4 behaves like the left-handed creation operators plus a singlet.

Thus $SO(10)$ has the interesting feature that it incorporates all the creation operators of a single family into a single irreducible representation. Of course, it also incorporates an extra singlet which does not correspond to any

DOI: 10.1201/9780429499210-25

of the particles that we have observed in the world. But we can hope that this extra particle will get a large mass in the spontaneous symmetry breaking process.

The extra singlet is also related to the particles that we actually observe in an interesting way. When we include it, we can restore the parity symmetry that is lost in $SU(5)$. To see what is happening, let us examine another subgroup of $SO(10)$.

The $SU(5)$ subalgebra of $SO(10)$ is regular, but not maximal. It is obtained by removing one of the circles from the Dynkin diagram. To find the regular maximal subalgebras, consider the extended Dynkin diagram:

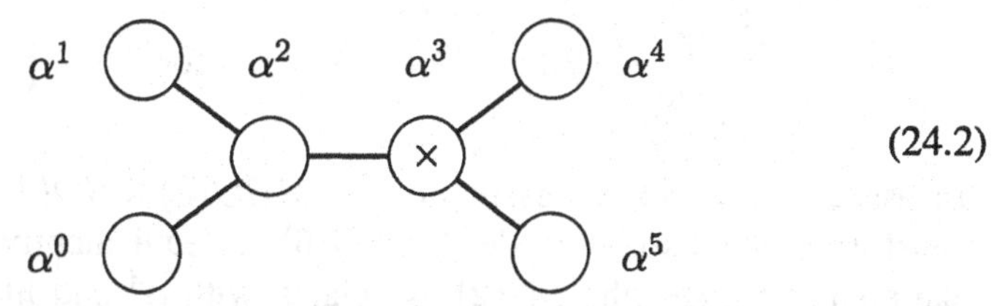

(24.2)

If we remove the circle labeled with the ×, the diagram falls apart into $SU(2) \times SU(2) \times SU(4)$. The $SU(2) \times SU(2)$ is the same algebra as $SO(4)$, and the $SU(4)$ is the same as $SO(6)$. Thus this is the subalgebra of the 10 dimensional rotation generators which are block diagonal on 4 and 6 dimensional subspaces.

We can use the Dynkin diagrams rather directly to see how the spinor representations transform under the subgroup. The weights of D^5 are

$$\sum_{i=1}^{5} \eta_i e^i/2\,, \quad \text{where} \quad \prod_{i=1}^{5} \eta_i = 1 \tag{24.3}$$

from (22.5)-(22.7). We know from the form of the extended Dynkin diagram that the $SU(4)$ (or $SO(6)$) subgroup has roots α^1, α^2, and α^0 (the lowest root). Explicitly

$$\alpha^1 = e^1 - e^2\,, \qquad \alpha^2 = e^2 - e^3\,, \qquad \alpha^0 = -e^1 - e^2\,. \tag{24.4}$$

Thus the roots of the $SU(4)$ subalgebra are entirely in the three dimensional subspace of the weight space formed by the first three components. The weights, (24.3), decompose into two copies of each of the two spinor representations of $SO(6)$, one for $\eta_1\eta_2\eta_3 = 1$, the other for $\eta_1\eta_2\eta_3 = -1$.

The $SU(2) \times SU(2)$ (or $SO(4)$) subgroup corresponds to the roots

$$\alpha^4 = e^4 - e^5\,, \qquad \alpha^5 = e^4 + e^5 \tag{24.5}$$

in the two dimensional subspace formed by the last two components of the weight space. α^4 generates an $SU(2)$ under which the weights

$$\pm(e^4 - e^5) \tag{24.6}$$

form a doublet. α^5 generates an $SU(2)$ under which the weights

$$\pm(e^4 + e^5) \tag{24.7}$$

form a doublet. Thus, the weights

$$\sum_{i=1}^{3} \left(\eta_i e^i\right)/2 \pm (e^4 + e^5)/2 \qquad \text{for} \qquad \prod_i \eta_i = 1 \tag{24.8}$$

are associated with a representation of the $SO(6) \times SO(4)$ subalgebra which transforms like a spinor under the $SO(6)$ (or equivalently, the 4 of $SU(4)$), like a singlet under the $SU(2)$ associated with α^4 and like a doublet under the $SU(2)$ associated with α^5 (we will call this $SU(2)'$ to distinguish it from the first $SU(2)$). Thus, under the $SU(4) \times SU(2) \times SU(2)'$ subgroup, (24.8) is a $(4, 1, 2)$.

The weights

$$\sum_{i=1}^{3} \left(\eta_i e^i\right)/2 \pm (e^4 - e^5)/2 \qquad \text{for} \qquad \prod_i \eta_i = -1 \tag{24.9}$$

transform like the complex conjugate spinor under the $SO(6)$ (or equivalently, the $\overline{4}$ of $SU(4)$), like a doublet under the $SU(2)$ associated with α^4 and like a singlet under the $SU(2)$ associated with α^5. Thus, under the $SU(4) \times SU(2) \times SU(2)'$ subgroup, (24.9) is a $(\overline{4}, 2, 1)$.

Thus

$$D^5 \to (4, 1, 2) \oplus (\overline{4}, 2, 1) \tag{24.10}$$

under $SU(4) \times SU(2) \times SU(2)'$. It then follows that the complex conjugate representation, D^4, is

$$D^4 \to (\overline{4}, 1, 2) \oplus (4, 2, 1) \tag{24.11}$$

The $SU(4)$ in $SU(4) \times SU(2) \times SU(2)$ contains color $SU(3)$. Each 4 is a $3 \oplus 1$ under $SU(3)$. Each $\overline{4}$ is a $\overline{3} \oplus 1$. Comparing (24.10) with (18.13), you can see in more detail how the $SO(10)$ unification works. The weak interaction $SU(2)$ in (18.13) must be identified with the $SU(2)$ subgroup of $SO(10)$. Under it, the creation operators for the right-handed antiquarks and the right-handed positron and antineutrino transform as doublets. Under the $SU(2)'$ of $SO(10)$, the creation operators for the right-handed quarks are a doublet, and there is another doublet comprising the creation operators for the right-handed electron and a neutral particle. This $SU(2)'$ is a mirror image of the weak interaction $SU(2)$. If we look at the creation operators for the left-handed fields, in the representation D^4, the antiquarks will be a doublet under $SU(2)'$, just as the right-handed antiquarks are a doublet under $SU(2)$ in D^5. Thus we should identify the neutral partner of the right-handed electron in the $SU(2)'$ doublet as a right-handed neutrino, the looking-glass version of the left-handed neutrino.

The $SO(10)$ unification thus restores the mirror symmetry of the theory that was lost completely in $SU(5)$. Of course, physics is not mirror symmetric. Parity is violated. The right-handed neutrino has never been seen. If $SO(10)$ unification is to describe the world, there must be some symmetry breaking that makes the right-handed neutrino and the particles associated with the $SU(2)'$ generators (and indeed, all the other $SO(10)$ generators except those of the $SU(3) \times SU(2) \times U(1)$ subgroup) very heavy.

24.2 * Spontaneous breaking of $SO(10)$

The spontaneous breaking of $SO(10)$ down to $SU(3) \times SU(2) \times U(1)$ can be considered in several steps. First one can ask, how can we break $SO(10)$ down to $SU(5)$? Then we can ask, what $SO(10)$ representation contains the 24 of $SU(5)$ that we use to break the $SU(5)$ subgroup down further to $SU(3) \times SU(2) \times U(1)$? Finally, we can ask what $SO(10)$ representation contains the 5 and 45 of $SU(5)$ that we use to break the $SU(3) \times SU(2) \times U(1)$ down to the $U(1)$ of electric charge?

24.3 * Breaking $SO(10) \to SU(5)$

The simplest representation that can break $SO(10)$ to $SU(5)$ is the 16 (or $\overline{16}$) This is evident from (24.1). Because the 16 has a component that transforms as a singlet under the $SU(5)$ subgroup, if that component has a vacuum value, the $SO(10)$ symmetry will be spontaneously broken down to $SU(5)$. In the first discussion of $SO(10)$, this representation was used. However, there is something slightly unattractive about it. It cannot be used (at least at the classical level) to give a large mass to the right- handed neutrino. It is important to have some mechanism for getting rid of the right-handed neutrino. It has not been seen, and if it were present, it would cause problems for both particle physics and for cosmology. The original discussion of $SO(10)$ assumed the existence of an $SO(10)$ singlet neutrino, in addition to the 16. Then the 16 Higgs representation could be used to put the right-handed neutrino together with this extra singlet into a massive particle, thus eliminating the unseen right-handed neutrino from the low-energy particle spectrum.

What is unattractive about this is that the addition of the extra singlet may be a step backwards. The $SU(5)$ model already unified the right-handed particles in two irreducible representations. The $SO(10)$ model without the extra singlet, with all the particles of the lightest family in a single irreducible representation, is perhaps more "unified," but with the extra singlet, it seems like an unnecessary complication of the original $SU(5)$ model. Is it possible to eliminate the right-handed neutrino without adding a singlet? The answer is yes, with a slightly peculiar twist. One can treat the right-handed neutrino as its own antiparticle, and find a Higgs in the tensor product of two 16's that is responsible for its mass. The vacuum value of the Higgs representation must be an $SU(5)$ singlet, as before, so that it can give mass to the $SU(5)$ singlet right-handed neutrino.

The tensor product of two 16s is

$$10 \oplus 120 \oplus 126 \tag{24.12}$$

where the 10 is the $SO(10)$ vector, the 120 is an antisymmetric three index tensor, and the 126 is a complex self-dual five index tensor. The 10 and 126 appear symmetrically in the tensor product of the two 16s (because the two 16s are the same representation, their tensor product can be classified according to representations of S_2), while the 120 appear antisymmetrically. Then we can easily figure out the $SU(5)$ content of each of these representations by doing the tensor product of D^5 in (24.1) with itself. The antisymmetric part is

$$\begin{aligned}
&(5+\overline{10}+1)\otimes(5+\overline{10}+1)_{AS}\\
&=(5\otimes 5)_{AS}\oplus(\overline{10}\otimes\overline{10})_{AS}\oplus(5\otimes\overline{10})\oplus(5\otimes 1)\oplus(\overline{10}\otimes 1)\\
&=10\oplus 45\oplus\overline{5}\oplus\overline{45}\oplus 5\oplus\overline{10}
\end{aligned}\tag{24.13}$$

Note that this is a real representation. The symmetric part is

$$\begin{aligned}
&(5+\overline{10}+1)\otimes(5+\overline{10}+1)_{S}\\
&=(5\otimes 5)_{S}\oplus(\overline{10}\otimes\overline{10})_{S}\oplus(1\otimes 1)_S\oplus(5\otimes\overline{10})\oplus(5\otimes 1)\oplus(\overline{10}\otimes 1)\\
&=15\oplus 5\oplus 50\oplus 1\oplus\overline{5}\oplus\overline{45}\oplus 5\oplus\overline{10}
\end{aligned}\tag{24.14}$$

Since the 10 is $5 + \overline{5}$, the 126 is

$$15 \oplus 50 \oplus 1 \oplus \overline{45} \oplus 5 \oplus \overline{10} \quad (24.15)$$

This representation is complex, and it contains the $SU(5)$ singlet component that we want to give the right-handed neutrino a mass. This is particularly useful, because an $SU(5)$ singlet vacuum value cannot give mass to any of the other particles, because their masses require a vacuum value that breaks $SU(2) \times U(1)$. Thus if the vacuum value of the $SU(5)$ singlet component of the 126 is very large, it will give a large mass to the right-handed neutrino, but all the other matter particles will remain massless until we turn on vacuum value for the Higgs that breaks $SU(2) \times U(1)$. This vacuum value can be much smaller (and indeed, as in $SU(5)$ is has to be if the couplings are to come out correctly), which explains why we don't see the right-handed neutrino — it is much heavier than all the other fermions.

Either with the 126 or the 16 and an extra singlet, we get rid of the right-handed neutrino and break the $SO(10)$ symmetry down to $SU(5)$, we can then discuss the further breaking of $SU(5)$ by asking what $SO(10)$ representations contain the $SU(5)$ representations that we discussed for the breaking of $SU(5)$.

24.4 * Breaking $SO(10) \to SU(3) \times SU(2) \times U(1)$

As in $SU(5)$, the obvious representation to consider is the adjoint representation, the 45 dimensional representation, D^2, which is a real two-index antisymmetric tensor. Let us ask how this $SO(10)$ representation transforms under $SU(5)$. We can find this by noting that the antisymmetric tensor product of two 10s is a 45. Since the 10 is $5 \oplus \overline{5}$, the 45 is

$$45 \to 24 + 10 + \overline{10} + 1 \quad (24.16)$$

This is encouraging. Because (24.16) contains an $SU(5)$ 24, we might expect to find a possible vacuum value that breaks $SO(10)$ to $SU(3) \times SU(2) \times U(1)$.

It is not quite that simple, however. The most general possible vacuum value for the antisymmetric tensor can be brought, by an $SO(10)$ transformation into the following canonical form:

$$\begin{pmatrix} 0 & a_1 & 0 & 0 & 0 & 0 & 0 & 0 & 0 & 0 \\ -a_1 & 0 & 0 & 0 & 0 & 0 & 0 & 0 & 0 & 0 \\ 0 & 0 & 0 & a_2 & 0 & 0 & 0 & 0 & 0 & 0 \\ 0 & 0 & -a_2 & 0 & 0 & 0 & 0 & 0 & 0 & 0 \\ 0 & 0 & 0 & 0 & 0 & a_3 & 0 & 0 & 0 & 0 \\ 0 & 0 & 0 & 0 & -a_3 & 0 & 0 & 0 & 0 & 0 \\ 0 & 0 & 0 & 0 & 0 & 0 & 0 & a_4 & 0 & 0 \\ 0 & 0 & 0 & 0 & 0 & 0 & -a_4 & 0 & 0 & 0 \\ 0 & 0 & 0 & 0 & 0 & 0 & 0 & 0 & 0 & a_5 \\ 0 & 0 & 0 & 0 & 0 & 0 & 0 & 0 & -a_5 & 0 \end{pmatrix} \tag{24.17}$$

The only way this can commute with $SU(3) \times SU(2) \times U(1)$ is if two of the as, say a_1 and a_2 are equal, and the other three as are equal, so that the vacuum value has the form

$$\begin{pmatrix} a_1\, I_2\, i\sigma_2 & 0 \\ 0 & a_3\, I_3\, i\sigma_2 \end{pmatrix} \tag{24.18}$$

where the first block is 4×4, written as a tensor product of two two- dimensional spaces, and the second is 6×6, written similarly as tensor product of two- and three-dimensional spaces. This commutes with matrices of the form

$$\begin{pmatrix} A_2 + \sigma_2\, S_2 & 0 \\ 0 & A_3 + \sigma_2\, S_3 \end{pmatrix} \tag{24.19}$$

where the As are antisymmetric and the Ss are symmetric matrices in the appropriate spaces. The $A_2 + \sigma_2\, S_2$ generate an $SU(2) \times U(1)$, while the $A_3 + \sigma_2\, S_3$ generate an $SU(3) \times U(1)$. There is an extra $U(1)$ here that we do not want, but it is broken by the vacuum value of the 16 or 126. Thus together with the 16 or 126, the 45 can do the required symmetry breaking down to $SU(3) \times SU(2) \times U(1)$.

There are other options for the $SO(10)$ breaking. We will just mention one. The symmetry can be broken by the vacuum value of a 54, a real traceless, symmetric tensor. This is also simple to analyze, because like the 45, it can be brought into a simple canonical form by an $SO(10)$ transformation.

In fact, this is even simpler — the 54 can be diagonalized

$$\begin{pmatrix} a_1 & 0 & 0 & 0 & 0 & 0 & 0 & 0 & 0 & 0 \\ 0 & a_2 & 0 & 0 & 0 & 0 & 0 & 0 & 0 & 0 \\ 0 & 0 & a_3 & 0 & 0 & 0 & 0 & 0 & 0 & 0 \\ 0 & 0 & 0 & a_4 & 0 & 0 & 0 & 0 & 0 & 0 \\ 0 & 0 & 0 & 0 & a_5 & 0 & 0 & 0 & 0 & 0 \\ 0 & 0 & 0 & 0 & 0 & a_6 & 0 & 0 & 0 & 0 \\ 0 & 0 & 0 & 0 & 0 & 0 & a_7 & 0 & 0 & 0 \\ 0 & 0 & 0 & 0 & 0 & 0 & 0 & a_8 & 0 & 0 \\ 0 & 0 & 0 & 0 & 0 & 0 & 0 & 0 & a_9 & 0 \\ 0 & 0 & 0 & 0 & 0 & 0 & 0 & 0 & 0 & a_{10} \end{pmatrix} \tag{24.20}$$

where

$$\sum_{j=1}^{10} a_j = 0 \tag{24.21}$$

The only way this can commute with $SU(3) \times SU(2) \times U(1)$ is if four of the as are equal, so that there is an unbroken $SO(4)$ that contains the $SU(2) \times U(1)$, and the other six as are equal, so that there is an unbroken $SO(6)$ that contains $SU(3)$. Thus a vacuum value for the 54 that preserves $SU(3) \times SU(2) \times U(1)$ automatically preserves a larger symmetry, $SO(6) \times SO(4)$, which is in fact just the $SU(2) \times SU(2) \times SU(4)$ subgroup that we discussed at the beginning of this section. But in fact, this is really all we need from the 54. The 16 or 126 vacuum value breaks the $SU(2) \times SU(2) \times SU(4)$ the rest of the way down to $SU(3) \times SU(2) \times U(1)$.

24.5 * Breaking $SO(10) \to SU(3) \times U(1)$

Finally, we can think about what $SO(10)$ representations can be responsible for breaking $SU(2) \times U(1)$ and giving mass to the quarks and leptons. We have seen that these must contain a 5 or 45 of $SU(5)$ and they should be contained in the tensor product of two 16s, (24.12). You can see that all three of the representations in 16×16 can do the job. Each gives a somewhat different pattern of mass relations among the particles of different charges and colors.

24.6 * Lepton number as a fourth color

There is one more thing that is worth mentioning about the $SU(2) \times SU(2)' \times SU(4)$ subgroup of $SO(10)$. This algebra is interesting because it is the smallest that contains the $SU(3) \times SU(2) \times U(1)$ symmetry in a **semi-simple** algebra. This is physically interesting, because it leads automatically to the quantization of electric charge. A simple algebra is not necessary for charge quantization, because it is only $U(1)$ factors that are not constrained by commutation relations. As we mentioned above, the $SU(2)'$ factor is a right-handed version of the electroweak $SU(2)$, so that in this model, the left-handed nature of the electroweak interactions is picked out not by the algebra, but by the vacuum — by whatever spontaneously breaks $SU(2)'$ much more strongly than $SU(2)$. But the $SU(4)$ is also quite interesting. One can think of it as treating lepton number as a fourth color. The neutrino is part of an $SU(4)$ 4 with the three colors of u quarks, and the electron is part of an $SU(4)$ 4 with the three colors of d quarks. It is only the spontaneous symmetry breaking, which leaves the color $SU(3)$ subgroup unbroken, that distinguishes leptons from quarks.[1]

Problems

24.A. Show that the matrices

$$\frac{1}{2}\vec{\sigma}\,,\ \frac{1}{2}\vec{\tau}\,,\ \frac{1}{2}\vec{\eta}\,,\ \frac{1}{2}\vec{\sigma}\rho_1\,,\ \frac{1}{2}\vec{\tau}\rho_2\,,\ \frac{1}{2}\vec{\eta}\rho_3\,,\ \frac{1}{2}\vec{\sigma}\vec{\tau}\rho_3\,,\ \frac{1}{2}\vec{\tau}\vec{\eta}\rho_1\,,\ \frac{1}{2}\vec{\eta}\vec{\sigma}\rho_2\,,$$

where $\vec{\sigma}$, $\vec{\tau}$, $\vec{\eta}$ and $\vec{\rho}$ are independent sets of Pauli matrices, generate a spinor representation of $SO(10)$. Find an $SU(2) \times SU(2) \times SU(4)$ subgroup in which one of the $SU(2)$ factors is generated by the subset $\vec{\eta}\,(1+\rho_3)/4$.

24.B. What is the dimension of the $SO(10)$ representation with highest weight $2\mu^5$. How do you know? Hint: consider $D^5 \otimes D^5$.

---

[1]This idea, along with the $SU(4) \times SU(2) \times SU(2)$ model, was an important precursor to unification, discussed in J. Pati and A. Salam, Phys. Rev. **D 10** (1974) 275.

Chapter 25

Automorphisms

An **automorphism** A of a group G is a mapping of the group onto itself which preserves the group multiplication rule:

$$A(g_1 g_2) = A(g_1)\, A(g_2) \tag{25.1}$$

For a Lie group, an automorphism of the group induces a mapping of the Lie algebra onto itself which preserves the commutation relations. Under an automorphism, the generators are mapped into linear combinations of generators

$$T_a \to A_{ab}\, T_b \tag{25.2}$$

such that

$$[T_a, T_b] = i\, f_{abc}\, T_c \tag{25.3}$$

implies

$$[A_{aa'} T_{a'}, A_{bb'} T_{b'}] = i\, f_{abc}\, A_{cc'} T_{c'} \tag{25.4}$$

25.1 Outer automorphisms

Some automorphisms are trivial in the sense that the mapping they induce on the generators is an equivalence:

$$A_{ab}\, T_b = R\, T_a\, R^{-1} \tag{25.5}$$

where $R = e^{i\theta_a T_a}$ is a group element. This is called an **inner automorphism**. But some of the Lie algebras have non-trivial or **outer automorphisms**.

For example, consider complex conjugation. If T_a are the generators of some representation, the mapping

$$T_a \to -T_a^* \tag{25.6}$$

 DOI: 10.1201/9780429499210-26

is an automorphism in which generators corresponding to imaginary antisymmetric matrices are unchanged, while generators corresponding to real symmetric matrices change sign (note that we can always choose the generators to be either antisymmetric or symmetric in the highest weight construction). Thus, an algebra can have complex representations only if it has some nontrivial automorphism.

We can identify the nontrivial automorphisms by looking at the symmetries of the Dynkin diagram. For example, consider $SU(4)$ with Dynkin diagram

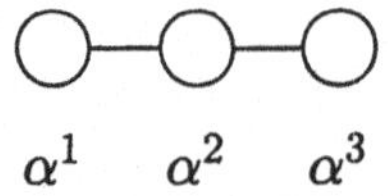

$$\alpha^1 \qquad \alpha^2 \qquad \alpha^3 \tag{25.7}$$

Since α^1 and α^3 appear symmetrically, the diagram with α^1 and α^3 interchanged has all the same lengths and angles and thus generates the same algebra. So there is an automorphism of $SU(4)$ in which the corresponding generators are interchanged,

$$E_{\alpha^1} \leftrightarrow , E_{\alpha^3} \tag{25.8}$$

along with all the other changes this induces in the explicit construction of the algebra from the diagram. This automorphism is nontrivial, since it interchanges the fundamental representations D^1 and D^3, which are inequivalent. In fact, since these representations are complex conjugates of one another, this is just the automorphism induced by complex conjugation, up to some trivial equivalence. All of the complex conjugation automorphisms are obtained in this way, associated with reflection symmetries of the Dynkin diagram.

But not all reflection symmetries correspond to complex conjugation. For example, the $SO(4n)$ groups have only real representations. Thus the reflection symmetries of their Dynkin diagrams correspond to nontrivial automorphisms that are not complex conjugations. The most interesting (and bizarre) example of this is the group $SO(8)$, with Dynkin diagram

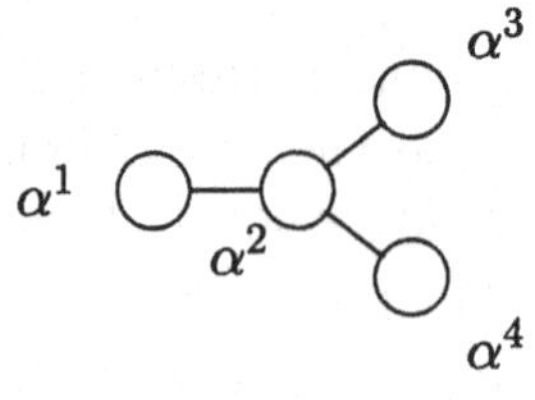

$$\alpha^3 \quad \alpha^1 \quad \alpha^2 \quad \alpha^4 \tag{25.9}$$

Here there is a separate automorphism for each permutation for α^1, α^3 and α^4.

25.2 Fun with $SO(8)$

The roots of the $SO(8)$ Dynkin diagram are

$$\alpha^j = e^j - e^{j+1} \text{ for } j = 1 \text{ to } 3\,, \quad \alpha^4 = e^3 + e^4 \tag{25.10}$$

The fundamental representations D^1, D^3, and D^4, with highest weights μ^1, μ^3, and μ^4, are each 8 dimensional. D^1 is the defining representation, an 8-vector. D^3 and D^4 are the two real spinor representations. The automorphisms map these three representations into one another in all possible ways.

First consider the mapping between D^3 and D^4. This corresponds to the symmetry of the Dynkin diagram that interchanges α^3 and α^4 with α^1 and α^2 held fixed. This is implemented by changing the sign of e^4. That means

$$H_4 \to -H_4 \quad \text{or} \quad M_{78} \to -M_{78} \tag{25.11}$$

and

$$E_{\eta e^j + e^4} \leftrightarrow E_{\eta e^j - e^4} \quad \text{or} \quad M_{j8} \to -M_{j8} \tag{25.12}$$

This, of course, is what we found when we explicitly constructed these representations. M_{jk} for $j, k = 1$ to 7 were the same for both representations, while

$$H_4 = M_{78} = \begin{cases} -\frac{1}{2}\sigma_3^1\,\sigma_3^2\,\sigma_3^3 & \text{for } D^3 \\ \frac{1}{2}\sigma_3^1\,\sigma_3^2\,\sigma_3^3 & \text{for } D^4 \end{cases} \tag{25.13}$$

The rest follow from commutation. Of course in this case, we got the result just from the symmetry of the Dynkin diagram, without constructing the representation explicitly.

Now consider the automorphism that interchanges α^1 and α^3 with α^2 and α^4 held fixed. This is rather weird, because it interchanges the vector representation D^1, generated by the antisymmetric matrices with only two non-zero elements with the spinor representation D^3, generated by products of Pauli matrices, whose square is proportional to the identity. The relevant mapping in terms of the weights is

$$\begin{aligned} e^1 - e^2 &\leftrightarrow e^3 - e^4 \\ e^2 - e^3 &\to e^2 - e^3 \\ e^3 + e^4 &\to e^3 + e^4 \end{aligned} \tag{25.14}$$

Solving this for the e^js we find

$$\begin{aligned} e^1 &\to \frac{1}{2}\left(e^1+e^2+e^3-e^4\right) \\ e^2 &\to \frac{1}{2}\left(e^1+e^2-e^3+e^4\right) \\ e^3 &\to \frac{1}{2}\left(e^1-e^2+e^3+e^4\right) \\ e^4 &\to \frac{1}{2}\left(-e^1+e^2+e^3+e^4\right) \end{aligned} \tag{25.15}$$

so the Cartan generators get mapped in the same way

$$\begin{aligned} H_1 &\to \frac{1}{2}\left(H_1+H_2+H_3-H_4\right) \\ H_2 &\to \frac{1}{2}\left(H_1+H_2-H_3+H_4\right) \\ H_3 &\to \frac{1}{2}\left(H_1-H_2+H_3+H_4\right) \\ H_4 &\to \frac{1}{2}\left(-H_1+H_2+H_3+H_4\right) \end{aligned} \tag{25.16}$$

or

$$\begin{aligned} M_{12} &\to \frac{1}{2}\left(M_{12}+M_{34}+M_{56}-M_{78}\right) \\ M_{34} &\to \frac{1}{2}\left(M_{12}+M_{34}-M_{56}+M_{78}\right) \\ M_{56} &\to \frac{1}{2}\left(M_{12}-M_{34}+M_{56}+M_{78}\right) \\ M_{78} &\to \frac{1}{2}\left(-M_{12}+M_{34}+M_{56}+M_{78}\right) \end{aligned} \tag{25.17}$$

Note how cleverly the theory has solved the mapping problem. The antisymmetric generators, $M_{2j-1,2j}$ of the vector representation get mapped into matrices of the form

$$\frac{1}{2}\begin{pmatrix} \pm\sigma_2 & 0 & 0 & 0 \\ 0 & \pm\sigma_2 & 0 & 0 \\ 0 & 0 & \pm\sigma_2 & 0 \\ 0 & 0 & 0 & \pm\sigma_2 \end{pmatrix} \tag{25.18}$$

whose square is 1/4.

The way the rest of the mappings go is this. The other generators break up into 6 sets of 4, each of which mix up among themselves like the four M_{12}, M_{34}, M_{56} and M_{78} that we just dealt with.

$SO(8)$ has a very interesting subgroup — $SU(2) \times SU(2) \times SU(2) \times SU(2)$. It is easy to see that this is a maximal subalgebra. It is just $SO(4) \times SO(4)$. But the way the automorphism works on the different $SU(2)$ factors is instructive. Let us look at the extended Dynkin diagram. It is

$$\begin{array}{ccccc} \alpha^1 \bigcirc & & \alpha^2 & & \bigcirc \alpha^3 \\ & \diagdown & & \diagup & \\ & & \bigcirc & & \\ & \diagup & & \diagdown & \\ \alpha^0 \bigcirc & & & & \bigcirc \alpha^4 \end{array} \tag{25.19}$$

where the lowest root, α^0 is $-e^1-e^2$. Evidently, we get the $SU(2)\times SU(2)\times SU(2)\times SU(2)$ subalgebra by removing α^2. The four $SU(2)$s are associated with the four mutually orthogonal roots,

$$\begin{aligned} \alpha^0 &= -e^1 - e^2 \qquad & \alpha^1 &= e^1 - e^2 \\ \alpha^3 &= e^3 - e^4 \qquad & \alpha^4 &= e^3 + e^4 \end{aligned} \tag{25.20}$$

Now consider the action of this subalgebra on the spinor representation D^3, with weights

$$\eta_j e^j/2 \qquad \text{for} \qquad \prod_j \eta_j = -1 \tag{25.21}$$

These break up into two sets transforming irreducibly under the subalgebra. The set

$$\begin{aligned} &\frac{1}{2}\left(e^1+e^2+e^3-e^4\right), \quad \frac{1}{2}\left(e^1+e^2-e^3+e^4\right), \\ &\frac{1}{2}\left(-e^1-e^2+e^3-e^4\right), \quad \frac{1}{2}\left(-e^1-e^2-e^3+e^4\right) \end{aligned} \tag{25.22}$$

is orthogonal to α^1 and α^4, and thus transform trivially, like singlets under the two $SU(2)$ associated with α^1 and α^4. But each weight is a component of doublet under the $SU(2)$ associated with α^0 and α^3. Similarly, the set

$$\begin{aligned} &\frac{1}{2}\left(e^1-e^2+e^3+e^4\right), \quad \frac{1}{2}\left(e^1-e^2-e^3-e^4\right), \\ &\frac{1}{2}\left(-e^1+e^2+e^3+e^4\right), \quad \frac{1}{2}\left(-e^1+e^2-e^3-e^4\right) \end{aligned} \tag{25.23}$$

is orthogonal to α^0 and α^3, and thus transform trivially, under these two $SU(2)$, and are components of doublet under the $SU(2)$ associated with α^1 and α^4. Thus we say that under the

$$SU(2)^0 \times SU(2)^1 \times SU(2)^3 \times SU(2)^4 \tag{25.24}$$

subgroup, the spinor representation D^3 transforms as

$$(2,1,2,1) \oplus (1,2,1,2) \tag{25.25}$$

There are two other similar possibilities for the tranformation property of an 8:

$$\begin{gathered} (2,2,1,1) \oplus (1,1,2,2) \\ \text{and} \\ (2,1,1,2) \oplus (1,2,2,1) \end{gathered} \tag{25.26}$$

These correspond to D^1 and D^4 respectively. How do we know which is which?

Problems

25.A. Carry through the argument discussed at the end of the chapter and determine which representation in the last equation in the chapter is D^1 and which is D^4.

25.B. Does $SO(8)$ have an $SO(5)$ subgroup under which one spinor (D^4, say) transforms like two $SO(5)$ spinors while the other spinor (D^3) transforms like an $SO(5)$ vector and three singlets? Explain.

Chapter 26

$Sp(2n)$

One reason that I want to discuss $Sp(2n)$ in more detail is that it is a good excuse to introduce another notation for $SU(n)$ which is often easier to work with.

26.1 Weights of $SU(n)$

The weights of the defining representation of $SU(n)$, ν^i for $i = 1$ to n, have the following properties

$$\nu^j \cdot \nu^k = \frac{1}{2}\delta_{jk} - \frac{1}{2n}, \qquad \sum_{j=1}^{n} \nu^j = 0 \tag{26.1}$$

The condition on the sum follows from the tracelessness of the generators. The n ν^j span an $n-1$ dimensional space, but we can make things much more obvious by embedding the ν^j's in an n dimensional space, as follows:

$$\nu^j = \frac{1}{\sqrt{2}}\left(e^j - \Sigma/n\right) \tag{26.2}$$

where the e^j are an orthonormal basis and the vector Σ is

$$\Sigma \equiv \sum_{k=1}^{n} e^k \tag{26.3}$$

In this notation, the simple roots of $SU(n)$ take the form

$$\alpha^j = \nu^j - \nu^{j+1} = \frac{1}{\sqrt{2}}\left(e^j - e^{j+1}\right), \text{ for } j = 1 \text{ to } n-1 \tag{26.4}$$

 DOI: 10.1201/9780429499210-27

These are also the first $n-1$ simple roots of $Sp(2n)$. The last simple root is

$$\alpha^n = 2\nu^n + \sqrt{\frac{2}{n}}\,\nu^{n+1} \tag{26.5}$$

where ν^{n+1} is a unit vector orthogonal to e^j for $j = 1$ to n. Now the point is that we can take

$$\nu^{n+1} = \frac{1}{\sqrt{n}}\,\Sigma \tag{26.6}$$

Then

$$\alpha^n = \sqrt{2}\,e^n \tag{26.7}$$

We could have guessed this from the Dynkin diagram by comparing with our form for the roots of $SO(2n+1)$. The only difference here is overall normalization of the roots and that the last root is longer than the others, rather than shorter. But you should now see the connection with the previous, somewhat cumbersome notation.

So the roots are

$$\frac{\pm e^j \pm e^k}{\sqrt{2}} \quad \text{for } j \neq k \quad \text{and} \quad \pm\sqrt{2}\,e^j \tag{26.8}$$

or equivalently

$$\frac{\pm e^j \pm e^k}{\sqrt{2}} \quad \text{for all } j, k \tag{26.9}$$

The weights of the $2n$ dimensional representation are

$$\pm e^j/\sqrt{2} \tag{26.10}$$

so D^1 has highest weight

$$\mu^1 = e^1/\sqrt{2} \tag{26.11}$$

It is easy to see that the other fundamental representations have highest weight

$$\mu^j = \sum_{k=1}^{j} e^k/\sqrt{2} \tag{26.12}$$

which is related to the antisymmetric tensor product of j D^1s.

26.2 Tensors for $Sp(2n)$

If we invent a tensor language in which the states of D^1 have a lower index (as in $SU(n)$), the tensor coefficients have upper indices and transform in the usual way

$$(T_a\, u)^\alpha = [T_a]^\alpha_\beta\, u^\beta \tag{26.13}$$

where α and β run from 1 to $2n$. It is most convenient to write these in the tensor product notation we introduced in which $\alpha = (j, x)$ where j runs from 1 to n and x from 1 to 2. Then T_a are the generators of the defining representation, with the form

$$\epsilon_a\, [T_a]^{jx}_{ky} = \left(\vec{S}^j_k\, \vec{\sigma}^x_y + A^j_k\, \delta^x_y\right) \tag{26.14}$$

An arbitrary linear combination of $Sp(2n)$ generators has this form where the $\vec{S}$ are three independent real symmetric $n \times n$ matrices and A is an antisymmetric $n \times n$ matrix. For example we could denote the generators as follows:

$$\begin{aligned}
[T_{\mu jj}]^{kx}_{\ell y} &= \frac{1}{\sqrt{2}}\, \delta_{jk}\delta_{j\ell}\, [\sigma_\mu]_{xy} \quad \text{for } \mu = 1 \text{ to } 3 \\
[T_{\mu ij}]^{kx}_{\ell y} &= \frac{1}{2}\left(\delta_{ik}\delta_{j\ell} + \delta_{jk}\delta_{i\ell}\right) [\sigma_\mu]_{xy} \\
&\qquad \text{for } \mu = 1 \text{ to } 3\,, i \neq j \\
[T_{0ij}]^{kx}_{\ell y} &= \frac{i}{2}\left(\delta_{ik}\delta_{j\ell} - \delta_{jk}\delta_{i\ell}\right) \delta_{xy}
\end{aligned} \tag{26.15}$$

The complex conjugate representation, with a lower index transforms as

$$(T_a\, u)_\alpha = -[T_a]^\beta_\alpha\, u_\beta = -[T^*_a]^\alpha_\beta\, u_\beta \tag{26.16}$$

But this representation is pseudoreal, because

$$-[R]^{\alpha\alpha'}\, [T^*_a]^{\beta'}_{\alpha'}\, [R^{-1}]_{\beta'\beta} = [T_a]^\alpha_\beta \tag{26.17}$$

where

$$R^{jxky} = \delta_{jk}\, [\sigma_2]_{xy} \tag{26.18}$$

R is antisymmetric, so the representation is pseudoreal. As we saw with the orthogonal groups, R is an invariant tensor with two upper indices and R^{-1} is an invariant tensor with two lower indices (they happen to have equal matrix elements in this case). Thus we need never consider tensors with lower indices, because we can always raise the lower indices and obtain tensors with the same information.

Now we can use this tensor notation to analyze the irreducible representations as tensor products of the defining representation. The point is that tensor analysis looks very much like that in $SU(2n)$, except that the existence of the invariant tensors R and R^{-1} allows us to get rid of lower indices and to reduce antisymmetric combinations. For example, consider the tensor product $D^1 \otimes D^1$, which explicitly looks like

$$u^\alpha \, v^\beta \tag{26.19}$$

The symmetric combination

$$S^{\alpha\beta} \equiv u^\alpha \, v^\beta + u^\beta \, v^\alpha \tag{26.20}$$

is irreducible because we cannot use R to decompose it further. This has highest weight $2\nu^1$, so this is that adjoint representation. Note that the number of components is right:

$$\frac{2n(2n+1)}{2} = 3 \times \frac{n(n+1)}{2} + \frac{n(n-1)}{2} \tag{26.21}$$

But the antisymmetric combination

$$u^\alpha \, v^\beta - u^\beta \, v^\alpha \tag{26.22}$$

can be reduced, because the combination

$$O \equiv [R^{-1}]_{\beta\alpha} \, u^\alpha \, v^\beta \tag{26.23}$$

is a non-zero singlet. The combination

$$A^{\alpha\beta} \equiv u^\alpha \, v^\beta - u^\beta \, v^\alpha - \frac{1}{n} [R]^{\alpha\beta} \, [R^{-1}]_{\beta'\alpha'} \, u^{\alpha'} \, v^{\beta'} \tag{26.24}$$

satisfies

$$[R^{-1}]_{\beta\alpha} \, A^{\alpha\beta} = 0 \tag{26.25}$$

because

$$[R^{-1}]_{\beta\alpha} \, R^{\alpha\beta} = \text{Tr}\left(R^{-1} R\right) = 2n \tag{26.26}$$

Therefore $A^{\alpha\beta}$ is the irreducible representation D^2 and we can write the product in terms of irreducible combination as

$$u^\alpha \, v^\beta = \frac{1}{2} \, S^{\alpha\beta} + \frac{1}{2} \, A^{\alpha\beta} + \frac{1}{n} \, R^{\alpha\beta} \, O \tag{26.27}$$

It is sometimes useful to see all this explicitly in the tensor product notation. Here the discussion should remind you of our treatment of symmetry and antisymmetry in the spinor representations of the orthogonal groups. The point is that the matrix generators of D^1 provide an invariant tensor that describes how the adjoint representation appears in $D^1 \otimes \overline{D^1}$. However, this tensor has one upper and one lower index. To find the invariant tensor in $D^1 \otimes D^1$, we must raise the lower index with R. That is, if we multiply the generators on the right by R which in the tensor product notation is just σ_2, we should get symmetric matrices with two upper indices. And this is right, because

$$\left(\vec{S} \cdot \vec{\sigma} + A\right) \sigma_2 \tag{26.28}$$

is symmetric because in each term the factors in the two spaces have the same symmetry. In the same notation, one sees that σ_2 itself is the singlet in $D^1 \otimes \overline{D^1}$, and that D^2 has the form

$$\left(\vec{A}' \cdot \vec{\sigma} + S'\right) \sigma_2 \tag{26.29}$$

where S' is constrained to be traceless.

Problems

26.A. Find a set of Cartan generators for $\vec{\sigma} \cdot \vec{S} + A$ that makes it obvious that this defining representation is the fundamental representation μ^1.

26.B. Find $D^1 \otimes D^2$ in $Sp(6)$ using tensor methods. Find the dimensions of each of the irreducible representations that appear in the tensor product.

Chapter 27

Odds and Ends

27.1 Exceptional algebras and octonians

The search for interesting unified theories, like the $SU(5)$ and $SO(10)$ theories, leads naturally to an interesting theory based on the algebra, E_6, one of the exceptional Lie algebras. The exceptional algebras are associated with the **octonians**, a peculiar set of objects of the form

$$a + b_\alpha i_\alpha \tag{27.1}$$

where a and the b_α for $\alpha = 1$ to 7 are real numbers. The i_α are a set of seven "imaginary units" which are generalizations of i and of the quaternions, $\vec{j}$. The i_α have the following multiplication law:

$$i_\alpha i_\beta = -\delta_{\alpha\beta} + g_{\alpha\beta\gamma}\, i_\gamma \tag{27.2}$$

where $g_{\alpha\beta\gamma}$ is completely antisymmetric. In some basis, g is

$$g_{123} = g_{247} = g_{451} = g_{562} = g_{634} = g_{375} = g_{716} = 1 \tag{27.3}$$

with all other components either zero or obtainable from (27.3) by antisymmetrization. This multiplication law can be obtained from the following pic-

DOI: 10.1201/9780429499210-28

ture:

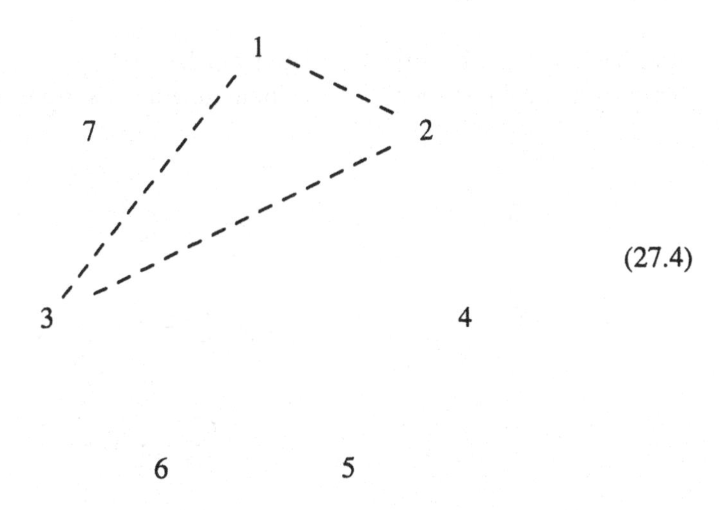

(27.4)

The seven sets $\{jk\ell\}$ for which $g_{jk\ell}$ is not zero are obtained by rotating the doted triangle, with the cyclic order of the indices maintained.

The algebra (27.2) shares with the real numbers, the complex numbers and the quaternions a nice property that was discussed in chapter 19. It is a division algebra with an absolute value that obeys a product rule. The absolute value,

$$|a + b_\alpha i_\alpha| \equiv \left(a^2 + b_\alpha^2\right)^{1/2} \tag{27.5}$$

is preserved under multiplication. If A and B are octonians, then

$$|AB| = |A||B|\,. \tag{27.6}$$

However, the octonian multiplication law is not associative! For example,

$$(i_1 i_2) i_7 = i_3 i_7 = i_5\,, \qquad i_1(i_2 i_7) = i_1(-i_4) = -i_5\,. \tag{27.7}$$

Because of the lack of associativity, octonionic matrices do not generally form groups (whose multiplication laws must be associative). However, the octonians are connected with the exceptional algebras in less direct ways. For example, G_2 is the subgroup of $SO(7)$ that leaves the object $g_{\alpha\beta\gamma}$ appearing in (27.2) invariant.

27.2 E_6 unification

Now back to E_6. Note first that to go from $E_8 \to E_7 \to E_6$, you remove one circle from the left branch of the Dynkin diagram, as shown below.

E_8

E_7

E_6 (27.8)

$E_5 = SO(10)$

$E_4 = SU(5)$

Continuing the same series, you see that $E_5 = SO(10)$ and $E_4 = SU(5)$. Because E_4 and E_5 both give sensible unified theories, E_6 is worth looking at. Of course, this argument is only suggestive, because it turns out that E_7 and E_8 are not useful for unification. But E_6 does turn out to be interesting.

Let us begin by considering the root system and some of the fundamental representations. Label the roots as shown below:

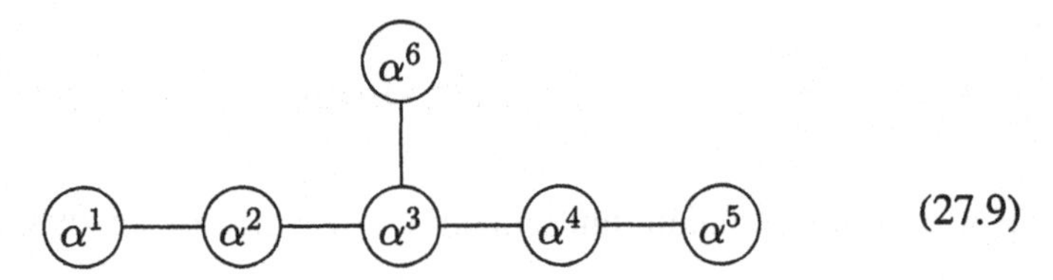

The Cartan matrix is then

$$\begin{pmatrix} 2 & -1 & 0 & 0 & 0 & 0 \\ -1 & 2 & -1 & 0 & 0 & 0 \\ 0 & -1 & 2 & -1 & 0 & -1 \\ 0 & 0 & -1 & 2 & -1 & 0 \\ 0 & 0 & 0 & -1 & 2 & 0 \\ 0 & 0 & -1 & 0 & 0 & 2 \end{pmatrix} \tag{27.10}$$

From this we can construct the weights of the representation D^1 using the Dynkin coefficients. The result (without all the lines — there were too many to draw) looks like this:

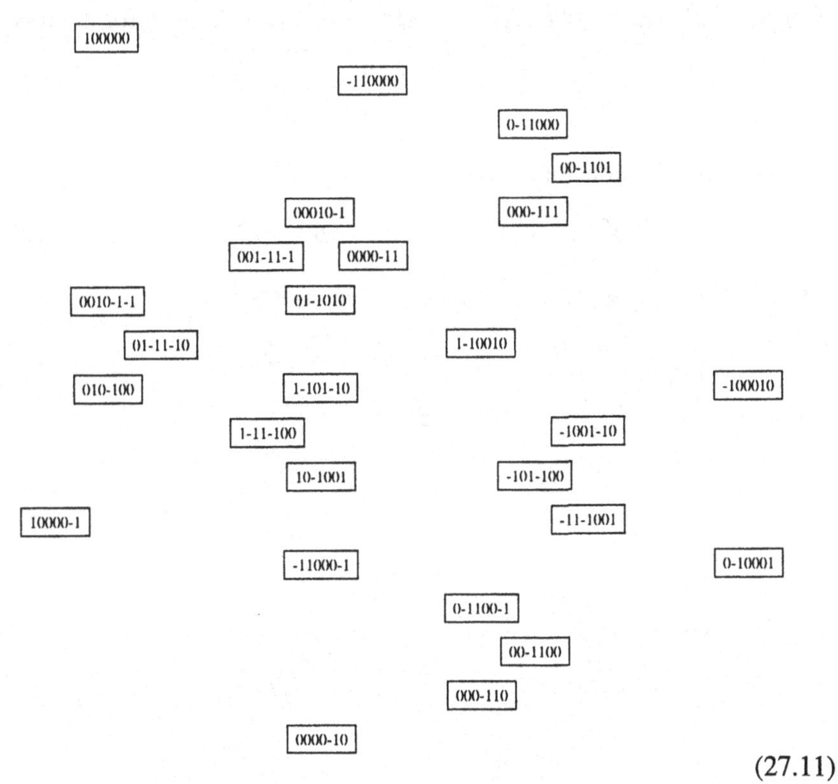

(27.11)

There are 27 weights here. This looks a little complicated, but one thing is clear. This is a complex representation— it is not symmetrical about the origin in weight space. To understand the structure of this object in terms of things we are more familiar with, let us see how this representation transforms under the two maximal regular subalgebras of E_6; $SU(3) \times SU(3) \times SU(3)$ and $SU(6) \times SU(2)$. We can do this by considering the extended Dynkin diagram and finding the Dynkin coefficients of the 27 under the subalgebras.

The extended Dynkin diagram is given by

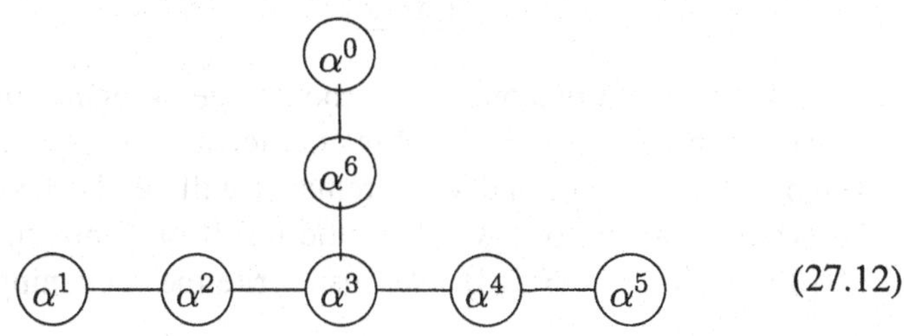

(27.12)

It follows from (20.15)-(a) that

$$\alpha^0 + \alpha^1 + 2\alpha^2 + 3\alpha^3 + 2\alpha^4 + \alpha^5 + 2\alpha^6 = 0 \tag{27.13}$$

Thus we can compute the Dynkin coefficient of the weight μ for the root α^0,

$$\ell^0 = \frac{2\alpha^0 \cdot \mu}{\alpha^{0\,2}} \tag{27.14}$$

as linear combinations of the 6 coefficients:

$$\ell^0 = -\ell^1 - 2\ell^2 - 3\ell^3 - 2\ell^4 - \ell^5 - 2\ell^6 \tag{27.15}$$

The result is shown below where the last column is ℓ^0.

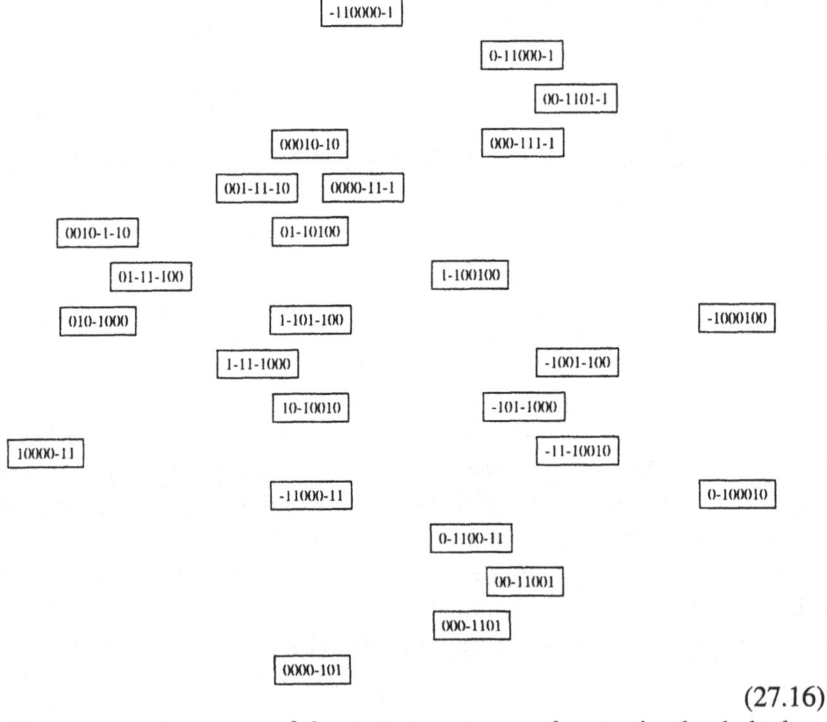

(27.16)

If we now remove one of the roots to get a regular maximal subalgebra, we can simply remove the Dynkin coefficients corresponding to the removed weight from (27.16), and what remains will be the Dynkin coefficients of the subalgebra. Below, we show the result of removing the root α^3 to get $SU(3) \times SU(3) \times SU(3)$. We also write the remaining coefficients in the order 120654. We do this so that in each of the three $SU(3)$ factors, the two roots appear together and the root on the outside of the extended Dynkin diagram appears first.

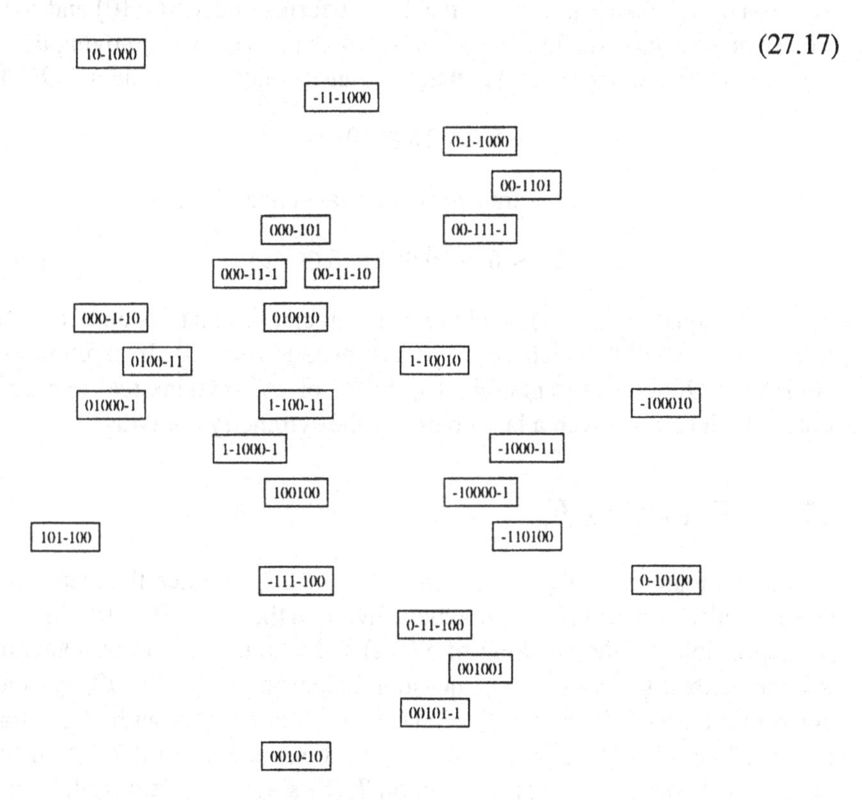

(27.17)

These can be organized into three sets of nine weights:

$$\boxed{\{10,\text{-}11,0\text{-}1\}\{01,1\text{-}1,\text{-}10\}00}\,,\ \boxed{\{01,1\text{-}1,\text{-}10\}00\{10,\text{-}11,0\text{-}1\}}\,,\quad \text{and}\ \boxed{00\{10\text{-}11,0\text{-}1\}\{01,1\text{-}1,\text{-}10\}} \tag{27.18}$$

In the set $\{10, -11, 0-1\}$, we recognize the Dynkin coefficients of the 3 of $SU(3)$. The set {01,1-1,-10} is the $\overline{3}$. Thus the 27 transforms as

$$27 \sim (3, \overline{3}, 1) \oplus (1, 3, \overline{3}) \oplus (\overline{3}, 1, 3) \tag{27.19}$$

An analogous argument, removing the root α^6, shows that the 27 transforms under the $SU(6) \times SU(2)$ subgroup as

$$27 \sim (6, 2) \oplus (\overline{15}, 1) \tag{27.20}$$

where the $\overline{15}$ is the fundamental representation D^4, antisymmetric in 4 upper (or two lower) indices.

Note that $SO(10)$ and $SU(5)$ are regular but not maximal subalgebras of E_6. You can probably guess how the 27 transforms under $SO(10)$ and $SU(5)$ from what we have already seen. But it is easy to check by removing the first Dynkin coefficient from (27.11) that the transformation law under $SO(10)$ is

$$27 \sim 16 \oplus 10 \oplus 1 \tag{27.21}$$

where the 16 is D^5 and the transformation law under $SU(5)$ is

$$27 \sim 5 \oplus \overline{10} \oplus 1 \oplus 5 \oplus \overline{5} \oplus 1 \tag{27.22}$$

From equations (27.21) and (27.22), you can see that E_6 unification can be related to $SO(10)$ much as $SO(10)$ is related to $SU(5)$. In addition to the complex $SO(10)$ representation, 16, the 27 of E_6 contains real representations which can be given a large mass by the symmetry breaking.

27.3 Breaking E_6

One of the features of E_6 unification is that the 27, in which the matter particles live, also contains $SU(2)$ doublets, living in the 10 of $SO(10)$, that could be responsible for the breaking of $SU(2) \times U(1)$ at low energy. One might ask the following group theory question. Can one break the E_6 symmetry down to the low energy $SU(3) \times SU(2) \times U(1)$ entirely with Higgs transforming like 27s. The answer is no. You can see this from (27.22). If there are 27s with vacuum values in the two $SU(5)$ singlet directions, that breaks the symmetry down to $SU(5)$. But there are no other $SU(3) \times SU(2) \times U(1)$ singlet components in the 27, because there are none in the $SU(5)$ 10, 5 or $\overline{5}$. Thus anything in the 27 that preserves $SU(3) \times SU(2) \times U(1)$ also preserves $SU(5)$. Something else is needed. Here we will simply state without proof that the additional breaking can be provided by a Higgs transforming like the 78 dimensional adjoint representation.

27.4 $SU(3) \times SU(3) \times SU(3)$ unification

In passing, it is worth noting that the $SU(3) \times SU(2) \times U(1)$ symmetry is actually completely contained in the $SU(3) \times SU(3) \times SU(3)$ subgroup of E_6. Observe further that the transformation properties of the 27 of E_6, from (27.19), are invariant under cyclic permutations of the three $SU(3)$ factors. This cyclic symmetry is an **inner** automorphism of the E_6 algebra, and thus a discrete subgroup of E_6. It is related to the symmetry of the extended Dynkin diagram of E_6, (27.12). But it is an inner, rather than an outer automorphism, because it is not a symmetry of the Dynkin diagram.

While $SU(3) \times SU(3) \times SU(3)$ is not simple, if it is supplemented by the cyclic discrete symmetry, it shares many of the properties of a unified theory. The 27 is irreducible under the combination of the continuous and discrete symmetry. In this model, one can do the symmetry breaking entirely with Higgs transforming like the 27.

27.5 Anomalies

There is a peculiar constraint on unified theories that follows from the structure of quantum field theory, the mathematical language in which all these theories are formulated. The constraint is that if the creation operators for all the right-handed spin 1/2 particles transform according to a representation generated by matrices T_a^R, then T_a^R must satisfy

$$\operatorname{Tr}\left(\{T_a^R, T_b^R\}\, T_c^R\right) = 0\,. \tag{27.23}$$

You can show that this symmetric trace of three generators vanishes for all simple Lie algebras except $SU(N)$ for $N \geq 3$ (and $SO(6)$ which is equivalent to $SU(4)$). In $SU(N)$, suppose that T_a^D generate the representation D of $SU(N)$. Then define the invariant tensor d^{abc} as follows:

$$\operatorname{Tr}\left(\{T_a^{D^1}, T_b^{D^1}\}\, T_c^{D^1}\right) \equiv d^{abc} \tag{27.24}$$

for the defining representation D^1. Then, for any representation, you can show that

$$\operatorname{Tr}\left(\{T_a^D, T_b^D\}\, T_c^D\right) = A(D)\, d^{abc}\,, \tag{27.25}$$

where $A(D)$ is an integer, which is called the **anomaly** of the representation D. Thus (27.23) is the statement that the creation operators for the right-handed particles transform according to an anomaly free representation of the unifying group.

You can easily derive the following properties of $A(D)$ (see problem 27.C):

$$A(\overline{D}) = -A(D) \tag{27.26}$$

$$A(D_1 \oplus D_2) = A(D_1) + A(D_2) \tag{27.27}$$

$$A(D_1 \otimes D_2) = \dim(D_1)\, A(D_2) + \dim(D_2)\, A(D_1) \tag{27.28}$$

For $SU(5)$, or $SU(2) \times U(1)$, the anomaly of the 10 is the same as the anomaly of the 5 (see problem 27.B), thus $5 \oplus \overline{10}$ is anomaly free. This in turn implies that the representation (18.12) of a family of right-handed particles under $SU(2) \times U(1)$ is anomaly free, because the generators of $SU(2) \times U(1)$ in this representation are just a subset of the generators of $SU(5)$ in the $5 \oplus \overline{10}$.

Problems

27.A. Find $A(D)$ for the 6, and 10 of $SU(3)$.

27.B. You can find the anomaly of the fundamental representations of $SU(n)$ by calculating the anomaly of the $SU(3)$ subalgebra of $SU(n)$ under which the n transforms like a single 3 and $n-3$ singlets. Use this to show that the anomaly of the 10 is the same as the anomaly of the 5 in $SU(5)$.

27.C. Prove (27.27) and (27.28).

Epilogue

Lie algebras, physics and mathematics

I hope that the reader has come to the end of this book with an enhanced appreciation for the mathematics of Lie algebras and its application to particle physics. This mathematics is a jewel — a crystalline treasure to appreciate for all time. The physics to which it is applied, however, is not such an unalloyed and eternal beauty. In physics, unlike mathematics, we are constantly pulled in opposite directions. At one pole, there is unification, simplicity and elegance — the Platonic ideal of Nature that is created by and creates mathematics. At the other, there is the marvelous chaos of this particular world — messy, contingent, and constantly evolving with our experimental ability to probe its richness. Good physics must embrace these antipodes. That is what makes it so much fun!

Index

Printed in the United States
by Baker & Taylor Publisher Services